Theory of Earth science

THEORY OF EARTH SCIENCE

WOLF VON ENGELHARDT

JÖRG ZIMMERMANN

translated by Lenore Fischer

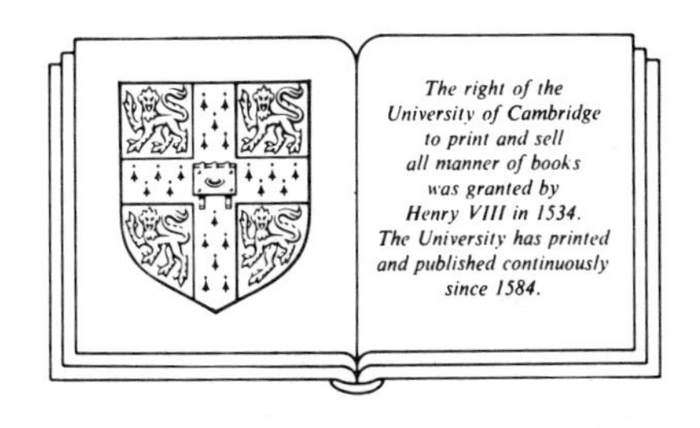

CAMBRIDGE UNIVERSITY PRESS

Cambridge

New York New Rochelle

Melbourne Sydney

Published by the Press Syndicate of the University of Cambridge
The Pitt Building, Trumpington Street, Cambridge CB2 1RP
32 East 57th Street, New York, NY 10022, USA
10 Stamford Road, Oakleigh, Melbourne 3166, Australia

Originally published in German as *Theorie der Geowissenschaft* by Ferdinand Schöningh, Paderborn, 1982 and © Ferdinand Schöningh 1982

First published in English by Cambridge University Press 1988 as *Theory of Earth science*

Printed in Great Britain at the University Press, Cambridge

British Library cataloguing in publication data

Engelhardt, Wolf von
Theory of earth science.
1. Earth sciences
I. Title II. Zimmermann, Jorg III. Theorie der Geowissenschaft. *English*
550 QE26.2

Library of Congress cataloguing in publication data

Engelhardt, Wolf von, 1910-
[Theorie der Geowissenschaft. English]
Theory of earth science/Wolf von Engelhardt and Jörg Zimmermann;
translated by Lenore Fischer.
p. cm.
Translation of: Theorie der Geowissenschaft.
Bibliography: p.
Includes index.
ISBN 0-521-25989-4
1. Earth sciences–Philosophy. I. Zimmermann, Jörg, 1946–
II. Title.
QE33.E5313 1988
550′.1–dc19 87-25631 CIP

ISBN 0 521 25989 4

SE

CONTENTS

PREFACE

Although historically derived from a single root, the natural sciences dealing with the current construction and history of the Earth developed in the course of the nineteenth and twentieth centuries into several independent sciences. In the universities these increasingly specialised disciplines developed their own chairs, institutes and teaching programs, and separate careers developed in industry. Many specialised journals were founded for communication as the number of empirical facts, theoretical discoveries and specialised methods which had to be known proliferated to such an extent that it became increasingly difficult for the individual to see beyond his own narrow discipline. Above all, a unifying theoretical concept was lacking that could have motivated specialists to discuss results or arguments with neighbouring disciplines or to ask such neighbouring disciplines for help.

Only in the last few decades has a new trend started to work against this splintering of disciplines. With the intensified efforts to understand complex geologic processes in the light of physical and chemical principles, and with the refinement of exact methods for observing and experimenting in all fields, the disciplines are now pulling together again. The observer of natural phenomena depends on the theoretician and experimenter in order to explain phenomena according to laws of nature, while at the same time a precise knowledge of the processes and conditions to be observed in nature is indispensable to sensible experimentation and to the development of fruitful theories. Thus progress in inquiry in each of the disciplines today depends more than ever on supra-disciplinary contacts. Geophysical investigations provide the foundation for theories of tectonics and mountain building; the geophysicist requires the help of petrology to interpret his measurements; laws of geochemistry and crystallography and the results of experimental mineral synthesis are used to explain the formation of igneous rocks, the deposition of various sediments in sea water, and their metamorphosis through temperature and pressure; the paleontologist uses mineralogical methods to identify the skeletal substance of fossil life forms, and sedimentological and geochemical findings to reconstruct the habitats of past epochs and the evolution of life.

Moreover the theory of plate tectonics has recently contributed marvellously to intense cooperation between the geological disciplines. Like no other theory before, it claims to solve the phenomena and problems of all disciplines according to a uniform, dynamic model of the evolution of the Earth's crust. Arguments from all disciplines have equal weight in the discussions for and against this theory, which is now fully fledged and is gaining proofs from an ever-wider sphere.

Despite specialisation, therefore, the discourse of current research (and consequently the practical application of scientific methods and results) in the main disciplines is merging into *one* geoscience. The main disciplines of this geoscience are: general, historical, and regional geology, oceanography, geological engineering and hydrology, sedimentology, soil science, geomorphology, palaeontology, mineralogy, crystallography (in part), petrology, geochemistry, economic geology and the geophysics of the solid Earth.

Externally this new unification is reflected in the various institutions and joint undertakings. In universities and research institutes of many countries the above disciplines are collectively administered under names such as '*Geowissenschaften*,' '*Erdwissenschaften*,' 'geosciences,' 'Earth sciences,' '*Sciences de la Terre*,' etc. Since the 1960s, international unions, particularly the International Union of Geological Sciences and the International Union of Geodesy and Geophysics, have been carrying out state-sponsored, long-term research projects involving the cooperation of all the geosciences. Many of the congresses held nationally and internationally have interdisciplinary programs. Not least, the exploration of planetary space, particularly through NASA, has promoted close cooperation between researchers of all geoscientific disciplines, and has moreover expanded the horizons of geoscience beyond the Earth to embrace other bodies of the planetary system. Earth science now belongs to the wider context of planetology.

We may now ask whether there are more than just pragmatic grounds for the rapprochement taking place between research and practice in these disciplines. Are the various disciplines held together simply by the fact that they study the same objects – the Earth with all its components, its history, the history of the life upon it, and the history of the Earth-like bodies of the planetary system? Or are there, over and above this, also similarities in the methods of inference, in the types of questions which research is to answer, and in the structure and function of theories? Are there characteristic features which justify Earth science's independence from the natural sciences?

In the present work we wish to help clarify this question by looking not at the currently recognised and 'confirmed' findings set down in textbooks and handbooks, but by attempting to investigate the science as an ongoing process of research. Our theme is therefore not the 'results' of research, but rather the methods through which the individual disciplines and geoscience as a whole, by observation, experiment, and argumentation, arrive at the results accepted as confirmed by the scientific community. Since we cannot accompany the scientist into the field, nor look over his shoulder in the laboratory or at his desk, the material for such an analysis must come primarily from the scientific journals in which researchers report on their framing of questions, on their observations and experiments, and on the founding and testing of hypotheses and theories. Besides their 'positive' information content, these reports contain the information we seek on the goals of current geoscientific research, on its methods and its objects.

On this basis we begin by turning to linguistic philosophy, to linguistics and semiotics. So far these concepts have rarely been applied to actual texts in the natural sciences. They are nevertheless fully suited for reconstructing geoscientific discourse, whether for problems of concept formulation, communicative functions, argumentation structures, or the use of graphic symbols in diagrams, tables, or maps. The subsequent analysis belongs to philosophy of science in its narrower sense, studying the geoscientific research process from the gathering of empirical facts, through the drawing of inductive generalisations, to the setting up of hypotheses and theories, which in turn depend on certain very general principles of research, such as actualism or uniformitarianism.

Philosophy of science, though a relatively young discipline within philosophy, has lately won increasing importance, reflected in its growing representation in university faculties. During its early development this discipline relied so heavily on the model of physics (and even then, largely on physics in its easily formalised logical and mathematical aspects), that questions of philosophy of science posed by other less 'mature' sciences were treated with scepticism or ignored entirely. The discrepancy between 'abstract' philosophy of science and 'concrete' scientific research has lessened slightly of late. Philosophers of science are increasingly aware of the historical and pragmatic aspects of scientific research, and therefore are dealing more precisely with the special problems involved. Nevertheless, attempts to reconstruct completely the research practice of a specific science have so far remained exceptional. In view of this, we hope that our investigation of the geosciences will help consolidate the position of philosophy of science as a philosophical discipline, and that it will also reduce the oft-deplored disinterest of the 'practitioners' in basic considerations of what they are really doing and hoping to achieve in their research.

We address ourselves above all to the geoscientists of all disciplines. It is our opinion that reflection on the language of science, on its non-linguistic modes of representation, on the fundamentals of concept formation and argumentation, on the structure of lawlike statements, and on the foundation of hypotheses and theories can be valuable on many counts. Such considerations can provide valid criteria (which may differ with viewpoint) for judging one's own and others' work and research aims. Such criteria are important today, when research priorities must be set repeatedly and in many contexts. A more self-critical view of hypotheses and theories will also be taken if one knows their function in science and realises that they can never be 'proved', but at best 'corroborated', thus feeding the hope of 'approaching the truth', even when this truth can never in fact be attained. Such an attitude shields one from identifying too uncritically with a theory and coming to believe in it as dogma. This is a trap into which geoscientists have frequently fallen – not only in the distant past – to the detriment of scientific progress. Finally, the realisation that there are methods and goals common to all geoscientific disciplines should deepen the consciousness of unity and help to clarify the special function of geoscience within the natural sciences.

This book is the result of many years of joint work between a geoscientist and a philosopher. It was made possible by the *Deutsche Forschungsgemeinschaft* through its program on 'philosophy of science'. We are thankful to the *Deutsche Forschungsgemeinschaft* for its generous support of our research. We would like to thank Mr Lothar Schäfer, of the *Philosophisches Seminar* of the University of Hamburg for reading through the manuscript and for his helpful discussions. We are grateful to Mrs Gabriele von Engelhardt for her help in evaluating the geological scientific literature.

August 1981

Jörg Zimmermann
Philosophisches Seminar
University of Hamburg

Wolf von Engelhardt
Mineralogisch-petrographisches
Institut, University of Tübingen

Translator's note

I wish to express my thanks to the authors for the support and help they have given me while translating this work. Most particularly I am grateful to Mr Nordmann of Hamburg University for his very painstaking reading of the manuscript. Needless to say, I accept full responsibility for any errors which may have arisen in the course of translation.

Lenore Fischer
Castleconnell, Co. Limerick

1

The structure of communications in Earth science

1.1 Introduction

Like all science, Earth science as a social institution is a form of collective practice,[1] that is, the scientists share a common attitude and behaviour-pattern which can be examined from various aspects. Of these we emphasise three, that dealing with the subject-matter, that with the instruments, and that with communications.

Geoscientists must behave communally *vis-à-vis* the subject matter of research, since the object of all the geosciences is in the final analysis a communal one: the study of the Earth, its components, its historical development, the history of life upon it, as well as all these in relation to other planetary bodies. Historically this uniform and shared nature of the research subject-matter has not always been so self-evident. The geognostic sciences developed differently in Central Germany (Thuringia and Saxony), France, England, and Italy, for each of the first geologists depended on his experiences of his own region, so that each developed different theories about the structure and history of the Earth, according to local factors. Geological, petrographical, paleontological, and geophysical observations made in various regions have since become generally available, so that all Earth scientists are in principle united by a common subject-matter. Traces of regionalism still survive in geology, however, a fact difficult for physicists and chemists to understand. This is because the geologist who has studied and worked in a particular region is best acquainted with a specific area of the Earth's crust. Thus a geologist working on the continental plate of Eastern Europe, which consists of more or less horizontal strata representing several hundred million years of Earth's history will be more likely to think in terms of permanent continental blocks than a colleague from a country surrounded or bordered by ocean. This colleague's need to explain the history of the ocean floor or similarities in continental structures across oceans will dispose him more towards a mobilistic theory, in which continental blocks have undergone considerable horizontal displacement.

As to the instrumental aspect,[2] all Earth scientists use the same means of dealing with their research object using communal operations of observation, measurement, and experimentation, i.e. applying all technological aids from the geology hammer to the mass spectrometer and computer.

In the communicative aspect,[3] there is communality not only in how Earth scientists exchange information, discuss research results and hypotheses, and criticise and test theories, but also in the methods of initiation into scientific practice (teaching and learning situations), as well as in representing the institution of Earth science in society, involving discussions of technological applications and future research priorities.

The instrumental and communicative attitudes and behaviour-patterns of Earth scientists with regard to their communal object is based on the acceptance (usually unspoken) of certain norms, embodied and maintained by the scientific community[4] to which the individual feels he belongs. This imaginary body becomes manifest at various levels. At the most general level, basic norms are accepted for scientific treatment of objects and for cooperation between scientists. Intermediate in level are communities resulting from membership in certain scientific disciplines: physicists, chemists, or geoscientists. Finally, groups formed at a lower level represent scientific 'schools', followers of a given theory, or member of an institution.[5] Depending on the analytical point of view, one and the same scientist can belong to different groups. The differentiation within the scientific community may often lead to conflicts or identity crises, the foundation of new subdisciplines or special groups, withdrawal and reentrance in other groups, or to changing allegiances. Furthermore, inevitable overlaps mean that no definitive boundaries can ever be drawn. This has of late become true, for instance, of the division between the physical and chemical disciplines on the one hand, and the Earth sciences on the other. Due to the increasing use of physical and chemical methods, the Earth scientist behaves instrumentally in the same way as the physicist or chemist. This is expressed particularly in names of such disciplines as 'geochemistry', 'cosmochemistry', or 'geophysics'. Again, the intermediate position of crystallography between Earth science and physics has meant that it is included in some German universities in the physics department, and in others in the Earth sciences. The special problems of such groups and their role in society are the subject of the sociology of science, and will not be discussed further here.[6]

Our discussion will elaborate on the obvious fact that Earth scientists show special traits in dealing with their subject as well as in their instrumental and communicative attitudes and behaviour-patterns, traits which differentiate them from other scientists. This is made clear even in the popular view that geologists 'go out into the field', collect rock samples and fossils in quarries, and use maps for orientation. True, laboratories are playing an increasing role in all branches of geoscientific research, yet contact and involvement with natural objects outside the laboratory remains a basic feature, one which distinguishes the Earth scientist's activity from that of the physicist or chemist. This is clearly related to a certain 'skill' in observing and dealing with the great complexities of geologic objects, a skill developed through study and practice, but probably based on the original predilection which caused the individual to enter this branch of science.

We are primarily interested in our context in the characteristics of the Earth scientist's communicative behaviour, for it is this which permits us to obtain a perspective on the particular interests and goals which determine the geoscientific dialogue.

The communications process we call scientific discourse.[7] Its course determines whether certain assertions made by scientists can be accepted by the scientific community (or by a representative group), how they are to be tested, and what relevance they have for application or in the context of the given problem. Continually arising differences of opinion lead in turn to further discussion, which can be carried out in oral or written form. In both cases specific communications media are necessary to permit contact between the members of the scientific community. Media for oral communication consist of universities, research institutes, conferences or workshops, a typical form for Earth scientists being the field trip. This is a didactic tool on the one hand, exposures being chosen on pedagogic grounds to show the student *in situ* how geoscientific research registers and explains given phenomena. On the other hand field trips are undertaken by scientists to demonstrate relevant empirical findings and to discuss certain hypotheses presented to explain them.

In the following we will, however, concern ourselves exclusively with the results of written exchanges of opinion, as these alone are generally accessible and form a reviewable record of interchanges. The media involved here are the various scientific publications. Scientific libraries are also important for written discourse, cataloguing and storing all published texts, so that these reference points of discussion are accessible at all times and places.

1.2 The clarity of discourse

Constant and complete communication between scientists is an ideal condition which has probably never been attained in the history of the sciences, and whose achievement has for various reasons become increasingly difficult in the last decades. The universal dissemination of the results of research in the Earth sciences has been hindered above all by the enormous increase in the production of such results. The members of the scientific community are hardly capable of working through all the texts relevant to them, and therefore depend not on the original works, but on such excerpts and summaries as appear in the various specialised publications. These reproduce the primary information in condensed and more or less incomplete form. Other researchers may from the start take into account only the publications of a limited group of scientists with whom they feel the common ties of similar methods, specialised interests, or shared schools of thought.

The large internationally circulating journals can only accept a limited number of works, and have only a limited amount of space available for each publication; they may even levy a high page charge, so that the author himself restricts the length of his contribution. This means that experimental procedures and theoretical conclusions are often given in such abbreviated form that the specialist alone can follow and critically evaluate the results. The overloading of journals and the consequent time lag between the delivery of a manuscript and its publication has a further consequence: it has become increasingly customary to publish research results in papers appearing as the proceedings of special congresses, as research reports from various research institutions or as private 'preprints' reproduced in limited numbers; these do not appear in the bookstores, and represent an 'apocryphal' literature available only to a limited circle.

Another hindrance which should not be underestimated is the language barrier obstructing universal communication. True, in the Earth sciences, as in the other natural sciences, English has won increasing recognition as a lingua franca. But beside all the publications in English, there are still many works printed in Spanish, Italian, French, German or Russian, and these tend to be ignored, particularly in the English-speaking world.

For these reasons, despite all efforts to produce translations, and despite the speedy compilation of international abstracts, clear and universal communication is impeded in many areas of Earth science. Even authors of reviews or monographs must admit that they can no longer cover everything. Thus it is stated in one review of the tectonics of Western Europe (1974):

> It cannot be doubted that difficulties have already arisen in Europe in evaluating and communicating the countless data. It was therefore impossible here to consider the entirety of all the existing contributions to the problem in question.[8]

The dangers of such a situation are obvious: publications which might be relevant to the progress of science are no longer noticed by their potential readers. On the other hand fruitless paths may be followed unhindered, due to the absence of critical control by the scientific public. The lack of communicative clarity may also mean that investigations may be pursued many times over, in ignorance of similar efforts being carried out by others, so that the principle of economy of intellectual and material resources is violated.

There are various ways of partially countering these dangers. Increasing specialisation in research subjects and methods has meant a corresponding narrowing of the readership addressed. The relevant scientific community is thus reduced to a specific group of researchers, so that the possibilities of discussing all the literature relevant to the solution of a group of problems, i.e. the possibilities of constant communication, are realised at least for a group of specialists. This has progressed very far in some cases, for example, in the proceedings of the annual conferences on lunar and planetary research held in Houston, Texas.[9] Between 1970 and 1980, these have grown into a series of 33 volumes of 35 800 pages, containing contributions by physicists, chemists, geophysicists, geochemists, geologists, petrologists and mineralogists on results obtained from lunar material and from observations made by the American and Soviet space missions, results therefore, concerning the planetary space near Earth and the composition and origin of the Moon and planets. Most of these papers are directed at a relatively small circle of interested people, e.g. at research groups dealing with radioactive dating or the properties of the solar wind, or at Earth scientists concerned with developing theories on the inner structure and geologic history of the Moon. Within these and other specialist circles, communication and mutual control is assured. A more comprehensive research programme, such as a planetology which would systematically and theoretically link the geology of the Earth with the now-developing 'geology' of the other bodies of the inner planetary system, requires a higher level at which particular communications processes could be coordinated and made accessible to a wider scientific community.

As in the area of planetology at present, other areas of geoscientific research lack efforts to promote clarity and remove barriers hindering communication between individual circles of specialists. In this context increasing weight should be given to synthesizing monographs which cannot be accommodated in normal journals, comprehensive papers held at conferences, which these days are overloaded by specialised short lectures, efforts to simplify and improve abstract journals, as well as international storage of geoscientific observation data and the making of such stored data accessible to the public.

1.3 Participation in discourse

Scientists participating in discourse function both as authors and as critical readers of texts. However, not all those who participate in discourse because they belong to the community of researchers have the same status. Some authors are privileged, either because they have better access to organs of publication, or because of a reputation gained through recognised achievement, as a consequence of which they receive more attention from the outset.[10] Also various organs of publication are ranked differently, as is the quality of research in certain laboratories, institutes, and even countries. Finally, barriers must be overcome in getting manuscripts accepted by journals. It used to be that acceptance standards were the personal responsibility of individual editors, but the editors have now largely been replaced by editorial boards. The individual works are read through by assigned referees who usually remain anonymous as far as the author is concerned, and who decide whether the work will be accepted as it is, must undergo certain changes, or will be rejected altogether.

Inquiring into the reasons motivating a scientist to publish papers, account must be taken not only of his objective or scientific interest in aiding inquiry by discovering new facts, improving a theory, or setting up a new hypothesis, but also his subjective interests external to science. Solla Price has suggested the provocative theory that the scientific paper, contrary to widespread opinion, is only secondarily intended to convey new information.

> First, scientific communication by way of the published paper is and always has been a means of settling priority conflicts by claim-staking rather than avoiding them by giving information. Second, claims to scientific property are vital to the make-up of the scientist and his institutions. For these reasons scientists have a strong urge to write papers but only a relatively mild one to read them.[11]

On this basis, scientific communication would be at least as personally orientated as it is materially orientated, in that it would serve not only to spread scientific knowledge, but also to establish, describe and reinforce a personal or group identity defined by one's membership in a subgroup of the scientific community.[12] This is especially true of publications meant to ensure access to the scientific community through the acquisition of a position at a university, institution, or in applied science. Such papers can sometimes be 'tactically' motivated, as expressed, for instance, in their adherence to a dominant theory or to the opinion of certain scientists whose good wishes the author hopes to gain or at least not to lose. Thus in the

United States in the 1920s, the overwhelming rejection of Alfred Wegener's hypothesis by the key geologists would have made it tactically very rash for any geologist seeking a secure position in an American university to express himself in print as supporting continental drift (see p. 306).

The reverse behaviour, that of holding fast to questionable theories which find little or no general acceptance, cannot wholly be explained as a quest for truth either. Here the critical factor may be a need for self-assertion, backed by a history of a difficult career.

In the Earth sciences, it may even be suggested that the lack of any compelling methods of determining the explanatory content of theories, and the practical difficulties of checking the reliability of some empirical findings (such as those made through extensive field work) may favour a communicative behaviour motivated by the need to ensure a personal identity (whether this be as an individual or as a member of a 'school').

A thorough study of the external motivations and interests of scientists as individual people or as members of social groups, while undoubtedly important, is a matter for the psychology and sociology of science, which we do not wish to discuss further here. We will, however, come back to the person of the author and researcher when we speak of the intentional and appellative functions of scientific texts (pp. 13, 16), and when we analyse the growth of geoscientific knowledge (p. 303).

1.4 Communicative functions of geoscientific texts

1.4.1 *Communication models*

The scholar is generally interested only in the informative content of a scientific text, and it is therefore generally tacitly or explicitly assumed that the imparting of substantive content constitutes the entire function of such texts. A closer look actually reveals that a scientific text exercises a far greater range of effects than that, however, each having its own influence on the communicative process of research. If we wish to understand Earth science as a collective practice, we must first attempt a general appraisal of the most important functions which are or may be carried out by the geoscientific text, beyond the mere imparting of information. Such an appraisal will serve firstly to set up a classificatory scheme of scientific texts, and secondly to form criteria on which the quality of a text may be judged with regard to its argumentational structure and linguistic form.

Based on concepts developed in communications theory and linguistic pragmatism to analyse the conditions under which texts are produced and received, we will look at geoscientific texts using the following model of the communications situation (modified after R. Jakobson[13]).

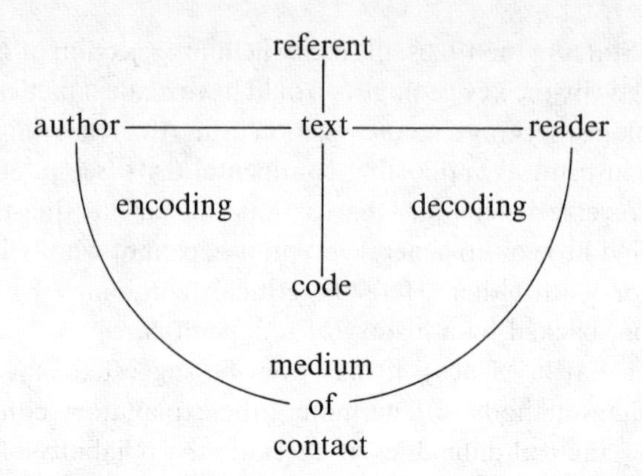

The *author*[14] produces a message which is received by the *reader*. The message refers to a certain *referent* or *state of affairs* (*Sachverhalt*), and is expressed in a *code* which must be partially or wholly shared by the encoding author and the decoding reader as the text could otherwise not be understood. Finally, a *medium* is required to bring about contact between the author and the reader; in our case this is the scientific publication or the library which provides access to the publication. This gives us six factors which influence the structure of every communication transmitted by signs: author, reader, referent, code, medium, and the message itself. All these factors can be 'addressed' by the content of the message, and each in its own way can affect what the reader understands. Jakobson therefore distinguishes *six communicative functions of linguistic expression*: the 'referential' relating to the referent or state of affairs, the 'emotive' relating to the author, the 'conative' relating to the reader, the 'metalinguistic' relating to the code, the 'poetic' relating to the message, and the 'phatic' relating to the medium.

We can ignore the latter two functions, as these play a minor role in scientific texts. The poetic function involves the text as an aesthetic form, existing as a simple linguistic structure 'for its own sake'. True, consideration of elegance, clarity or uniformity of presentation do play a role in scientific texts, yet stylistic criteria drawn from poetic considerations of clarity and intelligibility can be covered by the metalinguistic function. We will also not go into the phatic, medium-orientated function, which serves to attract the attention of the reader in the psychological sense, and to maintain and improve contact with the reader, as this effect may be achieved by expressions aimed at the reader. Finally, instead of the term 'emotive', we will use the broader term 'intentional' (i.e. referring to the intention of the author), and instead of 'conative', the more readily understood 'appellative', which emphasises the demands made upon the reader.

Basically all linguistic expressions can be understood in their referential, metalinguistic, intentional, and appellative aspects. They fulfill a referential function, in that they deal 'with something'; their metalinguistic function is fulfilled in that they choose certain elements from a pre-existing linguistic repertoire and recombine these such that a form of linguistic usage is at least implicitly imparted; they fulfill an intentional function in that they reveal the intention or desire of the author, if only that the reader should recognise the imparted content as being 'true', 'plausible', or 'relevant'; they have an appellative function in that they affect the reader in some manner, changing his consciousness or motivating him to action. A distinction should be made between those communicative functions which are only implicit and those which are explicit, even though the distinction is not always a clear one. Furthermore it should be kept in mind that psychologically interpretable clues may 'reveal' something to the reader which the author did not intend to convey when producing the text. This can be very informative in some contexts. We, however, will not deal with the implicit and hidden, psychologically interpretable functions, but confine ourselves to the explicit functions of geoscientific texts, dealing with each according to its *dominant* communicative function.

1.4.2 *The referential function*

The most obvious purpose for the majority of geoscientific publications, and the most important purpose for the growth of knowledge, is to present states of affairs *(Sachverhalte)* to the reader. This is carried out by the referential function of the text, but the mode of presentation may vary. Following Kant[15] we distinguish between assertoric, hypothetic, and apodictic modes of presenting states of affairs.

In the *assertoric mode* the factual existence of a state of affairs is asserted.[16] Such an assertion concerns things which, according to the empirical methods of natural science, are 'fact', i.e. what in the author's opinion any competent person can in principle check and confirm using recognised methods of observation. We say 'in principle', because in some cases the observations may concern process which are rare, temporary or not repeatable (such as a meteor's fall, or the course of a certain earthquake or volcanic eruption); the reader is in this case addressed as an imaginary witness. Factual claims appear in the form of measurement results, data tables, diagrams, experimental results, and observations of phenomena found in nature. Most geological and geophysical maps may be assigned primarily to this type of assertoric communication. Factual claims have a descriptive character. Taken together, they represent the *basis of empirical data* to which all further arguments of a hypothetical type must recur.

Testing factual assertions involves making decisions on how reliable the data communicated are and how adequate the description of the states of affairs is. The methods by which the data were obtained must also be checked, as must the terminology used to present them. Having passed these tests, states of affairs presented are accepted as *fact*[17] and are recognised as such by the scientific community.

In the *hypothetical mode*, the possible existence of certain states of affairs is asserted. Hypothetical assertions from the very start are made contingent upon further testing. The hypothetical mode is explicit when the truth of certain assertions is explicitly stated to depend on further discussion. The hypothetical character of these assertions is indicated by formulae such as: 'might be the case', or 'is probably the case', or limiting hypothetical conditions such as 'given that . . .' or 'if the theory is correct . . .'. Implicitly hypothetical claims resemble factual descriptions in that the hypothetical conditions and theories on whose validity the claim depends are not expressly named. Statements of this kind include, e.g. statements about processes or events in the geologic past, predictions of future events, or assertions as to conditions in the inaccessible interior of the Earth. A particularly common form of hypothetical claim consists of *interpretations*, statements in which certain empirical facts are understood and interpreted 'as something', e.g. when one unsorted breccia composed of various sizes of rock fragments and mineral grains is interpreted as a 'tillite' (glacial sediment) and another breccia as an 'impact breccia' (product of a large meteoritic impact), or when one microfold in sedimentary layers is interpreted as a 'synsedimentary slump', while another is understood to be the product of tectonic thrusting. Interpretations thus always anticipate alternative possibilities of seeing something 'as' something. A critical evaluation of hypothetical assertions involves firstly a decision as to which assertions in a given context are to be viewed as hypothetical at all. Certain interpretations of empirical phenomena (e.g. of a certain rock as being, based on its structure, a product of solidification out of a melt or magmatic rock) are founded on assumptions which by general consensus are no longer doubted, at least for the time being, and are therefore regarded as descriptions of fact. We will elaborate later on these conventional boundaries between empirical and hypothetical statements, when we come to the chapter on theoretical preconceptions about the empirical basis (p. 97). If the assertions of a text are to be viewed as hypothetical – and with the rise of a new theory even statements previously undisputed by the scientific community may come into this category – they must be viewed in the context of alternative interpretations. Generally the larger the number of alternative interpretations which may with reason be ruled out, the greater the gain in understanding achieved by the chosen interpretation. It will be shown in the course of our investigation that in the Earth sciences, presentation of states of affairs of a hypothetical type plays a very great role. That hypothetical assertions should be made and discussed against a background of many alternative possibilities is an important postulate in this science particularly, although it is not taken seriously enough even today.

The *apodictic mode* asserts the necessary existence of a certain state of affairs. It is always based on a formally logical inference, deducing one state of affairs from others. To test such an assertion, the logical inference must be examined for its correctness.

The referential content of a geoscientific text generally consists of these assertoric, hypothetic, and apodictic components. As these three components must each be evaluated according to differing standpoints, an overall evaluation of the entire text must necessarily involve recognising them separately and examining their interaction. In a later chapter we will investigate how these assertoric, hypothetic, and apodictic modes combine with the various types of inference (induction, abduction, and deduction) to produce a scientific explanation.

1.4.3 *The metalinguistic function*

The notion of metalinguistic function is based on a distinction made by Carnap[18] and other philosophers of science between two levels of scientific language. At the level of the 'object language', one speaks directly about states of affairs or the object of research, while at the level of 'metalanguage' on the other hand, one speaks about the terms and rules which make up the object language. Utterances at the metalinguistic level carry out three primary functions. By explicating the terms, firstly, they ensure communal understanding within the scientific community, by eliminating or reducing any possible discrepancies between the encoding of the text by the author and the decoding by the reader. Secondly, they extend and modify scientific language by defining and introducing new concepts. Thirdly they control the transfer of statements into other presentational forms.

The *explication* (explication, Latin, literally, 'to unfold') of scientific concepts or of complexes of statements, is predominantly conservative, that is, a scientific language is maintained and affirmed at a particular stage of its development by using existing conventions as a basis, whether these be generally accepted conventions or ones used only within a certain group.[19] On the one hand explications form bridges for understanding by clearly explaining the meaning and use of scientific concepts for the beginner or for a member of another discipline. This function is fulfilled particularly by text books and by articles introducing certain subdisciplines of Earth science or certain research areas. On the other hand explications are necessitated when it has been found that certain concepts are being used too broadly or ambiguously, so that more precise definitions or a breakdown into subordinate concepts is needed for research purposes.[20]

In evaluating the communicative function of explications, that is, determining whether they explain the use of certain terms to the reader and help avoid misunderstandings, our criterion must be that they are clear and understandable relative to how informed or 'initiated' the readership concerned may be. There are no absolute yardsticks for measuring understandability. The explication of a concept which is understood by one circle of specialists may need further amplification for members of another discipline. At any rate, the minimum criterion is that the explication must be understood by and be useful to that readership, large or small, to which the author addresses himself in his text. Besides being evaluated for their communicative value, explications must also be tested for their truth or falsity, for as clarifications, explanations, or refinements of terms already in use, they must adhere to the existing linguistic conventions.

The definition and introduction of new concepts, unlike explication, is predominantly innovative. These new concepts change or amplify the existing language of science. Each new theory introduces new concepts and conceptual systems into scientific language, and as old theories get cast aside, concepts in former use may often be rendered obsolete. Thus Darwin's theory of evolution introduced new concepts into paleontology such as 'line of descent', 'selection', and 'selection value'; the theory of plate tectonics introduced 'plate boundaries' and 'subduction zones' etc. while the impact theory brought into scientific language the new conceptual system of shock metamorphism. It might be said that progress in the Earth sciences is reflected in the changing and expanding terminology. During times of changes in theory, texts of predominantly metalinguistic function, concerning themselves 'only' with terminological and nomenclatural revision, become important.[21] New terms first become needed when new phenomena or objects are discovered or inferred from new theories. Thus hitherto unknown fossils or mineral substances may be discovered, which are named and described (i.e. defined) in accordance with the appropriate systematic nomenclature. During the Apollo Missions a new mineral was found in the lunar rock; it was named 'pyroxferroite', defined according to crystal structure and chemical composition (an iron–calcium silicate), and fitted into its place in the mineral system. An example of a term for a new mineral state discovered only through theory of impact is 'diaplectic' glass: a glassy (amorphous) substance produced not by rapid cooling of a melt, but in a solid state through the shock waves produced by meteoritic impacts.

A second case in which innovative changes in scientific language are undertaken is when entire nomenclatures, that is, terminological systems for the description of certain classes of objects, are recast, eliminating or redefining familiar categories and designations. If such a reorganisation is accepted by the scientific community, the scientist must not only learn new material, he must also relearn: he must forget the old familiar definitions and terms. As an example we might name the reorganisation of the nomenclature of igneous rocks (to be discussed in more detail in a later chapter, see p. 108) worked out in the last ten years by a commission of the International Geological Union and now largely accepted by the scientific community. The old, disparate and qualitative classification of rocks has been replaced by a system founded quantitatively on mineral composition. A number of old names have been entirely eliminated, yet at the same time only a few new terms have been introduced, and most of the old names retained. Long-used names such as granite, basalt, and andesite still occur in the new system, so that it might appear as though this were a simple case of explication. What is really involved, however, is a normalisation on an entirely new basis, the new terms being defined in a new way, so that, for instance, many rocks previously described as 'andesite' now fall into the basalt class.

Newly introduced concepts are evaluated according to their usefulness in research, and also according to how smoothly they fit into the accepted context of scientific language.

The metalinguistic level of a text finally includes *rules for transposition* of statements into structurally different methods of presentation, for example, transposition of verbal expressions into mathematical or graphic form, or processing of given material into a computer language. The criteria for evaluating such transpositions are freedom from contradiction, economy, and practicality.

1.4.4 *The intentional function*

Intentional functions are carried out by expressions in which the attitudes, evaluations, motives, and feelings of the author are expressed, and in which, therefore, the author expressly involves himself in the issue. This is usually done to sway the reader over to his side, and to evoke corresponding feelings in him. Most intentional expressions are therefore appellative in character. Four groups of intentional expressions are generally important.

Firstly, there are statements which relate to the emotional condition of the author. These may be called *statements of expression*. Negative statements of expression betray anger, worry, and the like. Feelings of dissatisfaction over such things as the dogmatic obtuseness of colleagues, lack of recognition of one's own achievements, 'trespassing' of other scientists into areas considered by the author to be his own special domain, or disregard of priority claims were all expressed much more drastically than today in publications of the nineteenth century and of the first decades of the twentieth century. The attentive reader, however, will find annoyance expressed in publications of today as well, when the author does not confine himself (at least not exclusively) to marshalling substantive arguments in his favour, but 'shortens' the operation by simply doubting the competence of the opposition to take part in discourse. Examples of such a non-communicative process of disqualifying the 'opponent' are especially frequent when a new comprehensive hypothesis is forming, as during the discussions about continental drift and plate tectonics (see p. 240), or in the dispute over whether the surfaces of the Earth and Moon were modified not solely by endogenous forces, but also by the impact of masses from planetary space. Personally motivated criticisms are also found in discussions of more specialised hypotheses, if only in hidden form, for instance in the tactic of leaving supposedly incompetent authors out of the discussion by simply not citing them.[22]

Instances of positive emotion also occur in geoscientific texts. This may relate to the work itself, as when the author conveys the feeling that he favours a certain solution to a problem, or that he feels himself bound to certain landscapes or to certain views and traditions associated with a school. The works of other authors may also be subjectively stressed by designating them as 'classic', 'outstanding', 'comprehensive', or 'seminal'. Non-verbal means may also be used to convey positive feelings, as when the author emphasises sentences with exclamation marks or italics.

Statements of hope or fear are also expressive. It is common for an author to state the hope that his investigations will help solve a certain problem, or that the research programme which he has encouraged or initiated will lead to the attainment of an ambitious goal. Rarer are statements doubting that a certain question can be solved, at any rate with the means available at the time.

Expressive statements are also found in the publications of other natural sciences, but members of other sciences have rightly remarked that they are more common in the publications of the Earth sciences, even in those of today. This is because of the nature of most geoscientific hypotheses. For one thing, they cannot experimentally be confirmed or refuted in the same way as physical and chemical claims, so that support for them is more often based on so-called subjective probability (see p. 157). Secondly, many of them are not presented in intangible, abstract, and mathematical form, but rather as complexes of states of affairs which are not observable, but which resemble observable facts. These can betray one into seeing as real the universe of the geologic past or of individual episodes posited by geoscientific theory. Personal identification with this may proceed to such an extent that a whole 'world view' collapses when, as research progresses, a new theory takes over the stage.[23]

The intentional function is secondly fulfilled by *statements which express the resolves, intentions, and expectations of the author*. Statements of this kind are a normal part of any scientific text. Every author will preface the results of his investigations by saying what he intended when he undertook the work, and what goals he had in mind, regardless of whether he fulfilled these wholly, in part, or not at all.

Thirdly, intentional statements include all *statements on the degree of certainty or uncertainty* with which an author regards certain states of affairs *(Sachverhalte)* as valid. This includes statements of doubt, conjecture, support, or conviction that something, be it one's own results or be it someone else's claim, is probable or improbable. Statements of this sort occur normally in scientific texts. Just as any elementary quantitative datum in physics is scientifically significant only when the margin of error of its numerical values is known, so any complex statement must be judged according to its degree of certainty or probability. The author personally guarantees this, even if only tacitly by publishing the text.

Fourthly and finally, statements with an intentional function include evaluations and estimations of whether something is good or bad, relevant or irrelevant. Such *evaluative forms of language-use* form a fundamental aspect of scientific discourse,[24] as they ultimately determine the future course of discourse, what is relevant, which claims should be accepted, which rejected. Evaluations relate either to 'things' or to propositions. 'Things' are seen as relevant to something else, particularly to certain contexts where they seem important within the framework of a problem or research programme. Evaluations of propositions are aimed reflexively at claims to validity raised implicitly or explicitly in scientific discourse. They decide whether a claim is 'true' or 'false', an interpretation 'plausible' or not, a premise 'probable', 'conceivable', or 'misleading', a presentation 'clear', 'thorough', 'one-sided', 'ambiguous', or 'unclear', a normative proposal 'useful' or not, an argument 'tenable' or not, or even whether an evaluative expression is 'well-founded' or not. Evaluations and estimates cannot be 'inferred' from simple determinations of fact; they are based on pre-existing theoretical contexts and goals of research, within whose framework it can in turn be determined how appropriate and well-founded they themselves are.

The critical evaluation of intentional statements must begin with the danger of their persuasive effect.[25] Such statements may persuade the reader or make him liable to accept the author's view without rational control. This effect is furthered by the tendency in scientific argument to formulate intentional statements impersonally, so as to emphasise their claim to objectivity. The individual author avoids the pronoun 'I', he speaks as 'we', anticipating the consent of the reader or of a larger public; or he may use formulae setting him in the role of speaker of the scientific community: 'every (competent) colleague will share my (our) view that', 'there can be no doubt that . . .', it is thereforc clcar that . . .', 'the conclusion therefore seems admissible that . . .'. It is up to the critical reader to recognise the intentional content of statements even when thus clothed, and to make sure of the grounds on which he should accept or reject imputations, anticipations, claims, or judgements of the author.

An evaluation of the various types of intentional statements, overcoming their persuasive effect, is based initially on the integrity or truthfulness of the author. For the particular types of intentional linguistic usage, more specialised criteria must be considered over and above this. Thus a declaration of intent calls for a corresponding fulfillment of that which was intended. It may be asked whether conjectures and assessments are well founded, and whether they are justified by the grounds given. It is more difficult to evaluate positive and negative expressive statements. They should be rejected when used to cover up inadequacies in the rational foundation of a hypothesis, or to play upon authority, and are only legitimate in scientific discourse where their content is objective and well-founded, regardless of how important their effect, detrimental or otherwise, has been on the actual history of science.

Despite the need for critical alertness, it can nevertheless be said that even the intentional component of a text, relating to the author and researcher, is of value to scientific discourse. Intentions, conjectures, and evaluations keep research going and imply a motivating personal conviction; as an expression of powers of judgement (in the Kantian sense), they cannot usually be eliminated and replaced by wholly impersonal criteria.

1.4.5 *The appellative function*

Generally speaking all scientific texts are appellative, in that they address themselves to the public for the purpose of gaining attention, informing others, evoking agreement or criticism, or even influencing future research.

The purest form of an appellative expression would be an imperative sentence. Since there are no command situations in scientific discourse, however, the appellative function is rather to be sought in expressions of encouragement, recommendation, or invitation to accept or reject certain claims, to test or to refute them. Invitations to continue work on the solution to certain problems, follow up questions arising from the author's investigations, participate in research programmes, or use methods which the author worked out or found useful during his work are all especially appellative. Texts of a dominantly appellative function include both those on the use of geoscientific methods and discoveries outside of academia (describing, for instance, geophysical and geochemical methods and apparatus to be used in prospecting or in more efficiently exploiting ore deposits), as well as those which, based on general theoretical considerations, suggest the likely occurrence of ores in certain areas of the Earth's crust (implying the recommendation that all explorations should take such considerations into account). Finally we should include in this group the investigations which have of late become so important regarding the endangerment of the geopotential so important for the future quality of life, for food and for technological civilisation. The threatened exhaustion of ore reserves, the wasting of water, the contamination of the air, waters, and soil, have all led to appeals to the government, economy, and the public.

Jackobson's 'phatic' expressions,[26] meant to capture or maintain the reader's interest, can be included under the appellative function. These embrace, among others, questions posed by the author to the reader, such as questions demanding an answer of the reader, or rhetorical questions asked only in order to obtain a specific answer or to show that the reader does not know the answer, so that the author may more effectively present his own solution, or show that no solution yet exists.[27]

Appellative expressions, like intentional ones, may exercise a persuasive effect on the reader. In general, however, pure persuasion, of which the reader is often not actually aware, is less the result of single statements, than of the argumentational style and linguistic usage of the overall text. The effect of appellative statements on the reader can only be judged by whether some reaction takes place or not. In the context of a publication, appellative statements must be evaluated according to how well they are objectively founded, if at all, remembering that authors tend to make their texts seem more important by appealing to the reader.

1.5 Types of geoscientific texts

Different types of geoscientific texts can be differentiated according to their dominant communicative function. They have been collated in Table 1.1, with titles of books and treatises as examples. Using this typology we may derive criteria for the appropriate evaluation of texts, for, as we have seen, scientific statements should be judged according to quite diverse standards, depending on the function which they fulfill in discourse. Normative metalinguistic statements may or may not be appropriate; assertoric

Table 1.1 *Types of geoscientific texts*

Dominant function	Text type	Sample title
1. **Referential**	Text books	Klockmann's *Lehrbuch der Mineralogie*[1]
General	Handbooks	*Handbook of Geochemistry*[2]
	Identification books	*Optische Bestimmung gesteinsbildender Mineralien*[3]
	Monographs	*Sedimentary Carbonate Minerals*[4]
Primarily assertoric	Specialised empirical works	Amorphous Copper & Zinc sulphides in the metalliferous sediments of the Red Sea[5]
	Reviews of empirical works	Comparative morphology of ancient and modern pillow lavas[6]
	Maps	*Geologische Karte des Rieses 1:50 000 mit Erläuterungen*[7]
Primarily hypothetical & apodictic	Specialised theoretical works	Magma genesis in the New Britain Island arc[8]
	Reviews of theoretical works	Origin of oolitic iron formations[9]
2 **Metalinguistic**	Dictionaries	Monolingual: *Glossary of Geology*[10]
		Multilingual: *Geological Nomenclature*[11]
	Terminological texts	*Zur Terminologie von Schieferungen*[12]
	Texts on nomenclature	Classification & nomenclature of volcanic rocks, lamprophyres, carbonatites and melilitic rocks[13]
	Computer programs	*Computerprogramme zur Berechnung von Mineral- und Gesteinsumbildungen bei der Einwirkung von Lösungen auf Kali- und Steinsalzlagerstätten*[14]
3. **Intentional-evaluative**	Book reviews	Book reviews in Tschermak's *Mineralogische und Petrographische Mitteilungen*
	Critical texts	*Kritik zur Plattentektonik*[15]
4. **Appellative**	Practical texts	*Geowissenschaftliche Probleme bei der Endlagerung radioaktiver Substanzen in Salzdiapiren Norddeutschlands*[16]

Literature cited in Table 1.1

[1] Ramdohr (1954)
[2] Wedepohl (1969–78)
[3] Tröger (1967, 1971)
[4] Lippmann (1973)
[5] Brockamp *et al.* (1978)
[6] Wells *et al.* (1979)
[7] Bayerisches Geologisches Landesamt (1977) and Gall *et al.* (1977)
[8] Paolo & Johnson (1979)
[9] Kimberley (1979)
[10] Gary *et al.* (1977)
[11] Schieferdecker (1959)
[12] Langheinrich (1977)
[13] Streckeisen (1980)
[14] Herrmann, Siebrasse & Könnecke (1978)
[15] Van Bemmelen (1975)
[16] Herrmann (1979)

referential statements may or may not be reliable; appellative statements may or may not be well-founded. Critical evaluation of an overall text will primarily employ the criteria appropriate to its dominant communicative function, but the criteria of other functions may also be brought into play, as any publication may, to varying degrees, serve other if not all communicative functions.

Among the texts serving a largely referential function we may distinguish the general group of textbooks, handbooks, identification books, and monographs. Textbooks primarily present facts, but in defining terms and nomenclatures, they serve both an explicative metalinguistic function and a didactic appellative function. They must therefore be evaluated according to criteria not only of referential, but also of metalinguistic and appellative functions. The referential character of handbooks and monographs is more marked, for these deal primarily with the scientific issues deemed valid at the time. Identification books do not describe just any features of fossils, rocks, or minerals, but only those which the user can employ with ease and certainty to identify the scientific names of natural objects. Since identification books place certain behavioural demands upon the user, outlining the processes he must use to name an object validly, they serve an appellative function as well.

Of the remaining predominantly referential texts, we may distinguish those which are largely empirical, characterised by assertoric statements, and those which are largely theoretical, in which hypothetical and apodictic assertions play a greater role. Since the gathering and description of nearly all empirical data presupposes theories and hypotheses, a work which imparts only empirical data without placing them in the context of some hypothetical conclusions could scarcely exist; and since every theoretical work must have and communicate its empirical basis, the boundary between these two groups cannot always be sharply drawn, and it can be difficult to assign a work to one or the other category. Still, a greater emphasis either on the empirical or the theoretical side of a text can usually be discerned.

Texts which serve a primarily metalinguistic function include dictionaries, which are largely explicative, publications on terminology and nomenclature, which are explicative and normative, and computer programs, which govern the transformation of data into computer languages.

Texts which critically evaluate the claims of other authors are specifically intentional in function. A typical case is the book review. Other such texts include those aimed primarily at judging whether given hypotheses and theories are well founded or not. This kind of criticism of hypothetical assumptions in texts which develop theories of their own becomes simply another device to emphasise the author's own referential claims as opposed to those of others.

The appellative function is especially important in texts which recommend methods, or which endeavour to apply geoscientific discoveries and methods to non-academic uses, especially to industrial purposes.

1.6 Text and discourse

1.6.1 *The structure of argumentation in geoscientific texts*

The role of geoscientific texts within the framework of scientific discourse is most adequately described if we regard science and research as activities aimed at the solution of problems;[28] contributions to this activity are made and published by individual researchers or research groups, and then become the subject of discourse in the scientific community. Such contributions to the problem-solving activity are primarily papers appearing in journals; these we have already described as being largely referential in function. We will now look at their structure in terms of their role as components of scientific discourse.

The scientific problem[29] whose solution is central to such a paper may be defined as a question demanding new definitions of a research subject, to be answered by carrying out a predetermined plan, the answer being supported by conclusions. The question is tentative, and changes in accordance with the state of scientific knowledge, for it arises out of a certain *problem situation* consisting of a complex of preconditions. These preconditions include such things as the theories and hypotheses which are assumed to hold; an explicitly formulated or previously-published and hence tacitly assumed programme of inquiry, within the framework of which the new investigation plays a certain role (e.g. as work on the central problem of the programme, work on a subsidiary problem; or the collection of facts for possible use in higher-level problem-solving). The problem situation also involves preconditions regarding, e.g. further problems generated by the solution, which may support or show the fruitfulness of the investigation; all known and relevant empirical facts; the instrumental methods and techniques available for the envisaged purpose; the accessibility of the material under investigation (e.g. in quarries, cores, from the sea floor, or from planetary space); and finally criteria of quality and value used to decide whether to investigate a given problem and how to reformulate it during the investigation.

The first thing a text should do in contributing to scientific discourse is to make the problem situation clear. The amount of detail required depends upon the author's audience: specialised journals may omit many components of the problem situation which would have to be explained in papers written for interdisciplinary purposes.

The second thing a scientific text should do is clearly formulate the problem which it intends to treat. Various kinds of *questions* underlie scientific problem solving. We distinguish firstly between open and closed problems or questions.[30]

Open questions are those for which there is no complete and exhaustive list of possible answers that could be given before commencing investigations. In particular they are questions which are answered by an indefinite number of statements, to which another statement can always be added, refining and supplementing, but not substantially changing the overall answer. The geologist who is the first to investigate an Antarctic mountain range's sedimentary and igneous rock structures, the paleontologist identifying the microfossils in a deep sea core, the geochemist (cosmochemist) analysing the trace elements in a sample of lunar rock, are all answering open questions. Clearly many works in geoscientific journals are concerned with such open questions. They constitute above all the empirical basis for geoscientific research.

Closed questions come with an exhaustive set of answers, or at least a framework for admissible answers. Such limitation obviously depends on theoretical (or hypothetical) presuppositions. The distinction between open and closed questions may therefore change with the growth of knowledge: it is relative and time-dependent. A supposedly closed question can become open, an open one closed. Within the category of closed questions, we follow Tondl[31] in distinguishing the following three main forms:

1 Whether-questions – decision-seeking questions,
2 which-questions – completion-seeking questions,
3 why-questions – explanation-seeking questions.

Decision-seeking questions suppose that a limited number of predicates are attributable to an object or class of objects. In the simplest case the answer is 'yes' or 'no', where the question concerns a single predicate, e.g. do lunar rocks contain aqueous minerals? In other cases two alternatives may be available ('whether–or questions') e.g. were the large craters on the Moon created by volcanic forces or by impact? Decision-seeking questions of the 'whether–or' type are typical of situations in which an *experimentum crucis* may decide between two competing and mutually exclusive hypotheses.[32] Thus, the trace element content may be used to decide whether a calcareous rock is of volcanic or sedimentary origin. It has been determined on the basis of their high content of rare earths that the Kaiserstuhl marbles are not sedimentary in origin, but volcanic (carbonatite).

In *completion-seeking questions* we may distinguish between those dealing with the subject and those dealing with the predicate. In the first case, we ask which particular objects permit the attribution of a given predicate, e.g. which crater structures of the Earth's surface have features suggesting that they were created by impact? In the second case we ask which predicates can be attributed to certain objects. Usually specific properties are involved, as in the question 'what minerals do the basalts collected in the Mare Imbrium of the Moon consist of?' This group also includes 'when' and 'where' questions, as in the question of 'when' a given lava was formed (to be determined radiometrically), or 'where' Precambrian ultrabasic volcanic rocks occur on the Earth's surface.

Every *explanation-seeking question* is based on a statement which is referred back to certain causes by placing it in an explanatory context. The statement may involve empirical, that is, observable facts, or it may involve states of affairs which have been inferred or assumed hypothetically. Explanation-seeking questions may be introduced by the word 'why', or by many other functionally equivalent formulae such as 'what is the cause of ...' or 'how can it be explained that ...'. The answers to explanation-seeking questions are rational justifications which in the Earth sciences usually take the form of hypothetical arguments. Thus the particular structure and mineral content of a rock are explained by suppositions regarding the conditions of its formation (which cannot be observed), or the inner and outer structure of a mountain range are explained by suppositions *re* tectonic processes in the Earth's past. Proposing explanations is the main activity of geoscientific research, and leads to the formulation of hypotheses and theories.

The various types of questions differ not only in the type and orientation of the question, but also in the way they are answered and justified. The proposing and testing of scientific explanations by inductive, abductive and deductive inferences are particularly significant for geoscientific research, as will be discussed in a later chapter.

The most important publications for scientific discourse are those which use data obtained by recognised methods for further argumentation, i.e. to answer scientifically relevant questions. The problem-solving context is crucial: like the type of question posed, it may be very general in nature. Knowledge of the geologic structure of the Earth's crust, for instance, is a basic prerequisite for developing hypotheses and theories in all branches of Earth science. The geological mapping of the Earth's surface, the simple description of the rocks in a certain region, the complete inventory of all fossils discovered in a formation, the investigation of the tectonic structure of a mountain range, are all descriptive reports which, as answers to open questions of a very general nature, are indispensable, even though the publication concerned (e.g. the notes to a geological map) may not state what further argumentational ends the data might serve. There is silent consensus that the expansion of the empirical basis through descriptive reports, e.g. through geological mapping or chemical analysis of many samples of a mineral from varying localities, is valuable even when the author cannot at the time of publication say to what supra-ordinate problem the data may relate. Accuracy and reliability of maps and data are of primary importance in evaluating such works, but at the same time the gathering and registering of data should be confined within reasonable limits: no one at present could justify geologically mapping the entire Earth's surface at a scale of 1:10 000, and there is no scientific value in adding 50 further chemical analyses of a mineral to the existing 500 ones, unless a certain problem-solving context should justify this, eliminating undesirable redundancy of data.

Works originating purely from the use of certain recognised instrumental procedures and consisting solely in reporting the application of these procedures should be evaluated particularly critically, especially those involving computer programs or physical apparatus that is newly introduced, not generally available, or difficult to run. These are usually only valuable as examples for the possible use, reliability, or accuracy of the methods.

Central to the study is the most significant contribution to scientific discourse, the argument[33] of the text. Here the states of affairs which have been described and worked up provide a basis from which to draw conclusions which should answer the questions formed at the outset of the investigation. In principle, these conclusions need justification, and must be exposed to the scrutiny of the discourse-community. Scientific discussion therefore arises because of and to the extent that claims are made which cannot be verified by immediate recourse to empirical fact.[34] Such claims, which, since they transcend empirical experience, we will call hypothetical, raise the work to the theoretical level of research, and raise the discipline to the theorizing stage; argumentation in discourse becomes a constitutive element of research, and every problem is then defined within the context of argumentation.[35]

The basic structure of geoscientific studies can be illustrated quite well with the aid of an old rhetorical scheme.[36] In the left column below we list the rhetorical terms derived from antiquity, on the right the translations used for our purposes:

A Exordium	Introduction
B Propositio	Presentation of subject
	(a) problem
	(b) methods
	(c) material

C Argumentatio	Presentation of hypothesis and reasons
(a) probatio	(a) supporting the hypothesis presented
(b) refutatio	(b) refuting other hypotheses
D Conclusio	Conclusion

The *introduction* is partly appellative in character. It sets up contact with the reader by designating the problem situation on which the investigation is based, and which makes the research significant to current scientific discourse in the discipline.

The *presentation of the subject* describes the course of the investigation and contains the background for the reasoning supporting the author's hypothesis. This section can be broken down into several parts. In presenting the *problem*, the author must give more detail than in the introduction concerning the question posited within the framework of the research situation; in particular he must refer to any previously published material on the matter. Secondly, where new empirical data will be given, the *methods* used must be explained. Thirdly, he must present the *material* to be dealt with.[37] In empirical investigations this consists of the natural objects on which the study is based, and of the data observed or measured by the described methods, along with any other relevant information from other authors. In theoretical works the material consists of states of affairs derived elsewhere and hypothetical statements whose sources are to be identified here.

In the *presentation of hypothesis and reasons* the author's argumentation proceeds from empirically derived data and other states of affairs not questioned in this context, to his hypothesis, which he regards as the answer to the initial question. Further recognised claims are generally also inserted, along with the basic data and states of affairs. The connective elements consist of inferences following the three modes of induction, abduction, and deduction (p. 80). The result is generally a complex structure which does not proceed certainly and automatically from premises to conclusion, but rather allows for further arguments which support or weaken the hypothesis, but whose number or relevance is not defined. The hypothesis developed from the basic material and supported by inferences is therefore not simply 'right' or 'wrong', but only more or less 'compelling'; we will come back to this in detail in later chapters (pp. 266, 274).

Aside from the reasons supporting the proposed hypothesis, the refutation of other suggested solutions to the problem is often important, for solutions in the Earth sciences particularly often involve considering various alternative models (p. 287).

The *conclusion* of a work should not merely repeat the claims developed and justified in the main body of the text; here above all the consequences which the solution under discussion may have for current research should be shown. This includes estimating both its internal and external value,[38] the internal value referring to the significance of the solution for research and the growth of knowledge, the external value referring to its practical relevance. Such an evaluation is often left out when the author cannot give any objective reasons for having undertaken to solve the problem.[39]

Further components of a geoscientific paper consist of the title, abstract, and references. The abstract briefly states the contents, describing the materials and methods and presenting the essentials of the results. The title indicates in even briefer form as much as possible about the subject, problem, and at times even the methods. The references fulfill two main functions.[40] Firstly, the citations indicate the sources for material and for theoretical discussions used in the context of the argument. The quality of the material and discussions need not be further tested, as it is assumed that this was done in the original publication. Secondly, the citation recognises the intellectual property of other authors. This property becomes the more valuable the more often a publication is cited and used, for this indicates that the author's work was fruitful.

The four components of the rhetorical scheme are not difficult to identify in most geoscientific papers, but their organisation and order is often modified. The textual divisions, therefore, do not always correspond with our outlined steps of argument. The introduction often contains not only a decription of the problem situation, but the formulation of the problem as well. When only the customary processes have been used to obtain data, a description of methods may often be left out. The description of the materials (empirical data and other states of affairs) on which the argument is based is not always in a special section, but is often worked into the description of the hypothesis and its justification. The hypothesis is not always developed out of the data, pre-existing subject matter, and assumptions, i.e. from the basis up, rather the theory or hypothesis is often stated first, and the empirically founded arguments then cited in its support. Most papers are intended to present and support a certain interpretation, hypothesis, or theory, without refuting any other hypotheses. Some papers, on the other hand, start off with arguments criticising or rejecting a certain hypothesis, with which they then contrast another. Less commonly, papers list a series of solutions to a problem and then weigh them. A section on conclusions is not always present.

As an illustration, finally, we will analyse the structure of argumentation of three sample geoscientific papers. All are taken from one issue of

Geologische Rundschau of the year 1979, dealing solely with works on diagenesis (the transformation of unconsolidated sediments into consolidated sedimentary rocks). Each of the three examples thus deals with a common domain, but each in a different aspect and with a different aim.

The first example is a specialised empirical study answering a narrowly-defined question on the basis of original (new) empirical findings. The second example is a specialised theoretical work using empirical data from the literature to set up a new partial hypothesis on the theory of diagenesis. The third example is a theoretical survey setting in context a complex of accepted hypotheses from the literature on the theory of diagenesis, and their justifications by empirical data. In our treatment of the three examples, the titles and subtitles of the various sections are given quotation marks.

Example 1: 'Diagenetic development and facies-dependent Na-distribution in carbonate rocks of Slovenia'[41]

'Introduction'

Presentation of the *problem-situation*: Known findings of other authors on sodium content in carbonate rocks, its dependence on facies (conditions of formation), and alteration through diagenesis.

Presentation of the *problem*: The question of the sodium content of the carbonate rocks (Upper Permian to Jurassic) in Slovenia (open question) and its origin (explanation-seeking question).

'Investigational Methods'

Presentation of the *methods* used to identify facies, mineral content, and sodium content of the rocks.

'Results'

Presentation of *material*: Findings regarding the formational situation (facies) of the limestones and dolomites investigated (shelf, coastal waters, reef, lagoon) extrapolated from their mineral content, fossils, and structures, and illustrated with sample photomicroscopy; analytically determined sodium content of the limestones and dolomites, ordered by facies; indications of diagenetic recrystallisation.

'Discussion'

Presentation of the *hypothesis and its supportive evidence*: The empirical data support the hypothesis that the sodium content of the investigated carbonate rocks in particular, and of marine carbonate rocks in general, originally increased with the salinity of the sea-water from which they were derived. Diagenetic recrystallisation under the influence of surficial waters of low salinity reduces the original sodium content.

Example 2: 'Geochemical aspects of the diagenesis of marine shales and carbonate rocks'[42]

'Introduction'

Presentation of the *problem-situation*: Conceptual distinction between early diagenesis (exchange in an open system between sea-water and unconsolidated sediment) and late diagenesis (exchange at depth between sedimentary rock and pore solutions).

'Examples of early diagenetic processes'

Presentation of the *problem* of material changes during the early diagenesis of clayey sediments, in the form of four questions:

1 What concentration gradients are found in pore water profiles of recent marine sediments?
2 What significance do compaction currents have (exit of pore water from the consolidating sedimentary layers into the sea-water)?
3 How rapidly do diffusion fronts move in weakly consolidated sediments?
4 What is the amount and directionality of the material exchanged between pore water and sea-water, with regard to certain important elements?

Since the author uses no measurements of his own, but only data published in the literature, a description of *methods* is not necessary.

Presentation of the *material* in answer to questions 1 to 3: Question 1 is answered on the basis of around 1000 pore water analyses of sediment samples from deep-sea cores; based on these data the author diagrammatically depicts what he conceives to be typical profiles of the way in which the potassium, magnesium, and calcium contents of pore solutions in clayey sediments vary with depth. The material for question 2 consists of the data available in the literature on the decreasing porosity with depth in recent marine sediments, and the rate of accumulation of clayey sediments on the ocean floor (sedimentation rate). From these data the author concludes that the amount of pore water produced by the compaction of the oceanic sediments has a negligible influence on the geochemical balance of the sea-water. The material for question 3 consists of the data in the literature on the diffusion constants of dissolved ions through the pore spaces of sediments, and of the average sedimentation rates. From these data the author concludes that the concentration gradients found in answer to question 1 indicate stationary conditions in the movement of material from the sediments into the sea-water and vice versa.

With this last statement the author has already moved on to the presentation of the *hypothesis and supportive evidence*, which is concluded in the answer to question 4. The measured concentration gradients and diffusion coefficients can, using physical and chemical laws of diffusion, be explained quantitatively by the hypothesis that (taking all oceans on an average) diffusion during the early diagenesis of clays causes transport of calcium, manganese and nickel to the sea-water from the pore water, while the pore water is at the same time enriched in potassium, magnesium, and sulphur from the sea-water. The calculated values of the material exchange are presented in a table in tonne/year and are compared with the introduction of material into the system from rivers.

'Examples of late diagenetic processes'

In this section, which we will not discuss in more detail, data from the literature on strontium and ^{18}O contents are used in support of the hypothesis of late diagenetic recrystallisation of carbonate rocks.

Example 3: 'Sandstone diagenesis in the light of recent literature'[43]

'1. Aspects and influences on diagenesis'

Introduction, in which the problem situation is presented by defining diagenesis and sketching the conditions of diagenetic change, particularly in sandstones.

Presentation of the theme as a formulation of the *problem*, namely the description and explanation of two phenomenological aspects of diagenesis in sandstones:

1 Reduction of porosity and permeability through change in grain fabric.
2 Change in mineral composition.

A description of the methods is not necessary, as all the data are taken from the literature. The material is not discussed separately, but is worked into the various sections in which the hypothesis is developed.

'*2. Mechanical compaction* (2.1 Hydrostatic pressure; 2.2 Superhydrostatic pressure). – *3. Cementation* (3.1 Pore cements, their formation and composition; 3.2 Origin and migration of the cementing substance; 3.3. Influences on cementation; 3.4 Influence of the depositional environment). – *4. Dissolution and displacement* (4.1 Dissolution of clastic minerals; 4.2 Dissolution of authigenic mineral substances; 4.3 Displacement)'.

In these sections, the *hypothesis and its supportive evidence* are presented. Following the current theories of diagenesis, the author discusses the component process of sandstone diagenesis in view of the two aspects of the problem indicated at the outset, and he describes the empirical data which are explained by and thus form the basis for these partial hypotheses.

'5. *Examples of diagenetic sequences. – 6. Progress in the last decades and future prospects. Conclusion.*'

In Section 5, examples from the literature are used to show the application of the described hypothesis for explaining diagenetic changes in sandstones from various geologic epochs. This is used to show the significance and fruitfulness of these hypotheses for current research. Section 6 sketches guidelines for future research.

1.6.2 *The collective evaluation of papers in scientific discourse*

The significance of a scientific text, particularly of a journal paper, for Earth science as a collective activity can be measured by the degree of attention paid to the publication by the community of Earth scientists.

In the worst, and probably most common case, a paper remains ignored and survives solely as a bibliographical reference in the large reference journals. In luckier cases, the text provokes reactions: it is either accepted and the information and insights recorded in it are used by later authors for further argument, or else it is regarded critically, and its results wholly or partially rejected. It then forms a background for rival interpretations or hypotheses. Devastating criticism can considerably shorten the life of a paper, particularly if this criticism comes from an authority in the discipline. Previously rejected opinions may sometimes be revived, as in the case of Alfred Wegener's hypothesis of continental drift, which, suggested in 1912, was almost universally rejected until resurrected in the 1960s (see p. 311).

The amount of attention which a text receives in the scientific community is indicated by the frequency and length of time for which it is cited in later papers. Over the years bibliographies have grown longer and longer, and geoscientific papers probably cite more works on the average than do physical and chemical publications. Solla Price (1963) suggests an average of ten citations per paper in physical and chemical publications. Fig. 1.1 shows an analysis of papers from two volumes of *Geologische Rundschau*. In 1961 most works cited 10–20 older papers, while in 1979 30–40 references were most common per paper. Most authors today must work through the results of 30–40 older publications before writing their own paper. Variations do of course occur. A particularly large number of older publications must be considered when writing general works, especially if of a regional character, such as an analysis of the structure of a mountain range, or a

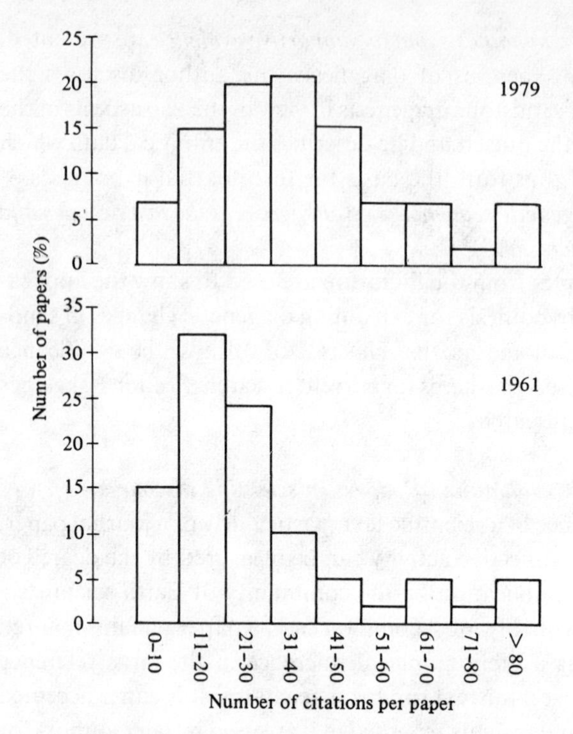

Fig. 1.1. Number of citations per paper for the years 1961 (maximum: 11–30 citations per paper), and 1979 (maximum: 31–40 citations per paper) of the Geologische Rundschau.

description of the geologic history of a continent. Fewer citations are found in papers having a greater physical or chemical character. Also, not all citations provide necessary background to the work concerned: other motives for citing papers include the desire to name influential and important scientists or befriended colleagues, to draw attention to one's own previous works, or by sheer quantity of references, to prove one's own scholarliness, knowledge, and thoroughness.[44]

It might be supposed that the considerable lengthening of the citation lists (shown between 1961 and 1979 for the papers published in the *Geologische Rundschau* and undoubtedly true of other journals as well) came about because the authors in 1979 cited firstly the works used by the authors in 1961 and then added further the publications which had appeared between 1961 and 1979. This is not the case, however, for the life-span of a paper in geoscientific discourse is astoundingly short. This is shown in Fig. 1.2, which shows the distribution of 2306 citations according to year of publication, cited in the papers of the *Geologische Rundschau* in 1979 (vol. 68). It is

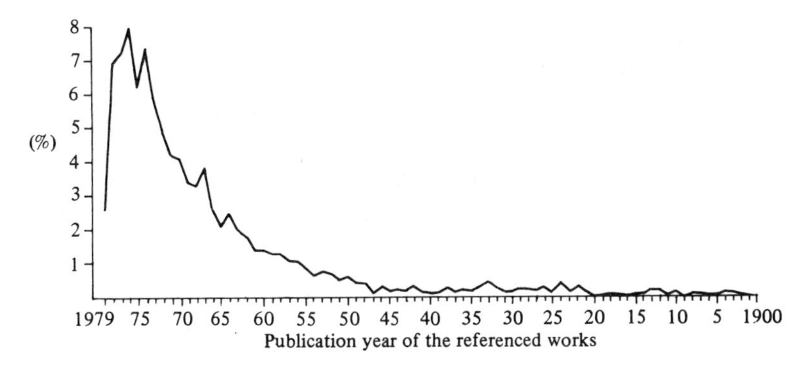

Fig. 1.2. Distribution of 2306 citations in the papers of Geologische Rundschau *(1979), shown according to their age.*

evident that as far as the scientific discourse being published in 1979 is concerned, the majority of participating authors were those whose publications were less than five years old. Works from before 1974 suffered a sharp drop in citation frequency with increasing age. This drop is roughly exponential: with every seven years increase in age, the number of citations fell by a factor of 2.[45] Half the works cited in 1979 were less than eight years old: only 15% were over 15 years old, and only 10% were over 20 years old. These figures demonstrate the extraordinarily short life-span of most publications in current geoscientific discourse. Anyone publishing a geoscientific paper today must hope that it will be 'talked about' within the next five years. Should the current trend continue, unless that paper proves in further discourse to be a fundamental contribution to the growth of knowledge, the chances that it will be cited or regarded after this time diminish rapidly. The reasons for the short-lived nature of most works and longer life-span of a very few others would be an interesting theme in the history of science, but we cannot follow it up here. We will limit ourselves to naming a few reasons for the short life-span of most texts.

Firstly, many discoveries become the general property of the scientific community by being incorporated into texts and handbooks, losing the rubric of the original author. Further, with the rapid development of new methods, data obtained by older processes lose their significance when new methods are developed which yield more precise and detailed results. Above all there is the development of new hypotheses, theories, or paradigms: observations which seemed relevant in the context of older theories, were interpreted in their light and seemed to support them, as well as explications of older theories and the conclusions deduced from them, seem uninteresting, obsolete, or simply false when a new theory takes over the stage, providing new patterns of explanation and directing attention towards phenomena which had hitherto remained obscure or ignored. Such developments are legitimate in themselves, but also carry the danger that theory-invariant facts which might be significant for current inquiry as well may remain buried in older works and become forgotten by the scientific community.

2

The language of geoscience

2.1 Introduction

The language of a science[1] in its totality is far from forming a closed system. On the one hand it is of necessity anchored in the vernacular, while on the other it overlaps to a greater or lesser extent with the languages of other sciences. Viewed historically, it is moreover subject to major changes, not only through the introduction of new terms, but also through major qualitative revisions. The interweaving both with ordinary language and with the language of other sciences is particularly evident in the case of geoscience.

Despite the growing influence of physical and chemical terms, the vernacular component is still very large as far as 'concrete' objects are concerned, objects which can be identified and described pre-scientifically. Such colloquial concepts, however, are usually used in a manner which has been normalised for geoscientific purposes, a process referred to as *terminological specification.*

This means that colloquial words are used either in a special or restricted sense (e.g. 'erratic', a transported rock fragment differing from the bedrock on which it lies, generally applied to rock fragments transported by ice; or 'fault', a fracture zone along which there has been displacement of the sides relative to one another parallel to the fracture), or vernacular elements are used to construct a new term (e.g. 'lag gravel', a residual accumulation of coarser particles from which the finer material has been blown away), or the meaning of the colloquial word is restricted by a more precise, often quantitative definition (e.g. 'clay', unconsolidated particles having a diameter of less than 0.002 mm).

Technical vocabularies[2] constitute an important source for terms in geoscientific language, a particularly large number coming from the language of mining (e.g. 'gangue', a mineral associated with an ore in a vein;

'lode', a mineral deposit consisting of a zone of veins; and many names for specific, petrographic types of rocks of specific age, such as 'greywacke', a kind of sandstone). Concepts from such technical contexts are often used in Earth science in a restricted or differently defined sense (e.g. 'marble' in the industrial sense refers to any dense limestone which will take a polish, while in the petrographic definition, 'marble' is a crystalline-granular rock produced from limestone through metamorphism.

Other elements have come from *other scientific languages*. Those from physics and chemistry[3] are used not only in geophysics and geochemistry, but with the growing use of physical and chemical laws as a basis for geoscientific explanations, they are increasingly entering the language of all the Earth sciences. Biological terms, of course, predominate in paleontology. Formulae, as components of mathematical language, are being used to an ever greater extent in geoscientific language due to the increasing introduction of physical concepts into geoscientific argument, and the increasing quantification of statements. The idea of a 'scientific language' has sometimes been used very strictly to mean a 'language of formulae'. This definition is not justified in the Earth sciences. Despite the continuing increase in the use of formulae, most geoscientific statements will always deal with qualitative matters which cannot be quantified, so that they can never be fully captured with mathematical tools. Furthermore, mathematically formulated statements can only constitute a 'language' when one, as Weizsäcker has stressed, 'lets the formula become a part of language, i.e. one uses it as it is always used conversationally.'[4]

Apart from the differences of linguistic usage due to the pragmatic communication situation (such as laboratory jargon or the style of popular scientific writing), the peculiarity of a scientific language consists above all in the *semantic structure* of its terminological vocabulary. A *term* shall designate any concept introduced by definition into the vocabulary of a science. *Terminology* refers to the totality of terms used in a discipline or in Earth science overall (e.g. petrological terminology includes all concepts used in describing rocks and in the theory of their formation). We speak of a *nomenclature* when referring to a closed set of terms dealing with a specific realm of systematically organised objects (e.g. the nomenclature of rocks, a subset of petrological terminology, consists of the systematically ordered and defined names of rocks).[5]

Along with terms deriving from the vocabulary of the given national language there are many *loan words* which derive from various languages and are incorporated into the international geoscientific vocabulary, as well as neologisms (newly formed words) which are constructed specifically for geoscientific purposes. Most loan words are of Greek or Latin origin but are used in a special and usually restricted sense. 'Facies', for example (Latin, face), designates 'a stratigraphic body as distinguished from other bodies of different appearance or composition'; 'metamorphism' (Greek, transformation) designates a 'process by which consolidated rocks are altered in composition, texture, or internal structure by pressure, heat, and the introduction of new chemical substances.' Other internationally used loan words derive from living languages, e.g. 'graben' (German), 'a block, generally long compared to its width, that has been downthrown along faults relative to the rocks on either side'; 'chernozem' (Russian), 'a black soil rich in humus and carbonates'; 'gyttja' (Swedish), 'a sapropelic black mud in which the organic matter is more or less determinable'; 'pahoehoe' (Hawaiian), 'a basaltic lava flow typified by a smooth, billowy, or ropy surface'. Loan words from English are particularly common nowadays among geoscientists internationally.

Greek and Latin words are preferred for the construction of *neologisms*, in order to ensure international applicability. Examples of this are 'diagenesis', 'process involving physical and chemical changes in sediment that converts it to consolidated rock'; 'anatexis', 'a high-temperature metamorphic process by which plutonic rock in the deeper levels of the crust is dissolved and regenerated as a magma'; 'coprolite', 'petrified excrement'. In addition to loan words and foreign words used internationally in a standard manner, the earlier regional development of some disciplines (in contrast, for example, to physics) has meant that scientific terms and even loan and foreign words may be exclusive to one country or be used differently in different countries. 'Anchimetamorphosis', for instance, 'metamorphosis at low temperatures', is used only in the German-speaking world and is incomprehensible in English. A particularly confusing example is the following: 'structure' (English) = *Textur* (German), 'larger features of a rock mass'; 'texture' (English) = *Struktur* (German), 'microscopic morphological features and size, shape, and arrangement of component particles of a rock'. Standardisation of terminologies and nomenclatures[6] therefore remains an important task of international commissions, such as the Commissions of the International Union of Geologic Sciences. The number of geoscientific terms is extraordinarily large and continues to grow. The *Glossary of Geology* (Gary *et al.*, 1977) in its third edition, 1973, contains *c.* 33 000 geoscientific terms while its second edition in 1960 contained only 17 000, a doubling in a period of around 13 years.

The development of scientific language is determined by various factors. New terms are necessitated by advances in theoretical knowledge and by the expansion of our domain of experience through technical or instrumental progress. Thus, in very recent times, geoscientific terminology has been greatly enriched by the theory of plate tectonics ('plate', 'subduction', 'convergent plate boundaries', 'divergent plate boundaries', 'obduction', 'transform faulting'[7]). The geological, petrographic, and mineralogical investigation of the Moon gave rise to the definition of many new terms, most of a descriptive kind. New terms must be defined when hitherto unknown or undifferentiated fossils, rocks, or minerals are described. Rock and mineral names are particularly revealing about the development of terminology and the principles of naming.[8]

Three principles should be observed when coining or developing terms:

1 maximum precision,
2 maximum economy, and
3 maximum practicality.

Depending on the predominant principle, various ways of coining terms present themselves. All three principles must probably be involved if the newly-proposed term is to be accepted, the most important one being that of precision, which can be measured relative to the state of theoretical knowledge. The basics of introducing, refining, and specifying new terms will be discussed in the following sections, along with their semantic organisation within word fields of certain sciences. A further chapter will deal with the non-verbal elements of geoscientific texts, that is with the role of diagrams, maps, and charts, as well as the function of schematic drawings, drawings, and photographs. These form a strikingly large component of geoscientific texts, and as it can be shown that such non-verbal notation and the principle of its analysis are often less clearly understood than is the use of verbal scientific language, we will have to discuss the basics of semiotics, the science of various notational systems.

2.2 The semantic structure of scientific propositions: identification and predication

Examining the semantic structure of a typical *proposition*[9] in science, that is, of a statement declaring that something is (was, will be, or might be) the case, we must first of all distinguish between the semantic referent of the entire sentence, and the semantic referent of its individual parts. The objective correlate of a scientific proposition independent of the conditions of its testability and acceptability, we call the *subject matter (Sachverhalt)*. 'As a rule, we only speak of a *fact* when dealing with something that is actually the case, in other words, when the sentence describing the fact is true. A state of affairs on the other hand can be something that is merely possible. A possible state of affairs can be described by any non-contradictory true or false sentence.'[10]

Scientific propositions whose truth must in principle be checked 'against reality', are, according to an old distinction, referred to as *synthetic statements*, as opposed to *analytic statements*.[11]

Analytic propositions are true independent of all empirical testing, as their predicates impart nothing new, but only elaborate what was already conceptually contained in the subject. Classic examples of such analytic propositions are generally taken from mathematics, and occur in the Earth sciences as well. Thus the statement from crystallography, 'in the cubic system there are five crystal classes', is an analytic proposition, for it can be deduced with the aid of group theory from the mathematical concept of crystal classes (a point group symmetry of finite order, containing only those symmetry operations admissible for crystals) and of the cubic crystal system (the class of all crystal classes relating to three axes of equal value placed at right angles to each other).

But the difference between analytic and synthetic propositions can be used not only to differentiate between analytic *formal sciences* (such as logic, mathematics, or formal crystallography) and synthetic *material sciences*, dealing largely with empirical states of affairs. Within such sciences as Earth science, a further distinction can be drawn between *analytically explicative* and *synthetically informative* statements. This makes the analytic–synthetic distinction a relative one. The set of analytically explicative statements includes all *definitions* generally accepted at a given stage of a science. The predicate of such a defining proposition therefore contains only something which, according to the existing conventions of linguistic usage in science, is already 'contained' in the subject of the sentence. Eco[12] has summarized this redefinition of the distinction between analytic and synthetic in the following manner: an analytic statement 'expresses that which is already "envisioned" in the code' (the accepted language of science), while a synthetic statement refers to something 'which is not envisioned in the code', and thus enriches the knowledge embodied in this language. Formerly synthetic propositions can become analytic when the predicate used to impart information becomes generally understood to be implied by the term's meaning and is therefore incorporated into the code. The boundary between the two types of propositions may therefore be historically regarded as fluid. But the process cannot be understood simply as an expansion of the analytical component, for the code may change, such that statements which at one time were considered analytic ('text book definitions') may lose this status, and in the course of the growth of scientific knowledge become eliminated as inappropriate.[13] Whether a particular statement in the concrete context of a scientific exchange is to be regarded as synthetically informative or not depends on whether alternatives to the proposed predicate are considered which would falsify the statement.

Analytically explicative statements (definitions) can also have alternatives, of course, but these are only variants in formulation, made for the sake of greater clarity, convenience, or economy of expression, and are not dependent on empirical review.

The statement found in older textbooks, that a 'granite' is invariably a rock produced by solidification of a mass 'emerging from the Earth's depths in a molten state', was in its time an analytically explicative statement, expressing only that which was implied at the time in the definition of the term 'granite'. This changed when the hypothesis arose in the 1930s and 40s, that 'granite' could also be produced during metamorphism through recrystallisation in a solid state, without involving a magma. This meant that there were now 'granites and granites', and that a statement such as 'the Bergell granite resulted from the solidification of a magma' was, given the alternative mode of formation, a synthetically informative statement, whose truth or falsehood had to be tested empirically.

Within a scientific proposition relating to a state of affairs, we distinguish between a *subject clause* and a *predicate clause*. Both can be more or less complex, and may include further attributive specifications. The subject clause represents the object 'about which' something is being stated, while the predicate clause represents 'what' is being said about this object. We further distinguish between *implicit* and an *explicit* predication as properties are attributed or their attribution withheld. Implicit predication relates to those aspects of the subject clause which not only *identify*, but also *characterise* an object, i.e. express 'what kind of object' an object is.

Only *proper names*[14] (singular terms) are used solely for purposes of identification. A proper name in its true sense cannot be used predicatively, it is used to identify a singular object. When speaking of the 'the Alps' or of 'Stromboli', what is meant is the one and only range of the Alps, or the single volcano, Stromboli. Since Earth science is again and again concerned with the description and explanation of singular phenomena, proper names play a much larger role in it than in other natural sciences such as physics or chemistry. Proper names are used when the singularity of the object concerned is to be emphasised, and when this can be done without ambiguity by use of a proper name. Identifying an object by means of a proper name and not by its properties, features, or through subsumation under a *Gattungsname* (general name or count term) is in principle a challenge to look for it where it can be observed directly in nature, or to inform oneself about it indirectly through descriptions or pictures.[15]

In contrast to proper names, *general names* (count terms, *Gattungsname*) and *substance names* (mass terms, *Stoffname*) in the subject clause of a statement are also implicitly predicating or 'characterising', since they could explicitly take the place of the predicate. The name of the ammonite genus *Perisphinctes* occurs as a subject in a statement concerning certain morphological characteristics of the shell. At the same time it is implicitly predicative, since it can appear as the predicate in a proposition identifying a particular fossil as a member of this genus. In statements on the mineral composition of the rock 'granite' (a substance name), 'granite' is the subject, but is implicitly predicative, since it can also be said that a particular mountain peak consists of 'granite'.

The difference in communicative terms between implicit and explicit predication is that using a subject clause to characterise an object serves only to clarify 'what' is being talked about, while explicit predication, by using a predicate clause, endeavours to inform the readership of something which is possibly of interest. In the case of defining propositions, explicit predication proposes a new linguistic convention or explicates an existing one.

A complex variant of subject clauses which can be used to identify a singular object is *typification.* This consists of a combination of the proper name and the general or substance name. Thus one may speak of 'the highest volcano in the Andes', or 'the only occurrence of carbonatite in the Upper Rhine Valley'. Typifications not only identify the particular object, but also characterise it by subsuming it under the general name ('volcano'), or substance name ('carbonatite').

Singular designation (identification of a particular object) must be distinguished from *universal designation* involving general and substance names. General names are also called *individuatives*, as their reference function may be '*divided*': Quine [16] speaks here of 'divided reference', to the point of enumerating the individual objects included in a general name. Substance names, on the other hand, are *cumulatives* as they can only fulfill their reference function globally. Substance names occupy a more or less intermediate position between proper and general names, though they are undoubtedly closer to the latter.

> Grammatically they are like singular terms in resisting pluralisation and articles. Semantically they are like singular terms in not dividing their reference. But semantically they do not go along with singular terms in purporting to name a unique object each.[17]

The totality of that designated by a substance name can also be expressed by a corresponding general name, in which case the reference can also be divided. For instance, instead of speaking of 'quartz', the totality of 'quartz crystals' may also be spoken of, and therefore also 'most', 'some', or 'three quartz crystals'. (Collective names or *collectives* must be distinguished from pure substance names; while behaving syntactically like substance names, semantically they are like generic names in being able to divide their reference function. Examples are 'the mineral content' of a rock, or 'the fossil content' of a stratum.)

The divided reference of generic names is regulated by quantifiers (*Quantoren*)[18] ('all', 'every', 'one', 'some', 'most', 'many'). Problems arising from generalisations will be discussed in more detail later. Here we will only stress that scientific assertions are to fulfill their reference function as precisely and specifically as possible, for the more precise the reference of a statement, the greater its testability, i.e. the easier it is to determine whether the objects actually possess the property attributed to them by the predicate. Inversely, unspecific reference ('some', 'many', 'most',) is often used as a shield against criticism.

It should be noted of general names that proper names are at times used atypically as predicates instead of solely for identification. The proper name is used as a general name, representing features which the object bearing the proper name exhibits in an 'exemplary' manner. Thus the proper name 'Stromboli' has given rise to the adjective 'strombolian', referring to a type of volcanic activity (rhythmic ejection of loose masses) typically observed at Stromboli. From the proper name 'the Alps' has been derived the term 'alpinotype', referring to a special structure (*decken* or nappe structure) first analysed in the Alps. Until recently it was common in petrography to name rocks after a *locus typicus*, that is, after a locality in which they could be observed particularly well. Today attempts are being made to replace such apparent proper names with generally defined general names, i.e. to provide general definitions for them. 'Larvikite' was the name of a rock occurring near Larvik in Norway, today called syenite; 'Kukkersite' was the name used for a bituminous rock found near Kukkers farmstead in Estonia, now considered an Ordovician oil shale; 'Suevite' ('Swabian rock') was the name for a breccia, regarded as unique, found in the Ries near Nördlingen: today this term is used generally for glassy breccias formed by meteoritic impact.

The use of proper, general and substance names to identify real phenomena permits the description of reality within the particular categorical order which lies at the basis of geoscientific observation and research. In a later chapter (p. 83) we will attempt to describe more precisely the categorical order underlying geoscientific disciplines. It differs from the categorical order of the universe of physics and chemistry in that individual objects and substances in physics, and individual objects in chemistry, are less important than in geoscience. It is for this reason that proper names, typifications, and substance names appear so frequently in geoscience texts.

When investigating the referential function of scientific expressions, we must differentiate between the extension and the intension of the expression.[19] The *extension* of an expression is the totality of objects for which the expression can be used. The intension of an expression is its meaning. The extension of an expression is unproblematic as long as we are dealing with the name for a quantity of empirically discernable and numerically countable objects. We speak in this case of *concretes*, that is, of names for factually existing concrete objects (stratovolcanoes, dunes, gas deposits). Where the subject clause of a scientific statement consists of an *abstract* (diagenesis, pseudomorphism, amphibolite facies), on the other hand, there are inherent difficulties in identifying the extension. This has led to elaborate controversies in philosophy of science and language, controversies which we cannot enter into here.[20] At least it is clear, however, that there are many concepts in scientific language which can be identified as more or less exclusively *inherent in a theory*, which are hence only very indirectly connected with concrete empirical objects, and which therefore cannot easily be determined extensionally. The concept 'subduction zone', for instance, has meaning only in the context of the plate tectonics theory: it refers to the phenomenon by which material from one plate will be thrust under the other along colliding plate boundaries (see p. 240). Which actual zones of the Earth's crust fall into this category is only explained in the context of the theory. The language of physics above all uses many such 'theoretical' objects that can be named only insofar as they are implicit in the presumed validity of entire theories. But many theory-inherent abstracts are used in geoscientific theories as well and in their application to explain real phenomena. Lorenzan and Schwemmer speak in this context of *ideata* which name 'ideal' objects such as 'mass points' in mechanics.[21] Such expressions which have no real extension in the sense of deictic discernability must be distinguished from expressions whose extensions may be empty in the here and now, but which could be extensionally fulfilled under certain conditions, e.g. 'the formation of protocontinents through major collisions' (four billion years ago) (Grieve, 1980). Such 'objects', in this case a hypothetical process and its possible result, are first constructed and then perhaps identified, e.g. by discovering corresponding structures in the geologically oldest parts of the Earth's crust. This problem of extensional determination also overlaps with that of confirmation or refutation of theoretical statements, a problem which we will discuss in detail later on (p. 266).

The *intension* of an expression, that is, its sense or meaning,[22] is identical to the conceptual features attributed to the object of a scientific statement by implicit or explicit *predication*. The totality of conceptual features is also referred to as the totality of properties. With regard to their complexity, predicates may be differentiated into *one-place* or *more-place predicates*. One-place predicates attribute a certain feature to the object designated by the concept, e.g. 'fluorite has a light refraction index of 1.434'. More-place predicates are also called relational predicates, as they define an object 'in relation' to others, e.g.: '*rutile* is less hard than *quartz*' (two-place relation); 'the deposition of *copper shales* in Thuringia took place after the deposition of the *Rotliegende* and before the deposition of the *Buntsandstein*' (three-place relation).

What is more important in our context than this differentiation (so indispensable for the logical formalisation of the semantic structure of predicates) are the principles of semantic ordering of predicates within the framework of scientific *conceptual fields*, i.e. of a collocation of terms which classify certain *fields of properties (Eigenschaftsbereiche)*.

2.3 Semantic analysis of geoscientific terms

2.3.1 *Colloquial word fields. Scientific conceptual fields*

Historically viewed, many scientific conceptual fields have evolved from colloquial word fields, while within scientific language itself, pre-theoretic and theoretic classifications can be semantically distinguished. Theoretic classifications are intensionally 'poorer', containing fewer conceptual features. Extensionally, however, they are far more precise, that is, they always permit a clear decision as to whether a particular object falls under a particular term or not.

Coseriu has summarized the differences between scientific conceptual fields and colloquial word fields in the following manner:

> Word fields provide no taxonomic classification of reality. A taxonomy usually attempts to be exhaustive, and to select the criteria for classification such that the resulting classes do not overlap. For example the discrete classes of zoology and botany rest on this principle. Word fields, on the other hand, are not discrete classes, and the classes which may be distinguished within them may fall in one word field or in two or more, that is, the language may classify part of reality through lexemes, and may leave another part unclassified.[23]

Oppositions within colloquial word fields are often not mutually exclusive: a subset of their semantic features may overlap, producing an intermediate area in which various words can refer to the same objects according to situation. Furthermore, the boundaries within word fields are not set by definitions, but may vary according to which of an open set of criteria are emphasised. Finally, the oppositions are nearly always of a qualitative nature, while in scientific conceptual fields, quantitative criteria are generally included at least as a subset of the semantic features stipulated by definition.

As examples of a pre-scientific word field, and of a pre-theoretic and theoretically-founded conceptual field, we will look at the object-class 'rocks', selecting out of this class the names 'basalt' and 'granite'. 'Basalt' was in use in German as early as the 16th century, as a pre-scientific word for homogeneous, hard, black rocks used as corner stones on houses and at street crossings. The word 'granite' has also been used in German since at least the 16th century, including at first the most varied of crystalline, granular rocks, even marble. In the 18th century its meaning in German became restricted to rocks composed of quartz, feldspar, and mica. These terms are still generally used in this vague manner by the stone-working industry today. During the development of petrography as a scientific discipline in the 19th century, the pre-scientific names were retained, but the attempt was made to restrict the name 'basalt' by definition to one rock-type differentiated from other dark-coloured, hard rocks, and the name 'granite' to one type of rock differing from similar granular crystalline rocks. This was particularly difficult, as rocks, unlike plants, animals, and minerals, form a continuous complex. Descriptive definitions of pre-theoretic concepts were produced based on qualitative features, such as the following formulation from a text book by Rosenbusch (1910).

> Basalts are rocks in part overtly, in part covertly porphyritic, almost to entirely granular, of a dark to black colour and generally lustreless lithoid appearance, with moderate to dense texture. They are characterised primarily by a combination of a basic calcium–sodium-feldspar with augite.
>
> All members of the granite family are characterised by a hypidomorph, granular structure, commonly tending to porphyritic, and by the mineral combination alkali-feldspar and quartz.

The pre-theoretic definitions of these and other rock names established 'types', but still left transitional areas between neighbouring types, where the nomenclature remained vague.

The stage of a largely theoretically-founded determination of the names 'basalt' and 'granite' has only been achieved in very recent times with definitions based on quantitative features within the framework of the modern systematic nomenclature of igneous rocks. This nomenclature will be described in a later chapter (p. 108). The boundaries of 'basalt' and 'granite' within the term field of igneous rocks have now become so precise through these definitions that it can unambiguously be decided whether a particular rock belongs to one of these classes or not.

Despite this development in petrographic terminology, pre-scientific words and pre-theoretic terms have not lost all significance. To identify a naturally occurring rock according to modern nomenclature, microscopic and chemical tests are needed which cannot always be carried out on the spot. The geologist involved in mapping therefore often uses macroscopic features to make a 'field identification' of the rock, using terms approximating the pre-theoretic or even pre-scientific names. More careful analysis may then show that the rock tentatively called a 'basalt' may actually be an 'andesite' or a 'melilithite' according to current nomenclatures, or that a tentative 'granite' may actually be a granodiorite or an alkali-granite.

2.3.2 *Synonymy, homonymy, and polysemy*[24]

In establishing the meaning of a scientific concept or term, the basic rule is that each concept should have only one meaning, while each meaning represented by a complex predicate should correspond to only one term. This ideal, however, cannot wholly be fulfilled in practice. Even scientific terms and their definitions depend as a rule on contingencies of national language, giving rise to translation problems which may cast doubt on the meanings of the terms that confront a communicating scientist. The literal meaning of many terms may be historically present, yet they acquire different meanings in the various disciplines. Sometimes competing definitions and hence competing meanings may exist even within one discipline, due to differing theoretical assumptions. Looked at semantically, these problems involve synonymy, homonymy, and polysemy of scientific terms.

We speak of *synonymy* when several words have the same meaning, e.g. the words 'eruptive rocks' and 'magmatite' as used today. Synonyms obscure the text and should be avoided, as they impede both the referential and appellative functions. International commissions such as the Commission for Mineral Names in the International Association of Mineralogy or the Stratigraphic Commission of the International Union of Geological Sciences have for years been attempting to eradicate synonymous terms and to bring about agreement between the researchers of all countries. There are various reasons for the existence of synonyms and for the tenacity with which they remain in scientific language. In most cases synonyms developed for historic reasons: old terms were replaced by new, better defined, and systematically more informative terms, yet the old ones remained in use simply through habit. Synonyms also arise when different authors assign different names to newly discovered phenomena at the same or nearly the same time. When the scientific community cannot agree unanimously on the one or the other term, whether because the question of priority cannot be resolved, or whether because different groups or schools adhere to one or the other author, the two terms may long remain in use side by side. As an example of this we cite two obviously synonymous terms from the American *Glossary of Geology* (Gary, *et al.*, 1973):

> Diaplectic Glass: (of quartz, feldspar, or other minerals) representing an amorphous phase produced by shock waves without melting.
>
> Thetomorph: a glass or glassy phase, commonly of quartz or feldspar composition, produced by solid-state alteration of an originally crystalline mineral by the action of shock waves.

We speak of *homonymy* when two or more meanings are associated with a single word form, that is, with words which are written or sound alike. These words are homonyms, and are taken to be separate words even though they cannot be differentiated by form. The meaning of a homonym can only be ascertained from its context. Since homonyms do not involve differences in meaning within one word, but rather separate words, they should be listed in dictionaries under separate entries.

We speak of *polysemy* when a single word has acquired various meanings through an extension of its range of meaning. This occurs especially by metaphoric extension, based on a similarity between its primary referent (the 'original' meaning) and its secondary referent (the 'metaphoric' meaning). The word 'milk', for instance, is metaphorically extended in meaning when used in the term 'glacier milk' for the milk-coloured meltwater emanating from a glacier. Here polysemy or multiple meanings are attached to the word 'milk'. Since the various meanings involved in polysemy all grew by extension of a word's original meaning, a relationship continues to exist between a word's different meanings. In dictionaries, the various meanings should therefore appear under a single entry, in contrast to homonyms. Another example of polysemy is the word 'hardness'. Three meanings for this word may be distinguished according to context:

'hardness' in the context of:

1. A human, the state or the police (psychology).
2. Water (in chemistry).
3. Solid bodies or plastic deformation (in physics or mineralogy).

In homonyms and polysemes the author's meaning emerges from context. Where this is not the case and misunderstandings are likely, polysemy or homonymy may be avoided by quoting a context in which the word is used as intended or by citing situations where the word is used in its undesired meaning. The third meaning of the word 'hardness' listed above might be stressed by quoting an example of this meaning such as 'diamond is the hardest mineral', or by citing the undesired meaning, as, 'ground water in karstic areas is hard.'

Since unambiguous scientific language is desirable, homonyms and polysemes may at first glance appear counter-productive, for each scientific term should have one and only one meaning, independent of context. Yet there is a reason for distinguishing between homonyms and polysemes. Homonyms should be avoided at all costs and eliminated entirely when possible, for words that are written alike but have different meanings can only lead to confusion, contributing nothing to information and communication. The case with polysemy is different, however. The use of words whose meaning has developed metaphorically from an original colloquial meaning may contribute to understanding and communication. Words such as 'mantle', 'glacier milk', 'vein gneiss', or 'ichor' (originally 'blood' or 'sweat', now a term for the mobile fluid phase in the deepest regions of the Earth's crust) are informative precisely because of the relationship between their scientific and colloquial meanings.

Homonyms and polysemes can only be spoken of in the true sense within the context of a single language, i.e. in relation to word forms in English, German, French, Russian, or another language. Scientific language contains many terms, however, which occur in the same form as loan words in all these languages. They should be used with the same meaning in all these languages, but this is not always the case. The plurality of the meanings of such terms developed during the time when, as already indicated above, the geoscientific disciplines were developing more or less independently within territorial boundaries across which there was still very little exchange.

We have already mentioned the example of the several meanings of the German and English terms 'structure' (*Struktur*) and 'texture' (*Textur*). Another example can be cited here. According to the American *Glossary of Geology* (Gary *et al.*, 1973), 'diabase' in the United States means an intrusive rock of plagioclase with ophitic texture (plagioclase crystals wholly or partially surrounded by pyroxine), whereas 'diabase' in England means an intrusive rock of plagioclase or pyroxene in which these minerals have been heavily altered.

2.3.3 *Types of definitions and classification of terms*

The *definition* of a term consists linguistically of two portions: the term to be defined (*definiendum*) and the complex predicate clause, with whose aid the meaning of the *definiendum* is established (*definiens*).[25]

A scientific definition is always in the form of a proposition in which the *definiendum* forms the subject clause and the *definiens* the predicate clause such that the predicate clause is semantically more explicit, that is, it lays out in separate predicates what is implied by convention in the subject clause. This presupposes that the individual components of the *definiens* are already semantically determined. By asking for the definitions of the predicates in the *definiens*, we arrive eventually at very elementary predicates, which simultaneously form the undoubted basis of scientific language, although their meaning can in principle be questioned. This 'foundation' in the construction of scientific terminology is a much-debated problem in philosophy of science which we will not enter into here.[26] Suffice it to say that the *ostensive* definitions often regarded as this 'foundation', by which the meaning of a term is supposedly explained by reference to the object for which it stands, do not constitute definitions in our sense, as they do not define the meaning intensionally through proposition, but only extensionally through the use of 'pertinent' examples. Such exemplification may be helpful in a learning situation, but does not replace a precise verbal definition.

A form of definition common in classical logic is the *definition according to* genus proximum *and* differentia specifica.[27] This is widespread in the various geoscientific disciplines, especially in paleontology (p. 118). The term to be defined is identified as an element of a class, indicated by the *genus proximum*. From this class the element is then differentiated by a specific semantic feature, the *differentia specifica*. We may name as an example:

Quartz porphyry: an anchimetamorphic rhyolite.

The class (*genus proximum*) is identified by the term 'rhyolite', which we assume to have been previously defined. The specific difference is indicated by the term 'anchimetamorphic', which must also have been predefined. In this example we see that two general tendencies may be distinguished in the way scientific terms are defined: one tendency towards general terms, and the other towards more specific ones.

This brings us to a *hierarchy of scientific conceptual fields* whose definitions are related to one another and which represent various levels of abstraction from 'lower' to 'higher'. The more abstract concepts are 'poorer' in intension, in that they are defined by a lesser number of predicates, but in extension they are 'richer', in that they refer to a greater number of objects. In more specialised concepts, matters are the reverse. The subsumation of a special concept under a more general umbrella concept is called hyponomy. All elements of one class falling under a common concept are regarded semantically as its hyponyms. Concepts at the same level of abstraction are *cohyponyms* with regard to each other, and can be grouped as a unit under the given class concept.[28] The feature which is picked out as constituting the *differentia specifica* of a particular cohyponym must be the feature which distinguishes it unambiguously from all other cohyponyms.

As an example of a fairly small conceptual field in Earth science characterised by semantic hyponymia, we shall give the basic classification of rocks. This involves a relatively high level of abstraction:

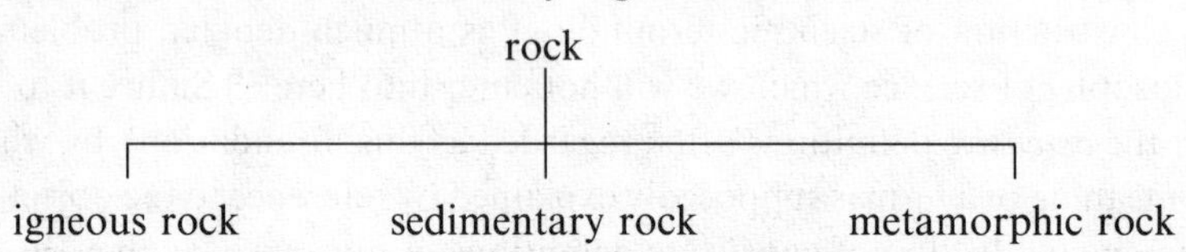

This diagram shows that the conceptual field represented by the cohyponyms consists of elements standing in *semantic opposition*[29] to each other. A rock which is igneous cannot at the same time be sedimentary or metamorphic.

A special case of semantic opposition involves *complementary* cohyponyms. Here we are concerned with hyponymic pairs of concepts semantically characterised by the fact that the negation of one concept implies the assertion of the other and vice versa. This is often formally expressed by an appropriate prefix (e.g. 'isotropic' v. 'anisotropic').

To be distinguished from this are *antonymic* oppositions such as 'large' v. 'small', which permit graduation and hence a continuous transition between the extremes. Antonymic semantic fields are common in the vernacular; in scientific conceptual fields they are usually replaced by quantitative graduations.

This brings us to a classification of different types of semantic relationships between cohyponyms of a scientific conceptual field, a classification which has been favoured from the first in philosophy of science, since, when viewed 'dynamically', it marks one of the decisive factors in the development of modern scientific languages: the transition from qualitative and comparative to quantitative classifications of terms.[30]

(a) *Qualitative:* cohyponyms not listed in any special order ('either (a), (b), or (c)').

(b) *Comparative:* cohyponyms ordered, but the 'distance' between neighbouring terms not quantitatively defined ('more (a) than (b); as much (a) as (b); less (a) than (b)').

(c) *Quantitative:* cohyponyms specifically ordered, and distances between them definitely determined by corresponding units of measurement.

The transition from qualitative to quantitative fields can be illustrated by the transition from vernacular word fields to scientific conceptual fields. One need only think of the transformation of the qualitative field of colour names into the quantitative field of colours ordered according to their wavelength. Even at the level of qualitative classification, the vernacular and scientific languages differ in that each opposition in the latter must be *exclusive*, that is, must be defined in each case by mutually exclusive semantic features, so that each possible element must be either (a) or (b) or (c).

Definitions and classifications based on qualitative features are unusually widespread in geoscience, and include the nomenclatures and systems of minerals, rocks, and fossils, to be discussed in a later section. In the diagram shown above for the definition of the major rock groups, the distinguishing qualitative feature consisted of the manner of formation, deduced from structural characteristics (i.e. a hypothetically inferred state of affairs). Comparatively-ordered conceptual fields are less common. The division of minerals according to Moh's Scale of Hardness is a comparative organisation, as are the geologic epochs, which in the days before absolute radiometric dating were defined as the following series of time divisions of unknown duration: Precambrian, Cambrian, Ordovician, Silurian, Devonian, Carboniferous, Permian, Triassic, Jurassic, Cretaceous, Tertiary, Quaternary.

With improved research procedures for producing quantitatively describable results, the importance of quantitatively ordered concepts in the geoscientific disciplines has increased. One example of this is the definition of the names of clastic sediments. The concepts 'clay', 'silt', and 'gravel' are cohyponyms under the overall term 'clastic sediment', differentiated by quantitative limits on the average diameter of their particles:

Clastic sediments

Clay	Silt	Sand	Gravel
(<0.002 mm)	(0.002–0.2 mm)	(0.2–2 mm)	(>2 mm)

The definition of a concept according to *genus proximum* and *differentia specifica* is only one of many possible types of definition. In physics, for example, we have many *functional definitions* where the concept to be defined is shown to be a function of other previously introduced concepts: 'velocity' as a function of 'distance' and 'time'; 'thermal conductivity' as a function of 'temperature gradient' and 'heat flow'. Such definitions are closely related to the formulation of corresponding laws, in which the relationships are generalised nomologically. In Earth science functional definitions are most often associated with states of affairs which are interpreted with the aid of physical laws. The 'permeability' of a rock, for instance, is defined as the proportion of the pressure gradient causing the flow of gases or fluids through the pores of a rock, to the quantity of gas or fluid streaming through a given unit of rock per time interval. Functional definitions occur in other contexts as well, e.g. in the '*N*/*S* quotient' important in the formation and development of soils, defined as the proportion of the average annual precipitation to the average atmospheric saturation deficit with respect to water in a given region.

Where concepts are defined by listing the experimental procedure used to produce them, we speak of operational definitions ('the manner of measuring a quantity defines it': Einstein). As in any natural science, these play an increasing role in those geoscientific disciplines using experimental means to determine features or behaviour.[31] One example consists of the use of the physical concept 'hardness' in mineralogy. The usual definition of the term is 'resistance to plastic deformation'. Several kinds of operationally defined concepts of 'hardness' may be distinguished depending on the procedure used. These consist in mineralogy mainly of the two following types:

Scratch hardness: scratching with certain standard minerals classifies the investigated mineral within a comparatively ordered 'scale of hardness'.

Microhardness: (after Vickers): measurement of the (microscopic) size of an indentation produced by a diamond pyramid of specific size and shape and employed with a given stress upon a flat surface of the investigated body indicates a quantitative unit of hardness.

Individual fields of Earth science as well as trends in research can be characterised according to the dominating or desired type of definition. Where the description of real objects is involved, the desired type of definition is always hyponymous. Progress in scientific inquiry is achieved, for instance, whenever simple definitions of the names of unrelated entities (e.g. the former names of minerals) can be organized into a hierarchic system. This is how systems of objects (e.g. of fossil life forms in palaeontology) or of matter (e.g. minerals and rocks) or of processes (e.g. of metamorphism) were constructed. We will discuss these in more detail in a subsequent chapter. In developing such systems the general tendency is to order the cohyponyms as quantitatively as possible.

As geologic phenomena are increasingly reduced to elementary principles of physical and chemical processes, the significance of functional definitions increases. This is true not only in geophysics and geochemistry, but in other disciplines as well. In paleontology, for instance, forms which had simply been morphologically described are now related to their biological functions, environmental conditions, and so on.

Operational definitions inevitably play a role where empirical facts are established not simply in their 'natural' state (phenomenological observation, see p. 126), but where the behaviour of objects is investigated under artificially controlled conditions (experimental observation, see p. 128). The establishment of physical and chemical parameters by means of recognised and reliable standard methods and the identification of technically important material characteristics through standardised tests imply operational definitions.

Finally we come to the distinction commonly used in the philosophical tradition between real and nominal definitions. *Real definitions* were understood as propositions on the 'nature' of an object, whereas *nominal definitions* implied the reduction of a concept to other, already known or predefined concepts. The nominal definition was taken to be a purely linguistic convention, by which a term was chosen more or less arbitrarily and could be replaced with great difficulty by a differently defined term. In the case of real definitions, on the other hand, the very 'nature of the object' should prevent such substitutions.[32] The distinction between real and nominal definitions is unimportant in semantic analysis. Semantically we deal only with nominal definitions, that is, with the conventional establishment of the meaning of a concept through a complex defining clause. To what extent we can proceed 'arbitrarily' in doing this, and in what areas any 'objective' or 'real' ordering of the objects exists to which our organisation would have to correspond is a matter of ontology. Indeed, when defining terms for some entities, such as mineral, plant, or animal species, we do not necessarily have a free hand but must somehow choose definitions corresponding to a 'natural' order. We will come back to this problem in the context of the so-called 'natural' and 'artificial' systems of classification (p. 95).

2.4 Non-verbal sign systems in Earth science

2.4.1 *Semiotic basics*

One of the peculiarities of geoscientific texts is that they are expressed not only in verbal language, but also to a great extent contain non-verbal representations such as photographs, drawings, graphs, and maps. The relationship of the non-verbal parts of the text to the verbal varies considerably, from the 'illustrated' paper where verbal expressions are explained, clarified, or merely adorned by non-verbal means, to the geologic map where information is expressed almost entirely in non-verbal form and which, given a knowledge of the legend, can be 'read' as if it were a verbal text without any need for transcription into verbal statements. An analysis of geoscientific texts would be incomplete if it were limited to an investigation of language in its narrower sense. We must therefore examine the special functions and structures of the non-verbal representational forms used in geoscientific texts. Clearly, communication between scientists and in wider contexts cannot do without non-verbal codes, as they help express something which can neither be directly expressed verbally, nor completely transcribed into verbal language.

In expanding into the area of non-verbal systems of signs, our investigation has passed from linguistics into the wider field of semiotics,[33] the subject of which is the investigation of the types and functions of all signs appearing within the communication process. Since in our context we will deal only with signs apprehended by the eye, we will confine ourselves to the semiotics of visual communication, that is, to codes which use visual, non-verbal signs. Non-verbal representations in principle fulfill the same function in the communication process as do verbal expressions. They each do this in different manners, however, so that in any given instance it may be asked what reason or advantage is involved in expressing the state of affairs non-verbally.

For non-verbal representations, it is again the *referential function* which is the most important, as the main purpose is to transmit information from the author to the reader. The character of the information conveyed varies according to the method of representation. A photograph imparts more vague, less interpeted information than a drawing, whereas a graph transmits abstract data and hypothetical inferences. Maps present simultaneous overviews and interpret what has been observed at many scattered points on the Earth's surface, etc.

Non-verbal representations may often carry out a beneficial *metalinguistic function*, since the definition of some geoscientific terms can be conveyed more easily and understandably with graphic than with verbal means. The diagram shown in Fig. 2.10, for instance, is used to define the names of sedimentary rocks composed of the three components 'sand', 'clay', and 'limestone'. The meaning of the term 'sandy, very clayey marl', obvious from the graphic representation, can only very cumbersomely be expressed in words. The definitions shown in Fig. 4.2 for the names of igneous rocks are another example.

An *intentional function* is expressed when graphic representations are used to sketch a hypothetical state of affairs in such a way as to overcome any remaining explanatory difficulties and to express a personal opinion more effectively than can be done in words. Fig. 2.3, showing one author's view on the explosive formation of the Ries near Nördlingen, could not be expressed so emphatically in words.

Non-verbal representations are often better suited than verbal texts to present certain states of affairs in a clear, easily understood, and apprehensible manner; they meet the reader halfway, and thus fulfill an *appellative function*. It is for this reason that such representations play a major role in didactic texts, and why they are always used when the author is interested in presenting his findings clearly to the reader, in attracting or maintaining the reader's attention, or in giving the reader instructions for performing certain operations.

Viewed semantically, i.e. according to their relationship to the referent, Peirce has distinguished three types of signs:[34]

Icon (pictographic sign) a sign similar to the referent, e.g. a photograph of a fossil or a drawing of the stratification of a cliff-face.

Index (indicative sign): a sign which is somehow related (e.g. causally), to the referent but which does not necessarily resemble it. A groove on a rock may be interpreted as the index of a glacier.

Symbol (conventional sign): a sign which neither resembles nor is in any other way connected with the referent, but is linked with it only by convention, e.g. the blue colour as a symbol for the Jurassic formation on geological maps.

Signs can only function as icons, indices, or symbols when they are understood as such on the basis of established coding and decoding rules. In particular, the representational character of signs is never 'purely' given, but is only a graduated approximation, established through interpretation.[35] When, for instance, a photograph is called an icon, because it resembles the ammonite or the mountain range depicted, this must be done with qualifications. A photograph never simply reproduces an object but resembles it only in certain aspects, being unlike it in others. Even the most perfect icon does not have all the properties of the natural object represented, but rather reproduces only some of the conditions expected by the scientifically trained or untrained perception; it selects certain stimuli to affect the viewer, permitting him to build up a certain structure having the same meaning as the perceptual event to which the sign relates.

The relationship between the components of the representation and the real event depicted are not governed simply by resemblence, but by convention.[36] It should therefore be noted that convention is involved not only in the meaning of symbols, but also in that of other iconic signs. Representational means can therefore be ranked according to their degree of iconicity. At the top of the list stands the 'faithful reproduction', such as a 3-dimensional model of a mountain landscape fashioned out of plastic, while at the bottom come symbols agreed on simply by convention, such as the crosses used to indicate granite on geological maps.

As in the case of verbal scientific language, the meaning of a non-verbal sign can either be bound by explicit definition, or governed by context with other signs. Following Bertin (1967), two types of sign systems may be distinguished. In *monosemantic signs and sign systems* the meaning of each sign is clearly defined. On the other hand the meaning of *polysemantic signs and sign systems* depends on the context of the signs; it allows room for interpretation, and is dependent on the attitude of the author or reader.[37] The meaning of polysemantic signs can thus be debated in principle.

Non-verbal sign systems occurring in geoscientific texts can be classified in the following types, differing in their degree of iconicity and polysemy:

1 photographs
2 drawings
3 graphs
4 diagrams
5 charts
6 maps

2.4.2 *Photographs*

Apart from multi-dimensional representational media such as films and models, neither of which occur in normal geoscientific texts, the photograph permits the most complete iconic representation of perceptual events. Despite all similarity to the reality represented, however, a photograph allows considerable interpretation, that is, it can be read in various ways, depending on the reader's previous knowledge or the author's verbal explanations. The individual sign elements, which in a black-and-white photograph occur as areas of various shades of grey (sometimes reproduced as various grids of dots), are polysemantic. Their meaning, that is their relationship to elements of the perceived world, depend on their context within the photograph of which they are a part. The photograph of a rocky highland, for instance, consists of shades of grey within a limited scale. None of these grey shades corresponds to what we call 'reality', namely our visual impression of the rock formations comprising the photographed mountains. The range of the scale depends on lighting and other conditions present at the moment the photograph was taken, as well as on photographic techniques, that is, on the lighting, grain of the film, method of developing the negative, and printing of the positive. Even the proportions between height and width of the objects shown in the picture are not dependent only on the 'real' dimensions, but also on the focal distance and aperture of the optic system. What is 'seen' on the photograph is to a large extent that which the author desires the reader to see, i.e. a specific selection of perceptual impressions conveyed to the reader by the author's explanations and by the reader's own previous knowledge. The author cannot, however control *selectivity* of impressions in a photograph to the same extent as he can in a drawing or in a verbal description, unless he *retouches*, and thereby *falsifies* the photograph.

The reader's previous knowledge is of crucial importance. A reader lacking in background will often find photographs of scientific objects incomprehensible, as an adequate understanding of the pictures presupposes the ability to differentiate relevant from irrelevant features on the basis of broad previous knowledge, and hence the ability to assimilate that information which the photograph is meant to convey.

Photographs may appear in geoscientific texts either as the objects of research, or as iconic supplements to verbal presentation. As objects of research, they substitute for the 'real' objects, and qualitative observations and quantitative measurements are carried out on them which would otherwise be performed on the real object. The photograph fulfills this function of replacing objects when the objects are too small or far away, or when they cannot be directly observed from any satisfactory standpoint. Topographic and even geological maps are prepared from photographs taken high up of the Earth, Moon, or other planets. Unique events, such as meteoric trajectories, atmospheric phenomena, or volcanic outbursts are photographically recorded; photographs of tiny objects taken with the light microscope or electron microscope are used for qualitative observations, counts, and various types of measurements. Due to the high iconicity of photographic representation, photographs can serve as direct objects of research if it has been ensured that interpretive leeway has been limited as far as possible. Such limitation (reducing polysemy) is achieved by verbally explaining all the relevant photographic conditions, permitting an exact reconstruction of the situation to which the photographic reproduction relates. Photographs made by the Apollo Mission of the Moon's surface, for instance, are accompanied by a specification of the exact position of the Sun, so that shadows may be used to infer the shape of objects, their surface composition, and so on. The more vague such explanations are, and the more blurred the relevant details on the photograph, the less suitable it is as an object of research.[38]

As iconic supplements to verbal presentations, photographs represent individual geoscientific states of affairs localised in time and space. Various photographic functions may be distinguished:

1 Photographs are useful to characterise an object, so that it can be recognised or found by anyone. A particular spot in the field can be unequivocably recorded photographically; a particular phenomenon (e.g. sandripples on a dune surface) can be depicted photographically in such a way that the reader will understand what is being discussed much more clearly than if words had been used.
2 Photography is useful to depict a singular object with many of its

visually perceptible features where this would be difficult or impossible verbally, or where the author does not wish to bind himself to a verbal description whose selectivity inevitably causes a certain interpretive perspective.

3 The meaning of a general term can easily be made clear using a photograph, by pointing out the given concept's characteristics on the object. A photograph of Etna, for instance, can be used to point out the characteristic features of all volcanoes belonging to the class of 'stratovolcanoes'; the morphological features of a fossil mussel species can be shown more clearly through a sample photograph than with words.

4 It is possible to use photographs to represent the variability of shapes or structures in objects belonging to one class, e.g. the different shapes of volcanic bombs.

As to their referential function, photographs are superfluous where they neither function as research objects nor, as descriptive media, say anything more than or other than what can be said in words or through graphic representations. When, for instance, in a report on the components of an ocean sediment, the presence of metallic spherules of a certain diameter are described, it is redundant to accompany this description with a microphotograph showing only a few blackened circles. Photographs are furthermore not useful when the leeway for possible interpretations is left too large, either due to poor technical quality of the photograph, or because the verbal explanations needed to limit polysemy are lacking.

2.4.3 *Drawings*

Drawings are less iconic than photographs, with many details of the concrete object left out. This straight away eliminates a series of readings which, though possible, are irrelevant in view of the state of affairs which the author wishes to portray. The author uses the medium of the drawing instead of a photograph when he wishes to suggest a certain, namely his own, interpretation of perceived objects. In contrast to the 'documentary' character of photography, which allows diverse interpretations (including the right one), the reader must always remember that the drawing falsifies a state of affairs right from the start.

In contrast to graphic representations,[39] drawings always leave a certain amount of room for interpretation. Drawings, like photographs, belong to the polysemantic sign system. What a certain line or area of a particular shade means, i.e. what the author wished to say about perceived objects with these signs, is not explicitly stated, but can only emerge from the context of the entire picture or the verbal explanations. A drawing can therefore also be interpreted in ways other than that intended by the author. Like photographs, drawings in the narrower sense refer to singular objects localized in time and space. Because of their reductive character, however, they are better suited than photographs to express the 'typical', i.e. to designate features of an individual object which indicate either its membership in a certain class of objects, or else its particular, aberrant character.

The *schematic drawing* frequently used in geoscientific texts is a drawing which has been reduced to few elements. The greater the reduction, the further removed it is from an empirical description of a singular fact, and the closer it is to a statement about a generalised or hypothetical state of affairs.

1 The schematic drawing may be used to generalise in representing an idealised object with characteristic properties on the basis of which actual individual objects may be assigned to a particular class. Fig. 2.1 shows an example of a schematic drawing of a conical folding system.
2 The schematic drawing is used to show a hypothetical state of affairs. Fig. 2.2 is a schematic representation of the hypothetical underground stratigraphy along a section from the northern edge of the Alps to the Bohemian Massiv in Lower Austria. Fig. 2.3 shows a (now outdated) hypothesis on the formation of the Nördlinger Ries through the explosive expansion of gases, imagined to have issued from a magma body which had penetrated at depth.

2.4.4 *Graphic representations*

Representational Methods and Functions

Graphic representations, unlike photographs, drawings, and schematic drawings are *monosemantic* systems as analysed in detail by Bertin (1967). The meaning of the signs from which they are formed is unambiguously established with no room for varying interpretations, as long as the signs are completely explained.

The essential difference between the transfer of information through verbal terms and graphic signs is that verbal and graphic representations are perceived by the reader in different ways. The understanding of a coherent verbal report depends on the sequential perception of a chronological succession of signs. Graphic representation, on the other hand, permits simultaneous perception of relationships between signs within the two dimensions of a plane. By varying the position and composition of the graphic signs, graphic representations permit information to be grasped and processed 'at a glance' which with verbal means could only be under-

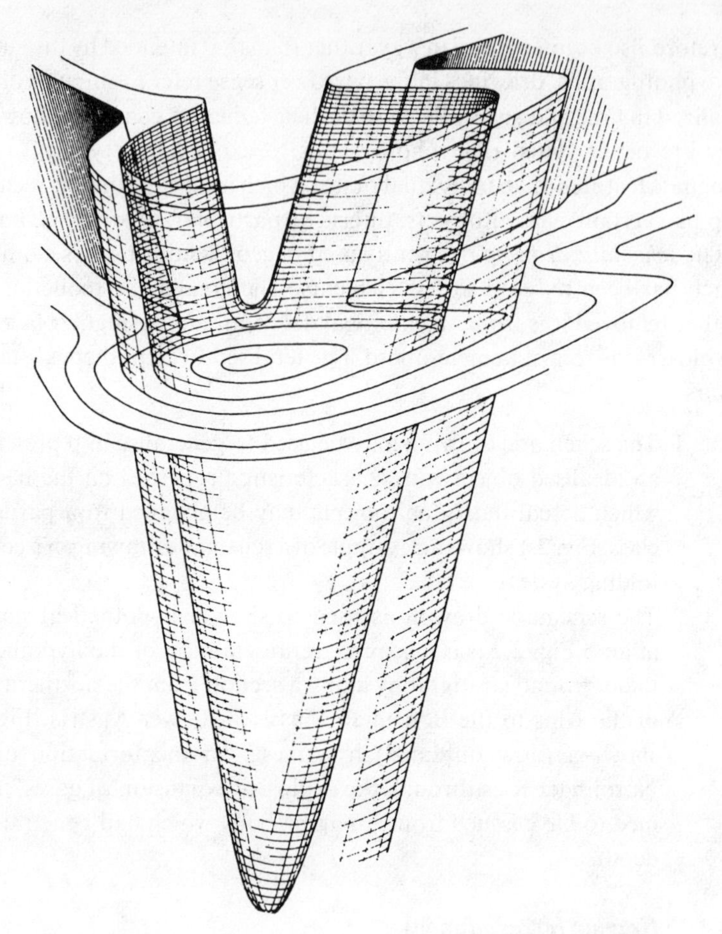

Fig. 2.1. Schematic drawing of a 'conical fold' observed in Norway. (Hansen, 1971, p. 78.)

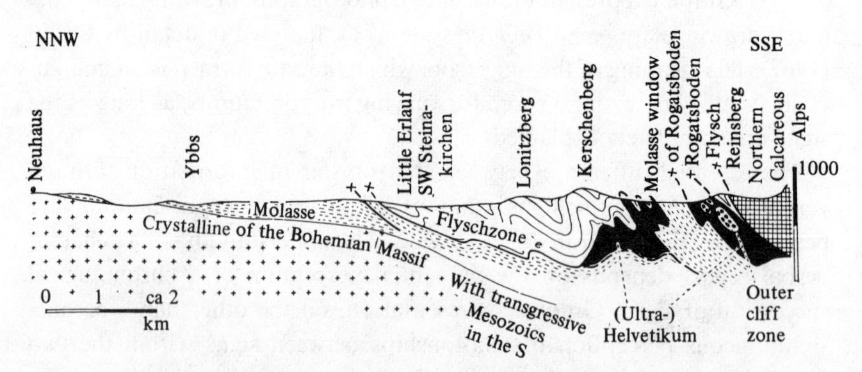

Fig. 2.2. Schematic profile from the northern edge of the Alps to the Bohemian Massif in Lower Austria (after Prey). (Gwinner, 1971, p. 341.)

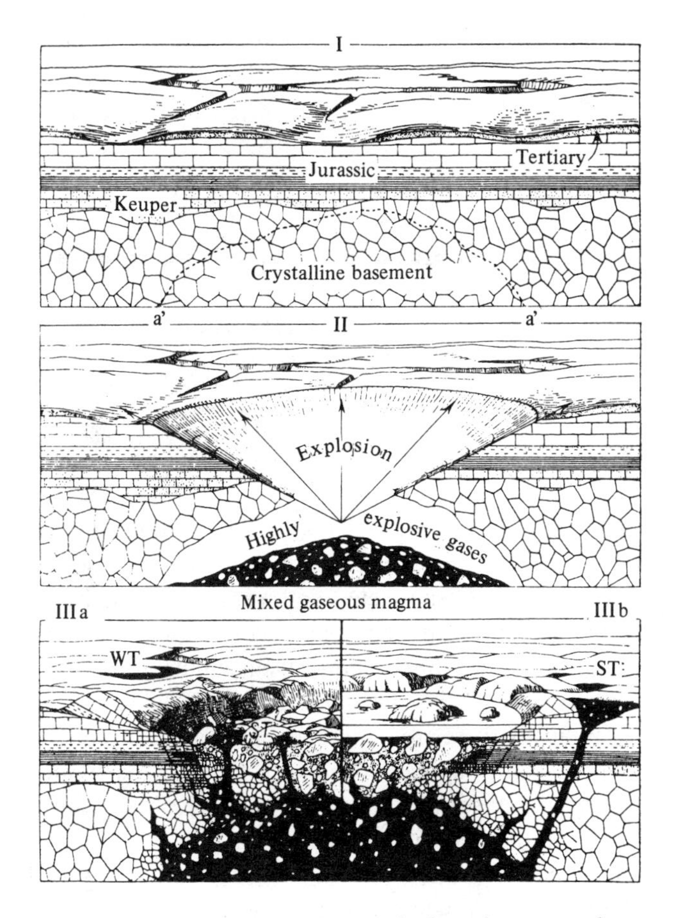

Fig. 2.3. Schematic representation of a hypothesis according to which the Nördlinger Ries was supposedly formed by the explosive expansion of gases emanating from a magma body which had penetrated at depth. (Schuster, 1926.)

stood 'bit by bit'. This is why graphically displayed information is easier to store in the memory. The instantaneous apprehension of such information, however, requires that no more than three *components* or complexes of the states of affairs be graphically represented. With more than three components, the information must either be presented as a *sequence* of several graphs, each to be instantaneously apprehensible (Fig. 2.7), or else through a complex *figuration*, which is no longer instantaneously understandable, but must be read in a series of stages. In order to read a graphic representation, it is necessary to make assignments which explicitly and unequivocably establish what the various graphic devices and their variations mean. Each complex of the state of affairs (the components) must correspond to one and

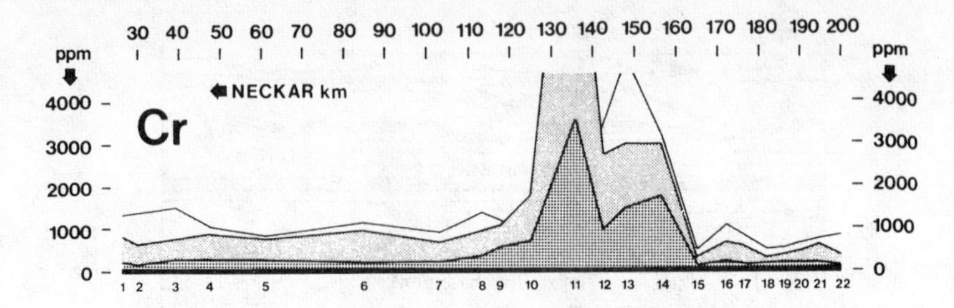

Fig. 2.4. Diagram of the chromium content (mg Cr/kg = ppm) of the clay sediments of the Neckar River along over 200 km of its course from Dezisau (10 km east of Stuttgart, point 22) to Heidelberg (point 1). The patterns indicate the fraction contents of various grain sizes (diameter of clay particles in micrometres); light grey: 0.06–0.2, dark grey: 0.2–0.6; dotted: 0.6–2.0 (Förstner & Müller, 1974, p. 120.)

only one *variable of representation*. Thus in Fig. 2.4, the component 'distance from the Neckar mouth at Heidelberg' is assigned to the representational variable '*x*-axis'. Components and variables must be organised analogously: the *elements* of the components correspond to the *levels* of the representational variables. The component 'chromium' in Fig. 2.4 is divided into three elements of the chromium content of different grain size functions. Graphically these are represented by gradations in the variable 'pattern' (shading of the area below the line). Here we must note that the available variables only permit the portrayal of a limited number of elements according to the kind of variable.

Semantic explanations ensure that a graphic representation can be read. Following Bertin, we distinguish external and internal identifications as two modes of semantic explication.

The graph is *externally identified* by its *title*. This must contain, firstly, the *invariant*, i.e. the complete characterisation of the entire state of affairs portrayed. This characterisation must be invariant with respect to all given concepts. Secondly, it must contain the individual portrayed components of this overall state of affairs.[40] Examples of external identification are given in the titles of Fig. 2.4 to 2.10.

Internal identification refers to the assignment of the individual components and their elements to the variously graded representational variables. This assignment is brought about largely through the labelling of the coordinate axes (if two components are to be represented via the dimensions of the plane) and the explication of additional variables by means of a *legend*.

The representational device of graphic systems is designated by Bertin in its most general sense as a *spot (Fleck)*. This first of all establishes a

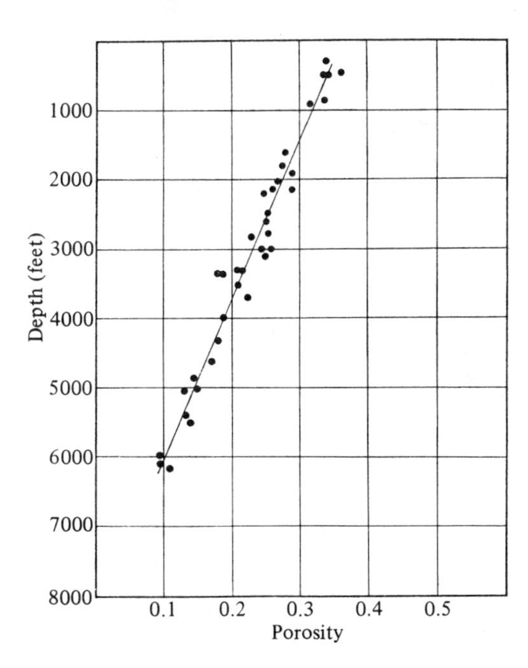

Fig. 2.5. Diagram of the porosity of clayey rocks from a deep core in Venezuela, after Hedberg (1936). (von Engelhardt, 1973, p. 288.)

relationship between the co-ordinates given through the two dimensions of the plane. Besides marking a particular point or area of the plane, the spot can also convey various gradations of information through its *size, light value, pattern, direction, shape* and *colour*. Including the two dimensions of the plane, graphic systems thus have more than eight visual variables to represent the components and their elements. The gradations of these variables, e.g. of direction or light value, are more or less limited, however, especially if the information is to be read quickly and unambiguously. Under these conditions, maximum differentiation is provided by coordinate axes, which for this reason must be regarded as the primary device of graphic representation.

Finally we must note that the spot can be used in three ways within the plane: as a point, line or area.

As a *point*, the meaning of the spot is independent of the area it takes up on the paper. Examples of such spots are representations on maps of trigonometric points, earthquake epicentres, or the points in Figs. 2.5, 2.6, 2.8, and 2.9.

As a *line*, the significance of the spot is independent of its breadth. Examples are contour lines and boundaries between the distribution areas of geological formations on a map, and the curves in Figs. 2.4, 2.7, and 2.8.

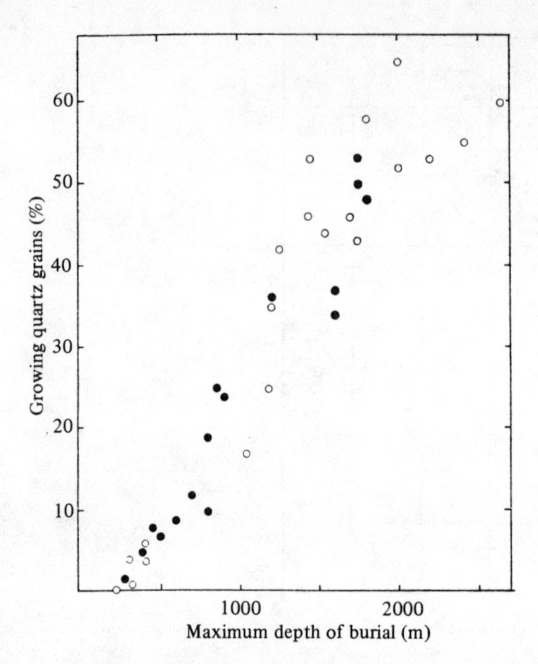

Fig. 2.6. Diagram of the percentage of quartz grains which continued to grow during diagenesis in oil and water-filled Jurassic sandstones of northwest Germany, shown in relation to their depth of burial. Points: oil-filled sandstones; circles: water-filled sandstones. After Philipp, 1963. (von Engelhardt, 1973, p. 261.)

As an *area*, the spot relates to the entire area bounded by a line. Examples are the distributions of geologic formations, waters, etc. on geological maps.

Each component and its elements correspond to a concept with its subsidiary concepts, which are usually cohyponyms. Thus the component 'chromium content' in the example of Fig. 2.4 forms a continuum whose elements are subsets measured in p.p.m. The component 'constituents of the lunar soil' in Fig. 2.8 is divided into four discrete elements: breccias, pyroxenes, plagioclase, crystalline rocks. As explained on p. 48, the cohyponyms of a conceptual field can be organised in three ways: *qualitatively*, *comparatively*, and *quantitatively*. The components may correspondingly also be organised qualitatively, comparatively, or quantitatively.[41]

The components of 'composition of the lunar soil' (Fig. 2.8) are qualitative, the components of 'Mohs' Scale' (Fig. 2.9) are comparative, and the components of 'Mole Ca/l solution' (Fig. 2.7) are quantitatively organised. Also the variables of the graphic representation must be chosen to permit the same type of organisation as the components which they represent.

Qualitative gradations may be realised through light value, colour, pat-

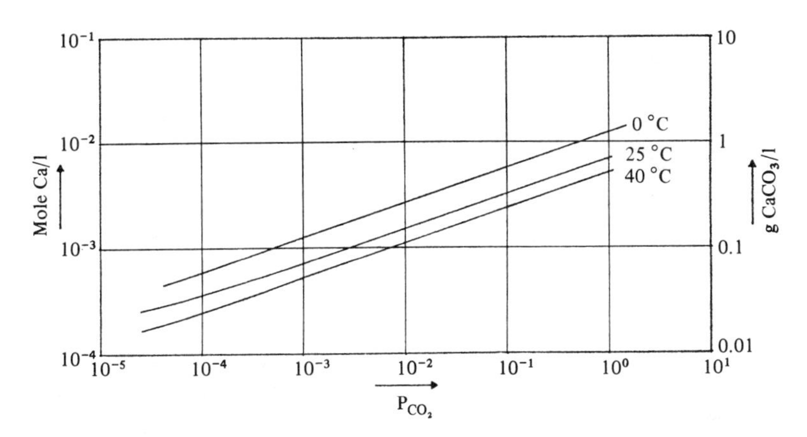

Fig. 2.7. Diagram of the solubility of calcite (as mole calcium per litre) in relation to the partial CO_2 pressure at various temperatures. (von Engelhardt, 1973, p. 184.)

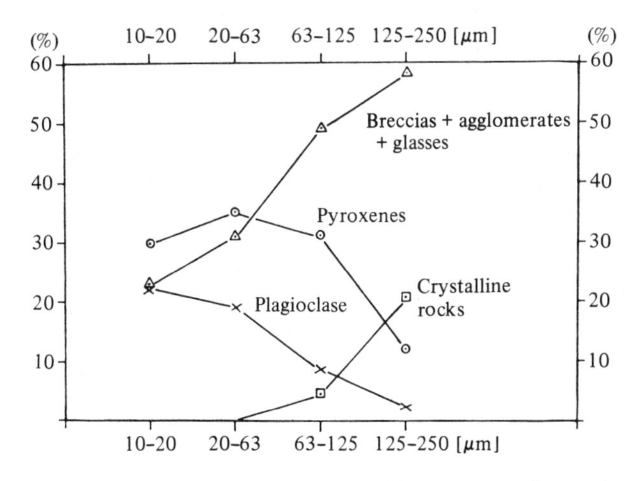

Fig. 2.8. Diagram of the percentage of breccias together with agglomerates and dark glasses, pyroxenes, plagioclase, and crystalline rocks (in grain number percentages) in a sample of lunar soil from the landing site of the Apollo 11 mission, in various grain size fractions. (von Engelhardt et al., *1971.)*

tern, direction, and shape of the spot or by discrete sites along the coordinate axes.

Comparative gradations can be realised through variations in the light value or size of the spot, and through the x and y values on the coordinate axes.

Quantitative gradations can be realized through variations in the size of the spot and through the x and y values of the coordinate axes.

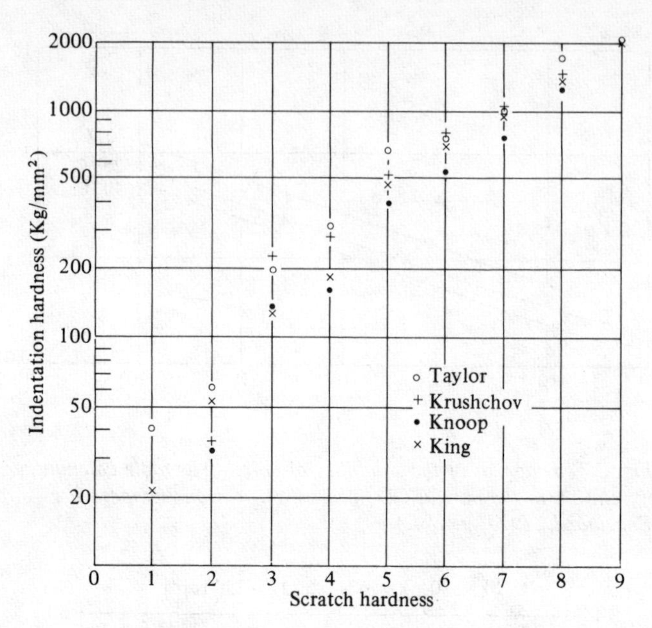

Fig. 2.9. Diagram on the relationship between indentation hardness in kg/mm² and Mohs' scratch hardness for various minerals in Mohs' Scale of Hardness, according to measurements by various authors. 1 = talc; 2 = gypsum; 3 = calcite; 4 = fluorite; 5 = apatite; 6 = orthoclase; 7 = quartz; 8 = topaz; 9 = corundum. (Barth, 1969, p. 148.)

In principle nearly all the representational variables are suited to represent qualitatively organised components, but it is convenient to choose those whose gradations can only be distinguished qualitatively. For this reason, the qualitatively distinguished constitutents of the lunar soil in Fig. 2.8 are represented by point-type spots which vary qualitatively in shape. Comparatively ordered elements, on the other hand, must be represented by comparatively ordered variables. Thus the comparative degrees of the hardness scale in Fig. 2.9 are represented by a sequence of sites on the x-axis. The primary device for representing quantitatively organised components are the coordinate axes, as quantitatively varying the size of a spot permits only a few gradations. An example of quantitative organisation is the representation of the component 'distance' through the x-axis in Fig. 2.4.

Graphic representations answer various kinds of questions. Following Bertin[42] we may distinguish the following types and levels of questions.

Every graphic representation answers as many questions as it has components. Thus in the example of Fig. 2.4, we may distinguish three *types of questions*: we can ask (a) what is the total chromium content in the clay sediments at a particular place? (b) where does the chromium content in the river sediments reach a particular value? and (c) how is the chromium content at a particular place distributed with respect to the various grain size fractions?

For each type of question, there are three levels of questioning:

I *Elementary level of questioning*, relating to one element of a component. (In Fig. 2.4, what is the chromium content of the grain size fraction 0.06–0.2 in the sediments at the mouth of the Neckar.)

II *Intermediate level of questioning*, relating to an extracted subgroup of elements, e.g. to maxima, minima, or concentrations. (In Fig. 2.4, where along the Neckar does the chromium content in the sediments reach its maximum.)

III *Highest level of questioning*, relating to an entire component, e.g. to the change in the component, described mathematically or perhaps simply qualitatively. (In Fig. 2.4, how does the chromium content in the Neckar sediments change along the course of the Neckar between Stuttgart and Heidelberg.)

If information is to be conveyed graphically, it must first be decided to which level of questioning the representation should correspond. The three levels correspond to three functions of representation, and require that certain things be taken into account in choosing the representational devices and forms. Following Bertin,[43] we distinguish:

I *Recording of information*, corresponding to the elementary level of questioning. This is achieved by representing an inventory of all elements of information which is exhaustive with respect to the state of affairs being portrayed. Such representations cannot be stored in the memory, and must fulfill the requirement that any given datum can be read from them at any time. In Fig. 2.5, the porosity of any rock sample taken at a given depth can be read off.

II *Processing of information*, corresponding to the intermediate level of questioning. This is achieved by representations permitting the discovery of previously unrecognised orders, groupings, and correlations. Used in this way, a graphic representation functions as an experiment, permitting the construction of various graphic images: the scientist 'plays' with these in order to find relevant correlations. In this very important function, used in many branches of Earth science, the graphic representation plays a specific role which cannot be replaced by verbal constructions, and which goes beyond the simple task of mere description. Many theoretical discoveries in Earth science were made possible only by the use of maps, an example being the theory of continental drift. Fig. 2.8 shows the processing of information with the aid of diagrams; in it, it may be ascertained that the pyroxenes in the lunar soils reach their frequency maximum in the grain size class of 20–63 μ.

III *Representation of information*, corresponding to the highest level of questioning. This is achieved by representations which are easily imprinted upon the memory, are therefore easily apprehended, and can be used to establish general patterns or lawlike statements. Here again, the graphic representation does not merely describe, but conveys insights which transcend direct experience. From Fig. 2.5, for example, it can be concluded that the porosity of clayey rocks generally decreases with depth in accordance with a linear function.

Graphic representations can be divided into diagrams, charts and maps, according to how the two dimensions of the plane are used.

The graphic construction is a *diagram* when the relationships between all elements of one component and all elements of at least one other component are expressed. The graphic construction is a *matrix* when the relationships between all the elements of one and the same component are expressed. The graphic construction is a *map* when the relationships between the elements of at least one geographic component are expressed, and are arranged according to their position on the Earth's surface.

Diagrams

The simplest type of flat diagram shows the relationship between the elements of two components. Six types of 2-component diagrams can be distinguished, according to whether they are divided up qualitatively (Ql), comparatively (C), or quantitatively (Qt): Q1–Ql, Qt–Qt, Ql–C, Ql–Qt, C–Qt, C–C.

Fig. 2.5 shows type Qt–Qt. Both components, depth (y-axis) and porosity (x-axis), are organized quantitatively. Each point represents one clay sample from a specified depth and of a specified porosity. The totality of the individual points on the diagram may be said to record individual data, and can replace a table of data. The derived line shows what the individual values yield by way of processing and transmitting information, namely the overall linear decrease in porosity with increasing depth.

Most diagrams use a right-angle coordinate axis for display in a plane, but other coordinate systems are also possible. When portraying the distribution of directions in space, e.g. the frequency of spatial orientation of quartz axes in a rock, the component 'direction in space' can be depicted in a stereographic projection of a spherical surface, while indicating the component 'number of quartz axes within a particular spatial angle' by coloured or patterned areas on the projected sphere.

Relationships between three components can best be shown on a flat diagram when the third component is ordered either qualitatively or comparatively. Some possible combinations are shown by the following examples.

Fig. 2.8 illustrates type Qt–C–Ql. The percentage amounts (y-axis, percentage of total amount) are quantitative, the grain size classes (x-axis) are comparative, while the four discrete elements of the component 'constituents of the lunar soil' are qualitatively organized. While the quantitative and comparative components are shown along the x- and y-axes, the four elements of the qualitative component are shown by variously shaped point symbols (triangles, circles, crosses, and squares). The diagram reveals a large number of discoveries concerning the state of affairs 'lunar soil 10084 of Apollo 11', characterised by the three components and their elements.

Fig. 2.9 also illustrates type Qt–C–Ql. In addition to the qualitatively graded indentation hardness and the comparatively graded Mohs' Scale, the qualitatively ordered complex of author also crops up, for each of the four authors carried out measurements using a slightly different system of measuring. The four qualitatively different elements are easily shown by differently shaped point symbols.

Figs. 2.4 and 2.6 illustrate type Qt–Qt–Ql. In Fig. 2.4 the components 'chromium content' and 'distance' are quantitatively organised, as are the components '% growth in quartz crystals' and 'depth of burial' in Fig. 2.6. The qualitatively ordered elements in Fig. 2.4 belong to the component 'grain size class', and in Fig. 2.6 to the component 'growing quartz grains' (namely in 'quartz grains in oil-filled and water-filled sandstones').

A particularly common way of graphing relationships between three quantitative components (Qt–Qt–Qt) is to portray them as a mixture of the elements belonging to the components A, B, and C. In this case they may be shown in a triangular coordinate system, as in Fig. 2.10. Each point within the triangle corresponds to a certain mixture containing proportions of A, B, and C. Along the sides of the triangle we may read off the proportions of mixtures containing two components each.

Matrices

A grid laid out in a plane may be used to show the relationships between all the elements of a single component. The plane of the drawing itself has no significance, so that the elements may be arranged in any way desired. To indicate the relation between any given two elements the connecting line symbol is used, or where this relation is non-commutative, going in only one direction (e.g. from father to son), this may be shown by an arrow. In general, the elements are so arranged that relationships intersect as seldom as possible.

Examples of matrices include the 'evolutionary trees' used in paleontology, such as Fig. 6.2 (p. 192) showing the phylogenetic evolution of amphibians and reptiles. The elements here are the individual animal taxa,

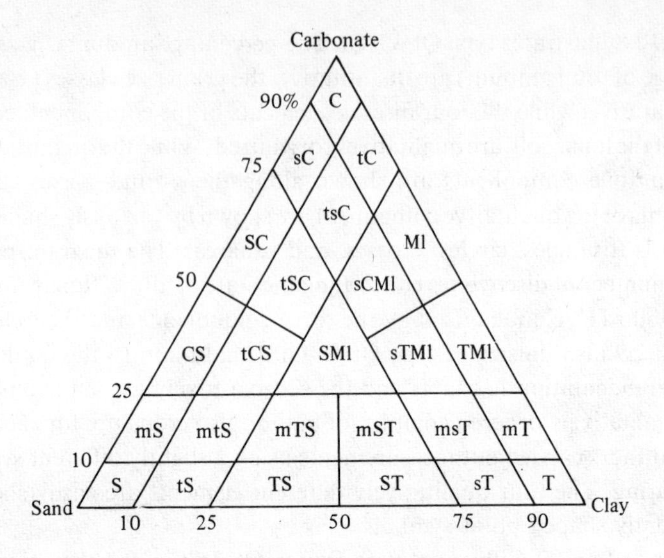

Fig. 2.10. Triangular diagram for classifying and naming sediments consisting of sand, carbonate, and clay.
S = sand or very sandy; s = sandy; T = clay or very clayey; C = carbonate or very carbonaceous; t = clayey; Ml = marl; m = marly. (Füchtbauer & Müller, 1970, p. 9.)

shown as areas; they are connected by narrow bands indicating their genetic relationships in such a way that descendant species are always shown above their ancestral species.

Another chart is shown in Fig. 2.11. Here the elements of the component consist of the various procedures and aims involved in investigating sedimentary samples. The relations between these elements consist of a chronological sequence, and are therefore shown as arrows leading from initial operations to subsequent ones, and finally to the desired information.

Maps

In all disciplines of geoscience, maps of various kinds form a particularly important type of non-verbal representation. They must be explained by verbal texts, but cannot be replaced by them.

A map conveys information on a complex of topographic–geographic states of affairs obtaining on the surface of the Earth, Moon, Mars, etc. These are transcribed from the curved surface area of reality onto the plane of the map sheet in a manner determined by the chosen form of projection and scale of reduction. This provides a fundamental topographic grid, onto which the geological subject matter is then entered by means of symbolic points (e.g. mountain peaks or trigonometric points), lines (e.g. contour lines or rivers), and areas (e.g. lakes or ice masses).

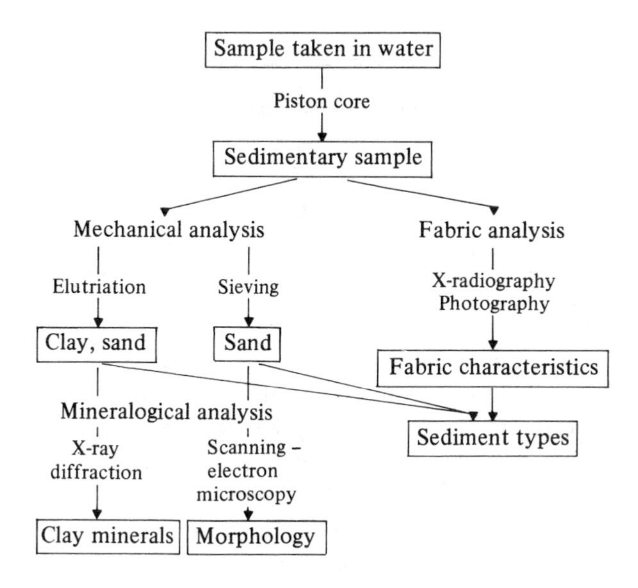

Fig. 2.11. Chart showing the sequence of field and laboratory operations involved in analysing sediments taken from the floors of water bodies. (Fürbringer, 1977.)

The only maps which can be read instantaneously are those which, along with the geographic components of latitude and longitude represented by the two dimensions of the flat sheet, admit no more than one other component by varying one of the additional visual variables, e.g. the component 'formational unit', shown by variation in the colour of the areas. To depict further components, the choice is either to produce a series of several maps, or to overlay several components in one map, producing a complex figuration which can be understood only by several stages of reading. Most geological maps are such complex figurations, using geographic information only as a background.

The complex of geoscientifically relevant subject matter located thus within the topographic grid can, like diagrams, be ordered qualitatively, comparatively, or quantitatively. Qualitatively ordered components include the formations and rock units occurring within the map region; these are shown by varying the colour, pattern, or light value of the spots. A comparative component would be, e.g. the sequence of rocks of various ages, shown by various shadings. Quantitatively organized components occur with particular frequency on geophysical maps, for instance, the distribution of the vertical components of the Earth's magnetic field, or of gravity anomalies. A wide-spread form of representation consists of lines connecting sites at which the variable involved has a constant value. Two

Table 2.1

	Photo-graphs	Drawings	Schematic Drawings	Diagrams	Matrices	Maps
Contributions	9	4	2	71	0	13
Geologische Rundschau	30	9	11	30	0.2	19
Total	20	7	7	50	0.1	16

such lines delimit an area within which the local values of the variable vary between an upper and a lower limit.

The map becomes a readable body of information only through its *legend*, which verbally defines the signs used for the elements of the qualitatively, comparatively, and quantitatively organized complexes of geoscientific subject matter. The legend thus fulfills a metalinguistic function: it translates the non-verbal signs into the concepts of the given scientific language. A fundamental requirement of every map, therefore, is that the interpretive legend be thorough and unambiguous.

We will discuss the problem involved in preparing and using maps in a later chapter.

2.4.5 *The frequency of usage of non-verbal codes in scientific papers*

As in all natural sciences, papers in the Earth sciences make ample use of non-verbal codes. Papers lacking photographs, drawings, or graphic presentations are rare in geoscientific journals. The preference for certain types of non-verbal presentations shows up characteristic differences, not only between Earth science and other natural sciences, but also between the various geoscientific disciplines.

The following frequency distributions given in Table 2.1 (in %) were found in a study of circa 700 photographs and figures taken from one year's issue of the journal *Contributions to Mineralogy and Petrology* (1978) and another 700 from *Geologische Rundschau* (1977).

Diagrams, photographs, and maps are the preferred methods of non-verbal presentation in geoscientific papers. The fact that maps are not only published as special geological and geophysical works, but also appear in many journal papers indicates that the empirical basis of geoscientific research consists to a large extent of descriptions of the phenomena to be observed on the surface of the Earth and other planets. The frequency of photographs reveals that the geoscientist must repeatedly deal with visually characterised singular objects, where adequate verbal description proves difficult. This is particularly true for disciplines such as paleontology. Photographs are less important in the more physically and chemically orientated disciplines of mineralogy, petrology, geophysics, and geochemistry than in historical and regional geology, hence the lower percentage of photographs in *Contributions to Mineralogy and Petrology* than in *Geologische Rundschau.* The high frequency of diagrams documents the important role of inductive research procedures in geoscience. Graphic methods can be used to assemble large amounts of data (information processing), to infer regularities heuristically from the data, or to use the data to confirm and strengthen hypotheses and theories (information conveyance).

3

The foundations of geoscientific research

3.1 Levels of knowledge (empirical and theoretical)

Various models have been proposed for the stages of scientific inquiry, all based on the distinction made by logical empiricists between propositions in an *observational language* and derivative propositions in a *theoretical language*.[1] One model which has proved useful for the analysis of geoscientific research texts is shown below. This consists of four levels of inquiry, the arrow indicating a progression towards propositions of increasing generality.[2]

↑ IV Level of regulative principles.
III Theoretical level.
II Level of systematisation and generalisation of empirical data.
I Empirical basis.

I. *Empirical basis.* The empirical basis consists of all propositions put forward as statements about facts ascertained by observation or experiment, the truth of which the author believes may be tested by any initiate. The boundary between the empirical basis and higher levels of knowledge is not based on the 'objective' content of propositions, but on their function in the context of a particular investigation. It must be noted that all statements of the empirical basis contain more than simply the observed or experimentally obtained sensory impressions, for intersubjectively testable facts are usually presented clothed in the garb of generally accepted interpretations. Thus, the statement that clayey rocks of the 'Lias-beta' formation were found at a certain locality belongs to the empirical basis of an investigation, even though this statement implies the hypothesis that these rocks were formed at a particular period in Earth's history. The empirical core of the hypothesis 'Lias-beta' consists of the intersubjectively testable fact that the rocks involved contain certain species of ammonites.

The statements of the empirical basis are characterised by two features. They relate to that which can be intersubjectively experienced, i.e. to singular facts localised in time and space. Secondly, their truth within the context of a specific study is not questioned.

II. *Level of systematisation and generalisation of empirical data.* This intermediate level consists of propositions summarising and systematising the results of many individual observations. It also includes generalisations evolved from these results, which, as regularities and ordering schemes, relate not only to the empirically obtained facts, but also to an infinite number of cases or to a finite number of unobserved cases, without as yet belonging to the next theoretical level.

III. *Theoretical level.* This consists of propositions and systems of propositions which transcend the statements of the empirical basis and level II generalisations; they either take the form of universal laws, or based on such laws, contain them as a necessary component. The theoretical level thus includes not only laws, but also explanations of facts based on laws, hypotheses about past, future, or unobservable states of affairs, and whole theories. Theories, being deductively ordered systems of partial hypotheses and laws, constitute the highest level in the hierarchy of knowledge.

IV. *Level of regulative principles.* In constructing hypotheses and theories, not only empirical facts and theoretical findings but also conventions, rules, or principles of non-empirical origin play a role.

Before looking in detail in the next chapters at the problems arising in geoscientific research at these four levels of scientific knowledge, we will analyse four examples from the literature, in order to illustrate the hierarchical structure and examine its usefulness. At the same time we will attempt to assign the statements in the texts to each of the four levels and to determine with which level each of the studies is concerned.

Example 1: 'Sediment transport at the mouth of the Alpine Rhine'[3]

Empirical basis. Around 500 water samples from the Alpine Rhine just above its mouth at Lake Constance were taken at regular intervals through the year at varying levels and velocities. The velocity and suspension load were measured in the expectation of finding a relationship between them.

Level of systematisation and generalisation of the data. A graph of type Qt_1–Qt_2 (Fig. 3.1) of the suspended load, shown in dependence on current velocity, is used to derive information and to find the mathematical function quantitatively describing the qualitative observation that suspended load increases with velocity. When velocity and suspended load are entered

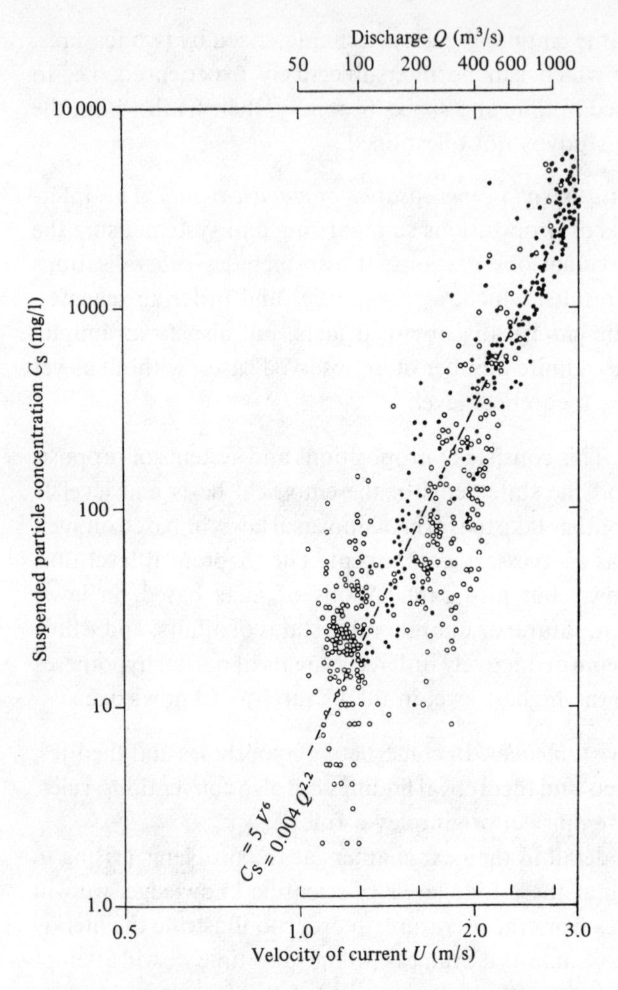

Fig. 3.1. Suspended load of the Rhine at its mouth at Lake Constance, shown as a function of stream velocity. (Müller & Förstner, 1968.)

into a logarithmic scale the measured points distribute themselves around a line corresponding to the equation

$$C_s = x(U^y)$$

where C_s equals the suspended load in mg/1, U equals velocity in m/s, $x = 5$ and $y = 6$. This function has the character of a quantitative, inductive systematisation of the measured data. The authors suppose that this is a general regularity applying to all rivers, different values for the constants x and y to be used for different rivers. This is corroborated by data from the literature for various rivers.

The *theoretical level* is not reached, as the inductively ascertained regularities are not attributed to any laws.

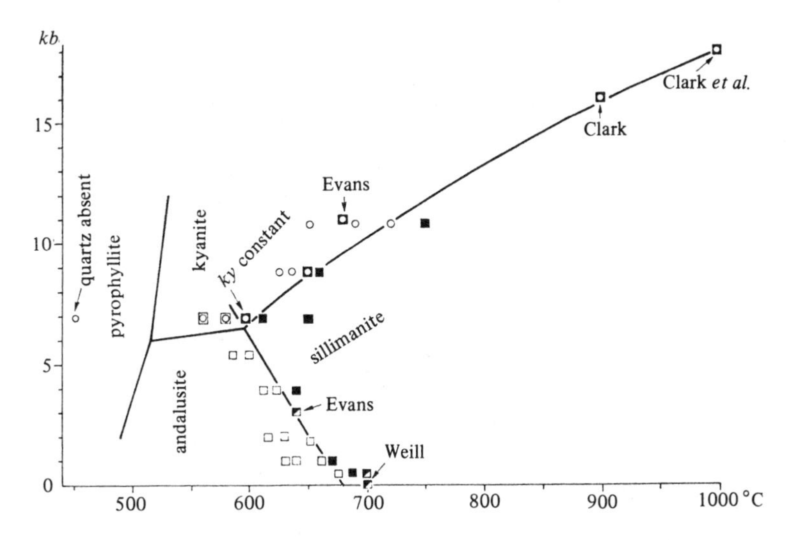

Fig. 3.2. Determination of the triple point of andalusite, sillimanite, and kyanite. (Althaus, 1967.)
Open squares: formation of andalusite; solid squares: formation of sillimanite; open circles: formation of kyanite; mixed symbols: equilibrium between the minerals or no reaction.

Example 2: 'The triple point andulusite-sillimanite-kyanite'[4]

Empirical basis. In order to identify the temperature and pressure-dependent stability field (important for the petrology of metamorphic rocks) of the minerals composed of Al_2SiO_5 (andalusite, sillimanite, and kyanite), powder of aluminium silicate minerals was exposed to pressures ranging from 1 to 11 atmospheres, at temperatures of between 400 and 700°C. The reaction products were identified by X-ray analysis. Andalusite, sillimanite, or kyanite were produced depending on pressure and temperature, as shown in the pressure–temperature diagram of Fig. 3.2, a diagram of type Qt_1–Qt_2–Q1.

Level of systematisation and generalisation of the empirical data. The lines drawn in the diagram graphically show the boundaries of the stability fields of the three phases in the pressure–temperature field. The triple point at which all three minerals coexist proves to have the values 6.5 ± 0.5 kbar, 595 ± 10°C.

Theoretical level: The author notes that the equilibrium points cannot be thermodynamically calculated at present as the necessary caloric data for the individual minerals are not available. Instead he shows that the position of the empirically determined triple point agrees with known phase equilibria in which Al_2SiO_5 minerals are involved.

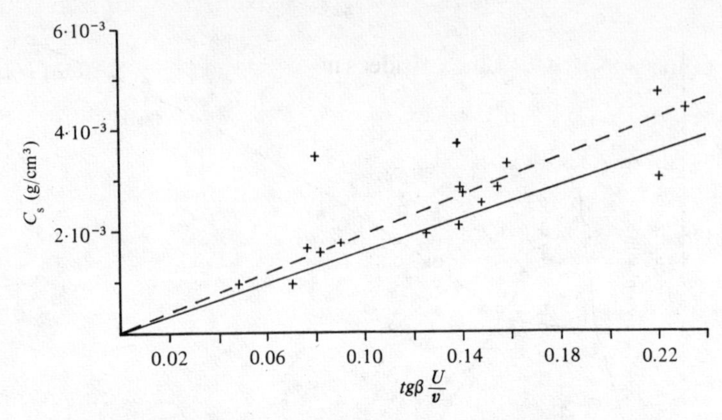

Fig. 3.3. Dependence of suspended load (C_s) *on slope* (tg *beta*)*, stream velocity* (U)*, and speed of sinking of suspended particles* (v) *in the Rio Grande. (von Engelhardt, 1973.)*

Example 3: 'Testing of a theoretically inferred law on transport of suspended matter in a river'[5]

Empirical basis: The following values were measured at various stations along the Rio Grande River: slope of river (*tg* beta), velocity (U in cm/s), amount of suspended load (C_s in g/cm³), and average speed at which the suspended particles sink (v in cm/s, calculated by grain size).

Level of systematisation and generalisation of the empirical data. Based on the assumption that the suspended load (C_S) will vary directly with slope (*tg* beta) and velocity (U) and inversely with speed of particle sinking (v), related values of C_s and (*tg* beta) (U/v) were entered as points in a diagram having the axes C_s and (*tg* beta) (U/v) (Fig. 3.3). The points are scattered around the dashed line

$$C_S = k' \ (tg \text{ beta}) \ (U/v), \text{ where } k' = 0.019.$$

Theoretical level. According to a theoretical relationship derived by Bagnold from physical laws the following equation should hold for suspended transport in a turbulent current:

$$C_S = k \ (tg \text{ beta}) \ (U/v), \text{ where } k = 0.016.$$

This is represented by the solid line in Fig. 3.3. The empirically found constant is 19% higher than the theoretically expected one.

The regulative principles guiding the researcher in generalising from the empirical data and in developing hypotheses and theories are not as a rule identified explicitly in papers; they belong to the assumptions of argumentation and theory-building which are tacitly accepted by members of the scientific community and which are not discussed. In the following Example

4 the author presents a particular interpretation of the principle of uniformitarianism, one of the key ideas in geological research (see p. 315).

Example 4: Investigations into the applicability of the model of plate tectonics to the evolution of the Earth's crust in the Precambrian[6]

Empirical basis: Observations on the distribution of sedimentary, plutonic, and volcanic rocks, and their radiometric age and magnetisation in the oldest parts of the continents (the Precambrian shields).

Systematisation of empirical data and theoretical level. Based on a reconstruction of the formational conditions of the oldest known sedimentary and igneous rocks and of tectonic crustal deformations during the early history of the Earth, the author develops the theory that since the Precambrian formation of the first continental cores about four billion years ago the continents have certainly drifted horizontally, due to convectional streaming in the mantle, but that plate tectonics in the sense of oceanic rifting and subduction of the ocean floor under drifting continental plates did not occur then in the form typical since the end of the Precambrian (se p. 240).

Level of regulative principles. The author points out that his theory contradicts the strict principles of uniformitarianism.

> The varing style in tectonic processes through earth history . . . leaves little doubt that global crustal evolution was non-uniformitarian.

The uniformitarian principle applies only to the temporal constancy of the forces causing crustal movement, but the effects produced changed during Earth's history:

> Uniformitarian causes (heat flow and mantle convection) have had non-uniformitarian effects (global crustal evolution) if sufficiently long periods of time are considered.

3.2 Paths towards scientific explanation

3.2.1 *Description and explanation*

The works cited in the previous section show how research progresses along the four levels of the empirical basis, inductive organisation, theoretical construction, and regulative principles. Before we begin in later chapters to examine which problems geoscientific research deals with on these levels, we shall look at the formal structure of the procedures which scientists employ to make the transition from elementary empirical findings to propositions of higher cognitive standing.

The fundamental propositions of the empirical basis rest on facts obtained through observation or experiment. They answer the question 'what is the case?'[7] The empirical descriptions may then be further subdivided, according to the particular specification sought. The first classification is provided by adverbial questions such as 'when' questions specifying time, or 'where' questions specifying place. Unlike in physics and chemistry, the 'when' and 'where' of empirical phenomena are often of great importance in geoscience. There are also 'is made of what' questions inquiring into the properties of things or substances (i.e. looking for predicates for given subjects), and 'what' questions inquiring into the thing or substance whose properties have been observed (i.e. inquiring as to the subject of predicates). Furthermore there are 'how much' and 'how large' questions, seeking quantitative specifications. A basic activity of researchers in all geoscientific disciplines consists of answering these kinds of questions in the form of describing facts. Preconceived categorical schemes are evident in the differentiation of questions and answers, schemes which also underlie the descriptive terminologies of the various geoscientific languages. These schemes and some of the terminologies will be dealt with in later chapters.

The desire for knowledge, however, will not be satisfied even by the most thorough and exact *description* of the world within the categories of geoscientific disciplines. We ask not only 'what is the case?' but also further 'why is this the case?' sometimes 'wherefore is this the case?' and finally 'what follows from this?' With these questions we demand *explanations*[8] for the existence and composition of observed facts of the present, states of affairs of the past, and events of the future. In the natural sciences we deal as a rule with 'why?' and 'what follows from this?' questions, and this is true for geoscience as well. 'Wherefore' questions demand teleological explanations and presuppose teleological laws which will be discussed in a later chapter (see p. 180).

The search for explanations governs the transition of research from the empirical basis to the higher levels of knowledge. To understand and evaluate the methods and results of geoscientific research, however, it is necessary to refine the concept of scientific explanation. This is not a superfluous task, for the word 'explanation' is used colloquially with various meanings which can hardly be reduced to one concept.[9]

Explanations may explain such things as the meaning of the phrase 'contact metamorphism', the best path to an outcrop of beryl in Bergell in the high Alps, how to use a polarised microscope, or the reasons why Wegener attempted the dangerous crossing of the Greenland ice sheet. Within the wide field of the word 'explain', however, certain characteristics distinguish the concept of *scientific explanation* which pertains to Earth science as to all the natural sciences. These characteristics we will illustrate in the following example.

A core drilled through sedimentary rocks reveals shales at a depth of 1000 m which, according to the evidence of the marine creatures preserved in them, were deposited as clayey mud on the floor of a marine basin. Measurement shows that this rock is considerably compacted relative to freshly deposited marine clays: the proportion of water-filled pore volume to solid mineral volume equals 0.25 in the rock, while in freshly deposited marine muds it equals 4.0 The question therefore arises, *why* this marine sediment has such a reduced porosity at a depth of 1000 m. The *explanation* of the observed fact results from the application of the law that pore volume in a shale will decrease when downwarped and buried by other sediments, according to the function

$$E_T = E_1 - b \log T$$

Where E_1 is the value of the pore volume at the start (4.0), T is the depth of burial (1000 m) and E_T the value of the pore volume at this depth (0.25). b is a constant whose value, when inserted in the equation, yields the observed pore value of the rock.

A complete scientific explanation consists of the logical connections seen in this example between three components, namely the statements describing the controlling state of affairs, the statements of law, and the statements describing the resulting state of affairs.

The structure of scientific explanations corresponds to the following outline:

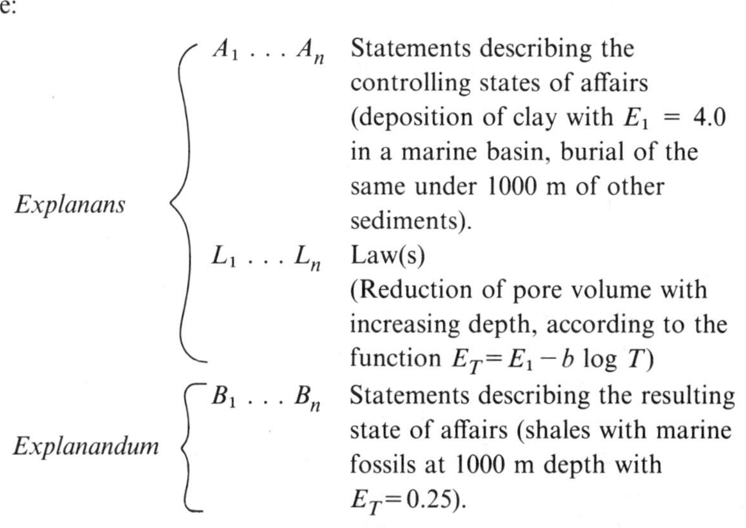

Scientific explanation is called *deductive-nomological explanation* (DN-explanation), because it contains a law (L) and draws conclusion B from A

and L in the logical form of deduction. If statement (A) describing the controlling state of affairs is true, and if (L) is a generally valid law applicable to (A), and if it is correct to infer (B) from (A) and (L), then (B) must also be true.[10]

Research proceeds by attempts to create scientific explanations, that is, by attempting to bring together controlling and resulting states of affairs with the aid of laws, adhering as closely as possible to the outline of deductive-nomological explanation. One difficulty consists in finding the appropriate laws to unite the controlling and resulting states of affairs. According to a view going back to Aristotle, explanation consists of relating the phenomenon to be explained to the already known and familiar.[11] This is certainly not a generally valid feature of scientific explanations, but is true of a type of explanation operating by analogy.[12] *Analogy models* are used to explain phenomena when the mechanisms bringing them about are not accessible to empirical examination. Thus the phenomenon of the flow of liquid or gaseous media through rocks where the flow processes through the microscopic pore systems cannot be studied in detail may be explained by analogy with electrical conductivity using laws corresponding formally to Ohm's Law. The controlling states of affairs here, pressure gradient and permeability of the rock, are analogous to electrical voltage and conductivity, while the resulting phenomenon (to be explained), the velocity of the medium, corresponds to the electrical strength of the current.

3.2.2 Forms of scientific inferences: induction, abduction, and deduction

The production of complete scientific explanations is a problem which must be solved by scientific methods, for the researcher does not at first have all three components of inference at his disposal. Three initial situations can be distinguished, each involving a different procedure for setting up an explanation best approximating the deductive-nomological ideal.

1 The controlling states of affairs (A) and the resulting states of affairs (B) are known. A law (L) is sought which will unite (A) and (B) in accordance with the deductive-nomological model. This is the process of *induction.*
2 Resulting states of affairs (B) and laws (L) are known. The controlling states of affairs (A) are sought which when united with (L) and (B) will satisfy the model of deductive-nomological explanation. This is the process of *abduction.*
3 Controlling states of affairs (A) and laws (L) are known. Resulting

states of affairs (*B*) are sought, which will be explained by (*A*) and (*L*) according to the deductive-nomological model. This is the process of *deduction*.

Only the process of deduction yields an inference in the strict sense of logic. Nevertheless it is common in Earth science, as in other sciences, to speak of 'inferences' in the cases of induction and abduction as well. It is called an 'inference' when the causes are inferred (abductively) from some phenomena. It is inferred, for instance, from the presence of marine fossils in a sedimentary rock, that it is of marine origin. Equally it is called an 'inference' when a law is 'derived' from many observations. From observations on the porosity of clayey rocks at various depths, the above-described law on compaction of clayey sediments is inferred. In our discussions we will follow this more lax word usage, and speak not only of deductive, but also of abductive and inductive inference, noting in all three cases the presence of two premises and a resulting inference. In doing this we follow the example of Peirce,[13] who developed and emphasised the specific roles of these three methods of research in the natural sciences.

We will illustrate induction, abduction, and deduction by means of three examples, each using or arriving at the same law.

Diamond is invariably produced if and only if carbon or carbon containing compounds are exposed to pressures of over 55 kbar at a temperature of at least 1000 °C in the absence of oxygen.

In the case of *induction*,[14] a law may be inferred as the result of 'inductive inference' from two premises, one describing a controlling and the other a resulting state of affairs.

First premise (controlling state of affairs): in various experiments, carbon was exposed to various pressures in the absence of oxygen at a temperature of 1000 °C.

Second premise (resulting state of affairs): in all experiments in which the pressure exceeded 55 kbar, and only under these conditions, diamond was produced.

Conclusion (law): if carbon is exposed to pressures of over 55 kbar in the absence of oxygen and at 1000 °C, diamond will be produced.

In the case of *abduction*[15] the premises consist of statements of a resulting state of affairs and a law. The result yielded by 'abductive inference' is the controlling state of affairs which caused the given resulting state of affairs.

First premise (resulting state of affairs): diamonds were found in volcanic pipes in South Africa.

Second premise (law): diamond is produced only from carbon and carbon compounds when the temperatures reach at least 1000 °C and when the pressures are at least 55 kbar.

Conclusion (controlling state of affairs): in the formation of the volcanic pipes, material was brought up from depths where pressures of at least 55 kbar obtain.

In the case of *deduction*,[16] the premises consist of a law and a statement about a controlling state of affairs. As the result of deductive inference, a statement regarding the resulting state of affairs is produced, which must follow from the controlling state because of the law.

First premise (law): at pressures of over 55 kbar and at temperatures of over 1000 °C, carbon in the absence of oxygen will change into diamond.

Second premise (controlling state of affairs): in an experiment, carbon is subjected to a pressure of 80 kbar and a temperature of 1200 °C.

Conclusion (resulting state of affairs): In the experiment diamond will be produced.

Using the abbreviations A = controlling state of affairs, B = resulting state of affairs, and L = law, we present the following formal table of the three 'modes of inference' which are used in research to arrive at scientific cxplanations.

	Induction	*Abduction*	*Deduction*
Premises	A	B	L
	B	L	A
	—	—	—
Conclusions	L	A	B

Induction, abduction, and deduction each offer their own opportunities for going beyond the limits of the world of empirical facts[17] and constructing hypotheses and theories. Each of the three procedures, however, has its own limitations and risks which inevitably render all hypothetical constructs open and subject to review. We will discuss the particular applications of induction, abduction and deduction in later chapters. We will merely note in advance here that, due to its aim of unravelling the history of the Earth and other planets, Earth science relies heavily on the abductive method of inferring unknown causes from known effects. As Peirce[18] noted, Earth science differs significantly in this respect from experimental sciences such as physics and chemistry, where induction and deduction are the predominant methods.

3.3 Categorical schemes

The diversity of scientific disciplines shows that 'the world' may be viewed with different eyes, described in various manners, and interpreted and explained according to differing models. The languages of the sciences are based on differing *categorical schemes or frameworks* used to describe theoretical models and observational results.[19] The vocabulary of the descriptive language within Earth science also reveals specific categorical concepts, which determine what type of entities and relationships can be recognized in geoscientific research. These may be identical with or differ from those of other natural sciences.

Körner[20] has worked out the categorical frameworks existing in the language of physics with regard to mutable and movable things in space and to things in fields. The universe of geoscientific research contains things, configurations, and substances in space which change through time. The categorical concepts of geoscientific language therefore relate to the entities of *things (Dinge)*, *configurations (Gestaltungen)*, and *substances (Stoffe)*, and to their organisation in the dimensions of *space* and *time*.

3.3.1 *Things*

The thing-concept constitutes the basis of the categorical ordering of the world as described by the Earth sciences.[21] The scientific thing-concept has been abstracted from the thing-category which underlies all our daily experience and language. According to this everyday category, things are material objects which in some way can be isolated from their surroundings and which preserve their identity and can be recognised through all changes in time. These characteristics are more precisely specified in the scientific thing-concept, which can be formulated as follows:

1 At every moment of its existence a thing takes up a certain, three-dimensionally bounded *thing-space (Dingraum)* which can be localised with respect to other things, and which cannot be occupied at the same time by any other thing, unless the one is a part of the other.

2 Each thing displays *properties* expressed linguistically through predication. Of these we may stress the *necessary properties (notwendige Eigenschaften)*, upon which the identity of the individual thing depends. The necessary properties are those attributable to the thing throughout its existence so that it may always be recognised by them. Those properties of a thing which may change during time without affecting the identity of the thing, we call *contingent properties (kontingente Eigenschaften)*.

3 Each thing consists of a *substance* defined by certain properties, or of several such substances. Its material composition is often one of the necessary properties of a thing.

4 Things have *temporal duration (zeitliche Dauer)*, that is, they preserve their necessary properties by which they can be recognised through time. During the duration of their existence their contingent properties may change either qualitatively or through movement (change of location, i.e. of spatial relations to other things). The temporal duration of a thing is usually limited: things can be created at certain times, and they can finally perish.

Some examples of things dealt with in the Earth sciences are small bodies such as a sand grain, quartz crystal, fossilised sea shell, pebble, or meteorite, as well as larger bodies with defined boundaries, such as a lava flow, granite massif, sandstone bed, or volcano.

There are various ways of classifying individual things into genera or *classes*,[22] defined by certain necessary properties shared by all the elements of the class.

(a) Classification according to location in space, e.g.: all volcanoes of the Andes; all barchans of the Sahara; all quartz crystals in the crevices of the Gotthard Massif.

(b) Classification according to shape and extension of the thing-space, e.g.: all calcite crystals of prismatic habit, or all grains of a sand whose diameter falls within certain specified intervals.

(c) Classification according to substance, e.g.: all grains of a sand consisting of feldspar; all volcanic structures composed of tuffs and basaltic lavas.

(d) Classification according to non-material properties, e.g.: all fossilised sea shells of the genus *Pecten*; all quartz crystals which twist the plane of polarised light to the right (dextral quartz).

(e) Classification according to genesis, e.g.: division of the grains of a sand according to sedimentary, metamorphic, or igneous origin (type of genesis); granitic plutons in the Alps of Tertiary age (time of genesis); granite erratics of Finnish origin in an Ice Age ground moraine along the Baltic Sea coast (place of genesis).

The Earth sciences deal not only with things described as members or elements of defined genera or classes, but also very often with *individual things* which can only be identified by proper names (see p. 38). This, as we have already mentioned, is in typical contrast to the physical, chemical, and biological sciences, so that proper names play a far greater role in geoscience than in other natural sciences.

Examples of proper names of things include the names of the planets, individual units of the Earth's crust (the Canadian Shield, the Antarctic Plate), of volcanoes (Mauna Loa, Etna), of rock masses (Bergell granite), the description of individual meteorites according to their place of landing (the Stannern meteorite) or of discovery (iron from Canyon Diablo).

One notable feature of the thing-concept is that a thing is not a permanently-fixed ontological unit. What is defined as a thing in the context of a discipline depends upon the research goals, and may change with these goals; in particular it does not need to correspond to daily usage. The same space–time continuum will be structured into different thing-units by different sciences or disciplines. In the folded sedimentary rocks of an outcrop, for instance, the paleontologist will see fossilised shells and ammonites as disparate things within the rocky medium; the stratigrapher will differentiate the individual folds as thing-units; the sedimentary petrologist will see separate strata demarcated chemically and mineralogically by sharp boundaries. The boundaries and definitions of a thing, of course, may also change with time as knowledge within the discipline advances.

3.3.2 *Configurations*

'Configurations' are the formations and structures imposed upon material substrates. They are classified spatially or genetically, according to the forces or processes which produced them.[23] There is no sharp boundary between a configuration and a thing: one and the same object may be considered as a thing in one context, but a configuration in another. A dune, for instance, viewed as a three-dimensionally bounded body of sand, is a thing; it is a configuration, however, when interpreted as a form imposed upon the desert sand by the action of the wind. Some configurations, however, could not under any circumstances be thought of as things: these are forms which have no closed three-dimensional boundaries.

Examples of configurations include stratification in sediment and sedimentary rock masses, soils forming structurally differentiated, surficial formations covering large areas, and geomorphological formations such as valleys, craters, or mountain forms.

Morphological classes of configurations include types of folding in stratified rocks, rock textures (porphyritic, evenly granular, etc.), and types of bedding.

Genetic classes of configurations include tectonic folds, synsedimentary slump folding, and volcanic or impact craters.

3.3.3 *Substances*

Substances are the material of which a thing consists, and form the substrate of configurations. Among the substances one can construct hierarchical series where each member forms the substance of the following member, as in the following sequence:

chemical element – mineral – rock.

Substances can be defined and classified in very different ways, as, according to necessary properties (e.g. chemical elements, minerals), according to the more elementary substances of which they are composed (e.g. minerals, rocks), or according to their manner of formation (e.g. rocks).

3.3.4 *Categorical concepts of space*

The categorical concepts of space include the *coordinate systems* of the surface and bodies of the Earth, other planets, and extra-planetary space.[24] These systems govern the organisation of things, configurations, and substances in space. The definition of such geometrical ordering systems and the identification of locations in them are topics of geodesy and astronomy. Geological maps, which we discuss in another context (p. 131), also present observations and their interpretations according to their distribution in space.

But space can be organised not only geometrically in terms of the distribution patterns of things and substances; it can also be differentiated and specified in terms of the behaviour of things. It is to this that the concept of fields relates: *fields* are unbounded or poorly bounded spaces distinguished by the fact that things and even substances behave in a special way within them.[25] Fields are characterised by the spatial distribution of physical and chemical parameters and play a large role in geophysics and geochemistry. We distinguish scalar or vector fields according to whether the parameters involved are simple numerical values or directional quantities. In contrast to things, different kinds of fields may occupy the same space-area.

Examples of physical scalar fields include temperature distributions on areas of the Earth's crust or porosity values in a sedimentary body of rock; chemical scalar fields include the spatial distribution of the uranium content of a granitic pluton or the distribution of pH (acidity) in a soil profile.

Examples of physical vector fields include gravity acceleration fields in a given area of the Earth's surface, temperature gradients around a hot body of magma, or values for the directional permeability of a sandstone body; chemical vector fields include reduced water content with increasing depth in rocks of the Earth's crust, or increase in alkali content with greater proximity to the contact zone around a magma body.

3.3.5 *Categorical concepts of time*

The dimension of time[26] is of particular importance in the universe of Earth science, since all geoscientific disciplines are ultimately involved in explaining the phenomena observed in nature as a result of processes that combine into a global Earth or planet history. Geoscientific theories provide models for rationally reproducing this geohistory in whole and in part, which in turn permit the deduction of predictions for the future. (*Geohistory* is used broadly here and below to include the history of the planets near Earth.)

The scientific aim of explaining the course of time and of constituting a scientific geohistory is hampered by the fact that very few temporal processes in nature can actually be observed in research, and those investigated experimentally are usually much shorter-term than geohistoric events. The subjects of geohistoric research usually consist of the long-lasting spatial and material forms and configurations which make up the solid Earth and planets. The basic problem of geoscientific research is that its actual theme, geohistoric time, cannot be confronted directly, but is only apprehended indirectly through the physical and chemical states of affairs, configurations, and forms brought about and left behind by geohistoric events.

The significance and function in research of categorical concepts of geohistoric time can only be understood in light of the non-temporal contexts from which they must first be developed.[27]

Time points and time intervals

Minerals and rocks, their textures and depositional conditions represent short-lived events or longer-lasting processes, and therefore, moments (time points) or intervals of geologic time. A few characteristic examples are:

(a) Each spatial element of a rock represents the process involved in the first formation of the rock it contains, and, therefore, represents either a specific *time point* in geohistory (e.g. the final solidification of a magma, the deposition of sedimentary particles, or chemical precipitation from a solution) or a certain *time interval* in which a sequence of partial events took place. The time interval in which a magma cooled, for instance, is revealed in its texture, the first minerals to crystallise out being surrounded by later ones.

(b) In the spatial elements of nearly every rock, changes can be found in the original composition representing processes which occurred after the original formation of the rock, and therefore indicating later intervals of geohistoric time. Such phenomena consist, for instance, of diagenetic compaction and cementation of sediments deposited as loose, porous masses (clays, sands, and calcareous muds), or the 'growth' of metamorphic minerals (e.g. garnet) through the effects of higher temperatures and pressures (metamorphosis).

(c) In a spatial element of an igneous rock or in a mineral grain the number of atoms formed by radioactive disintegration since the time of formation (where no loss has occurred) represents the time interval from the instant of formation to the present.

(d) The thickness of a homogeneous bed of a sedimentary rock represents the process of steady sedimentation on this site, and thus a certain interval in geohistoric time.

(e) In each (undisturbed) spatial sequence of sedimentary rocks, each stratigraphic boundary, being a discontinuous change in material, represents a punctuated event in the process of sedimentation on this site, and thus a time point in geohistoric time.

(f) In an originally horizontal and evenly-deposited sedimentary rock, each uptilting or folding of a stratum represents a deforming event, and thus a time point or time interval in the course of geohistoric time since the formation of the rock concerned.

States and processes

Minerals and rocks, their textures and depositional conditions, show more than that geohistoric time is divided up into the moments and intervals discussed in the last section. For geoscientific research, they are important indices (see p. 53) of certain states which obtained in Earth's past and of processes and events which changed these states.

We describe as a *state* the totality of parameters which characterise a given system at a given point in time.[28] There are 'inner parameters', that is, values of properties and spatial relationships between material components (things, configurations, and substances) of the system, and 'outer' parameters, or conditions (e.g. uniform pressure or temperature) characterising the field in which the material components are embedded and measuring the influences acting upon them. Current states can be observed by measurements in nature or in the laboratory. Past states, on the other hand, can only hypothetically be inferred from the configurations of chemical substances and minerals found, e.g. in a spatial element of rock.

The concept of state refers in the first instance to a point in time. Should the parameters characterising the state remain unchanged for a time interval, then a *stationary* state of a given span obtains. The stationary state is *stable* if the system persists under the given external conditions in accordance with physical and chemical laws of nature. The state is *unstable* if, firstly, the external conditions should correspond, according to physical and chemical laws, to inner parameters other than those actually observed, and if, secondly, it can be observed or expected that the state will change and move towards a new, stable state. This may occur spontaneously or may be triggered either chemically (catalysis), or by energy impulses. The state of the system is *metastable* if it exists for longer periods of time despite being unstable according to chemical and physical laws, and if catalytic or energy impulses can, if at all, only slowly and incompletely force it towards a stable state.

Example of stable states: the mineral assemblage of a metamorphic rock corresponding to a certain pressure/temperature range, as long as these pressures and temperatures continue to obtain; the mineral assemblage of 'lateritic' soils (aluminium and iron oxyhydroxides and kaolinite) produced from various original rock types under tropical rainforest conditions, as long as these conditions continue to exist (high precipitation, high mean annual temperatures, low pH values).

Examples of unstable states: landforms in the high mountains; heavy metal sulphide minerals produced under reducing conditions, where these come into contact with atmospheric oxyen when exposed along a vein.

Examples of metastable states: bituminous sediments and organic rocks in contact with the atmosphere; the exposure at the Earth's surface of diamond created under high pressures and temperatures; glassy rocks produced by the rapid cooling of melts.

Processes[29] are changes in the states of systems, observed either directly (current geology) or reconstructed hypothetically. The reconstruction of processes of the Earth's past is a major task of the Earth sciences, and the concept of a process is, therefore, a very commonly used category. Processes are named and distinguished according to the parameter which changes during the process and by which the stage of alteration can be recognized.

Chemical processes express themselves through alterations of chemical composition. Oxidation alters pyrite (iron sulphide) into iron oxides, ferric oxyhydrates, and sulphuric acids; desilicification signifies the disintegration of various silicates in the soil and the formation of kaolinite and aluminium oxyhydrates in tropical soils.

Physical processes may alter the shape and size of things, and create or destroy configurations. Examples are: increase in rounding and decrease in size of bedload during transport by a river, creation of asymmetric ripples on an even sand floor by flowing water, and disappearance of these when the water exceeds a certain critical velocity.

There are also equivalent *mineralogical processes* in the creation, alteration, and destruction of mineral substances, *morphological processes* in the creation, alteration, and destruction of landforms, and so on.

Processes can furthermore be divided up according to their temporal structure: there are *stationary* processes which run at constant speed, *delayed*, and *accelerated* processes.

A *stationary process* consists, e.g. of the transport of terrestrial clays into a sinking marine basin, proceeding at a steady rate over a long period of time and leading to the formation of a thick, uniform packet of clayey rock. The speed at which suspended matter is carried along a river course forms a delayed process.

Secular processes are usually stationary in kind, proceeding very slowly and taking effect only over geologic spans of time. Secular processes include slow crustal movements such as the uplifting of Scandinavia since the melting of the Pleistocene continental ice sheets or the drifting apart of North America and Europe, which, though theoretically assumed, has never yet been measured directly.

Cyclical processes in Earth science are not circular processes, leading back to the initial state, but repeatedly observed sequences of typical changes or events which appear to evolve from an initial state to an asymptotically achieved end state. Cycles play a role in the geomorphological description of landscape evolution: depending on the dominant agent of erosion one may differentiate fluvial, glacial, arid, and marine cycles, each divided into stages of 'youth', 'maturity', and 'old age'.

Circular processes are those which repeatedly reproduce the same state. Circular processes include the recycling of elements in geochemistry, where the path of individual chemical elements is followed, e.g. the 'exogenous' recycling of substances involved in the geologic processes of weathering, transport into the sea, sedimentation, formation of sedimentary rock, uplift of the rock, and renewed weathering.

Rhythmic processes consist of regular or irregular sequences of similar changes. The so-called 'strombolian' activity of some volcanoes is rhythmic, where fine ash and coarse material, ejected in more or less regular sequences, are deposited as tuff layers whose fine stratification reflects the rhythmic events. The rhythmic alteration of quantitatively and qualitatively differing summer and winter sediments in a lake, leading to the formation of so-called banded clays or varves, is a rhythmic process useful for chronometry.

Events are momentary changes such as the onset or cessation of a process, or processes of very short duration.

An event can for instance be the transgression of the sea over previously continental areas, caused by crustal movements. This can be recognised in the stratigraphic sequence as the sudden appearance of marine beds. All stratigraphic boundaries, i.e. discontinuous changes in deposited material (clay over sand, chalk over clay, etc.) document events in Earth's history. The same is true for tectonic phenomena.

Catastrophes are events which in some way cause spectacular effects, e.g. volcanic eruptions, sudden turbidity currents on the submarine continental slopes, and major meteoritic impacts.

The directionality of geohistoric time

The dimension of geohistoric time is reflected in processes. It therefore has direction, an arrow of time. In geohistoric time 'later' follows upon 'earlier' such that on the one hand the 'later' outstrips and continues the 'earlier', while on the other, the 'earlier' contains the germ of the 'later' within it. In order to arrange the processes and states of the geohistoric past as reflected in rocks and minerals according to their temporal succession, i.e. in accordance with the arrow of time, criteria for 'earlier' and 'later' must be established. The most important criteria for the direction of geohistoric events are (a) superposition, (b) inclusion, and (c) evolution.

(a) According to the principle of *superposition*, the spatial overlapping of rock beds reflects the temporal succession of geohistoric events. The time interval represented by homogeneous bed A should be historically earlier than that represented by homogeneous bed B, where B lies above A in the spatial sequence of beds. As early as 1669, Steno used this principle in the first attempt at an Earth history based on observation. It is still a useful instrument for determining the direction of geohistoric events, although it is not always reliable, for the following reasons. Strata do not always develop one after the other, but may form simultaneously (an example of this is the stratum-like A, B, and C horizons which form during soil development). The present spatial arrangement of strata may not always correspond to that which existed at the time of their formation: packets of beds may be tilted or even turned upside down by tectonic events. Tectonic influences can, as in the nappe structure of the Alps, thrust younger beds over older ones. In all these cases, spatial arrangement cannot be used to determine which is 'younger' and which is 'older', and in order to determine the direction of geohistoric time in the rocks other criteria such as evolution must be considered.

An example of an inappropriate use of the principle of superposition is the opinion expressed around 1800 by many geologists that the granites and gneisses of mountains such as the Saxonian Erzgebirge, the Harz, and the Alps, must be older than the sedimentary rocks there, as the sedimentary rocks overlay them.

(b) An *inclusion* is another form of spatial relation which provides a key to the temporal sequence of events. Where structure A is surrounded by structure B the formation of structure A *can* have taken place before that of B. We cite three examples.

A quartz porphyry contains inclusions of uniformly well-formed quartz crystals in a fine-grained matrix. One infers that the quartz crystals formed first, during the cooling of the magma, and the minerals of the matrix crystallised later.

A piece of stratified rock is found swimming in an igneously produced granite. One infers that this stratified rock is a fragment of a formation which formed before the intrusion of the granitic magma took place.

In a sedimentary conglomerate we find a rounded cobble of porphyry. Three successive events may be inferred from this: solidification of a magma into porphyry; fluvial transport and rounding of a fragment of this porphyry; deposition of the cobble in a conglomerate.

The principle of inclusion, however, is not wholly reliable on its own: totally enclosed crystals, concretions, and similar structures can form in a solid state during metamorphism or diagenesis, that is, they can develop later than the surrounding rock.

(c) *Evolution*. Processes manifesting themselves in minerals and rocks can indicate the arrow of geohistoric time when they show an evolution. Such an evolution is evident when a sequence of events reflects a trend showing the directionality of time. The features to be stressed in such an evolution are increasing randomness (increase in entropy) but also increasing differentiation and complexity.

Examples of increasing randomness: mineralogically heterogeneous primary rock complexes are, according to their grain size and mineral composition, reduced by physical and chemical weathering to very homogeneous masses of clayey rocks. A morphologically richly varied landscape is reduced by erosion and weathering to a featureless plain.

Examples of increase in differentiation: the process of fractionated crystallisation (see p. 236) produces a number of different igneous rocks from a homogeneous magma. A packet of stratified sedimentary rock is folded, faulted, overthrust, etc. during mountain building into a complex system of tectonic structures; individual events may be isolated out of this whose chronological sequence represents increasing complexity. The remains of organisms of an animal phylum embedded in sedimentary rocks are interpreted as stages of progressive evolution through geohistoric time if they can be arranged into a series showing increasing differentiation.

Evolution is the most important time-concept of geoscientific research. One speaks of the evolution of planets, the evolution of the oceans and continents, the evolution of mountains, of landscapes, of soils, or of plant and animal species. All these uses of the concept indicate the intention of geoscientific research to reconstruct the direction of geohistoric time out of the present composition of things, configurations, and substances.[30]

Age

The age of a rock or mineral or of any event fixed in the rock is the time interval which has elapsed till the present (e.g. till the year 1988) since the formation of the rock or mineral or since the inception of another event.

Phenomenological age is qualitative (relative) in kind and is reckoned according to features of superposition, inclusion, or phenomenological criteria of evolution. It will be established, for instance, that event A is older than event B or that the time interval between events A and B is greater than that between C and D.

Chronometric age is defined quantitatively and expressed in time units (e.g. in years). It presupposes the existence of a clock of hourglass type, i.e. of cumulative mechanisms in which rhythmic or steady accretion of material occurs per time unit, which accretion can be measured or counted and converted into time units.

Rhythmic clocks include tree growth rings and seasonally differentiated sediment sequences (e.g. banded clays). The slow sedimentation of bedded rocks on the sea floor was formerly used as a clock of the evenly accreting type. Today radiometric clocks are used based on the decay and decay constants of radioactive atoms. Where these constants are known, the proportional quantities of mother to daughter elements can under certain conditions be used to determine the amount of time which has elapsed since the time when the given rock or mineral contained only the mother element. One condition is that no loss of the daughter substance has taken place since then.

Simultaneity

To set up a global geohistory it must be possible to correlate chronologically the sequence of events found to have occurred on various parts of the Earth's surface, and indeed, on other planets as well. It is particularly important to determine the simultaneous occurrence of events reflected in the minerals or rocks of different sites.

Chronometrically the simultaneity of events (particularly of the formation of minerals or rocks) can be determined by radiometric dating. Not all rock series, however, contain the appropriate minerals and rocks for this.

Phenomenological correlation of rock formations presupposes that events occurred which took place simultaneously or coetaneously at spatially distant places. Such correlations are based, for instance, on the reversals of the magnetic field, marine transgressions and regressions, regional mountain formation (tectonic 'phases'), climatic changes (registered in climatically indicative sediments, faunas, or floras), phases of biological evolution (marked by so-called 'key fossils'), or the deposition of products of major volcanic eruptions.

The various methods of phenomenological correlation have different ranges of application and specific degrees of uncertainty and inaccuracy; these depend to a large extent on the state of geologic knowledge and on the theories underlying the correlation methods.

Time scales

The ordering of the geoscientific universe requires not only a spatial coordinate system for the Earth, planets, and planetary space, but also a global time scale into which all geohistoric events and developments may be fitted. The definition of such scales and the standardisation of their divisions is one of the most important tasks of research and of international communication, and is not by any means complete. Two time scales claiming to have global validity are in use today.

The *geochronometric time scale* is based on radiometric determinations of age and divides geohistoric time into absolute time units of set magnitude. A hierarchy of 10^3, 10^6 and 10^9 years is usual.

The *geochronostratigraphic time scale* is based on the time intervals during which the stratigraphic units of sedimentary rocks (so-called formations) were formed. These divisions have been internationally agreed upon and their boundaries are defined by fossils and other phenomenological criteria contained within their beds.

The geochronometric and geochronostratigraphic time scales are related to one another. The geochronostratigraphic scale is necessary, for it is not possible to date most sedimentary rocks absolutely.

3.3.6 *Ordering into complexes*

All disciplines of geoscience deal to a far greater extent than, e.g. physics, with wide complexes of individual entities. Things, substances, configurations, processes, and states must be named. In order to describe them and most especially to form explanations, hypotheses, or theories, definitions must be found such that individual objects which are defined as homogeneous can be subsumed under conceptual classes or genera, which in turn must dovetail into a lucid system. In the various fields of natural history, 'natural' and 'artificial' systems used to be differentiated; the term 'natural' system referred to a real, ontologically founded ordering of an object-realm based on 'real definitions' (see p. 51), while an 'artificial' system was a more arbitrary and conventional construct based on 'nominal definitions'. An example would be the contrast between the 'artificial' system of Linné, and the 'natural' system of the plants.[31]

Many geoscientific complexes are conceived of as 'natural' orders whose structuring is not determined by convention alone. No matter how the concepts 'calcite' and 'quartz' are defined, the existing minerals are different. Nevertheless it must be stressed that, as the history of the discipline shows, different ways exist and have existed for ordering the various complexes. Definitions of terms, classes, and systems have each been developed to serve the needs and goals of research. With this in mind, we will look in a later chapter at some examples of terminologies and nomenclatures in use today.

Following a suggestion of Hubaux,[32] the following types of geoscientific complexes can be formally distinguished.

1. *Discrete complexes.* These consist of members between which there are no transitions as regards characteristic properties. Some examples are chemical elements, mineral substances, and plant and animal species. In such complexes there are various ways of distinguishing higher-level concepts depending on the choice of distinguishing features, such as the mineral class 'phyllosilicates', which includes many separate minerals such as biotite, muscovite, or kaolinite as cohyponyms at a lower level. Such higher-level concepts can be defined by convention. The lower classes of minerals, however, occur in nature as discrete entities, with as few transitions or intermediates as exist between chemical elements. Ordering at this lowest level of discrete complexes is ontologically predetermined, the only thing left to convention being the definition of the natural differences.

2. *Continuous complexes with known coordinate axes.* These consist of elements whose characteristic properties grade into one another so that there are transitional members between any two elements. The elements can graphically be visualised as points in a spatial continuum whose coordinate axes correspond to known, measurable properties. Examples are individual loci in fields, each characterised by given spatial coordinates and by one physical or chemical parameter each, such as temperature or content of a chemical substance. Another example is the one-dimensional ordering of points in time (e.g. a series of volcanic eruptions) and of time intervals (e.g. of global ice ages) on the scale of geohistoric time.

In complexes of this kind the qualities by which individual elements are distinguished from one another are naturally present. It is convention, however, which draws the boundaries between 'individuals', which then no longer appear as points in the spatial image, but rather as elementary fields. Convention must then establish further groupings which provide for internal differentiation.

3. *Continuous complexes with unknown coordinate axes*. These consist of elements having transitions and intermediate forms with regard to all of their properties, and for which it is furthermore not clear at the outset which parameters should be used to distinguish and subdivide them. Such objects can graphically be visualised as points in a multi-dimensional spatial continuum where neither the number nor the kind of coordinate axes are known or 'naturally' defined. In this case the coordinate axes must be decided upon before delineating elementary fields and internal groupings.

Many sets of geoscientific objects belong to this kind of complex. As some examples we may name the continua of rocks (p. 108), tectonic forms and units, geomorphological configurations (p. 121), and soils. But in nearly every specialised investigation, the researcher encounters continua which must be rationally ordered, such as the textures of genetically related rocks, the continuum of chemical composition in rock types from one area, or the various morphological and compositional parameters characterising the rounded pebbles in a conglomerate bed.

Discrete complexes are the easiest sets of homogeneous objects to deal with in research. This explains the vain attempts to discover or construct discrete groupings in complexes which in nature are continuous: it was at one time believed, for instance, that rocks, like minerals, formed discrete 'rock types'. Empirical review has shattered this hope: rocks form a continuous complex.

Unlike discrete complexes, continuous ones must first be ordered even at the lowest level. Such a start depends on two provisos. Firstly, the creation of a scale according to which each element can quantitatively be characterised and distinguished from other elements. Secondly, it requires the constant awareness that the aim is not simply to create an inventory, but one that fulfills the interests of research, and that the terms must therefore be defined in accordance with research goals. In complexes of type (2) this relates largely to the delineation of boundaries within the continuum. In complexes of type (3) one must also determine the coordinates by which the continuum will be divided up and the objects ordered. Thus, given the intention of explaining the formation of igneous rocks according to chemical laws, the continuum of igneous rocks must be organised according to chemical parameters, not, e.g. according to textural features.

4

Problems at the empirical basis

4.1 Introduction

In our analysis of the structure of argumentation of geoscientific papers (p. 20) we distinguished two major parts: the presentation of the theme (*propositio*) and the presentation of the hypothesis (*argumentatio*). The most important component of the first part, namely all the findings obtained through observation and experiment as material for discourse, we call the 'empirical basis'. These are presented by the author to the reader as facts, and it is on them that the presentation of the second major part (the argumentation) is based. The second part develops propositions which transcend the assertions about facts in the form of inductive generalisations, explanations, laws, hypotheses, theories, and so on.

To distinguish informative or descriptive statements of the empirical basis (those which the author does not question within the framework of his text) from statements considered problematic on higher levels of discourse, we do not use features of their content. That is, we do not identify descriptive statements because they speak only of that which can be directly perceived. Rather we attribute to the empirical basis all statements of a text which the author intends should fulfill a certain function: they are statements presented to the reader as reports of observed or experimentally obtained, intersubjectively testable singular facts.

This pragmatic, functional definition of the empirical basis of geoscientific research contrasts with some current views. We start by noting that the statements of the empirical basis are not simply reports on subjective experiences of the author, nor are they descriptions of elementary sensory impressions and natural phenomena stated in words of daily discourse. The scientific experience proceeds from daily experience; it is not a continuation of it, but develops out of it through selection and abstraction, i.e. through processes which destroy or at least transform the objects of naïve experience. 'A scientific experience is therefore an experience which contradicts the common experience'.[1] What an Earth scientist notes with regard to certain objects and phenomena differs fundamentally from what the layman notes. The rocks which a hiker gathers while walking through the mountains are generally quite different from the hand specimens which a geologist gathers along the same trail. Selective acts of observation and experiment establish and produce only such facts as are relevant for the theoretical goals of the individual disciplines. These facts, the results of observations and experiments, are recorded and imparted not in every day vocabularies, or at least not only in every day language, but by means of descriptive terminologies and nomenclatures.

Statements at the empirical basis differ, therefore, from non-scientific accounts both in origin and in form. They are obtained by observation and experiment according to standardised methods of dealing with nature and they are expressed in a standardised form of scientific language as developed in each separate discipline. Standardised methods and standardised language are essential for that intersubjective testability required of all statements at the level of the empirical basis. The empirical basis must include only information which can be acquired, not perhaps by everybody, but by any member of the scientific community familiar with the methods of observation and experiment and with the terms of the descriptive language.

Since both the descriptive methods and the terms used to obtain and describe empirical data rely on theoretical concepts, explanations, and goals, it is impossible to distinguish statements of the empirical basis from those of the higher level of knowledge using the Two-Language-Model proposed by R. Carnap and other philosophers of the school of logical empiricism.[2] According to this model it should be possible to assign any scientific statement, independent of context, either to the level of theory-neutral observation language, or to the theoretical level of language. To ascertain such a thing, however, it would have to be possible to distinguish the vocabulary of a theory-free observation language from that of theory-laden language. Moreover, one would have to determine what is to be designated as 'observable' and what as 'non-observable'. Even Carnap has to admit that there is no sharp boundary between these two concepts:

> There is a continuum of meaning (namely of the concept 'observable') which starts with direct sensory observations and proceeds to enormously complex, indirect methods of observation. Obviously no sharp line can be drawn across this continuum.[3]

In order to uphold the concept of two languages, Carnap suggests that the 'observable', i.e. theory-neutral values and concepts include anything which can either be directly perceived with the senses or measured with 'relatively simple methods'.[4]

Criticism of the Two-Stage-Model is based on the recognition that there are no 'pure' observational statements in science, since any linguistic formulation of observational results presupposes the acceptance of certain theories, hypotheses, or generalisations transcending the realm of what is empirical in the narrower sense. As Schäfer stresses,[5] Duhem showed that the popular view in which a physical experiment consists of instrumental observation uninfluenced by theories is wrong. A physical experiment, rather, involves two components.

> Firstly it consists of the observation of certain facts. In order to carry out these observations, it is enough to be attentive and to have receptive senses. A knowledge of physics is not needed; the director of a laboratory may be less adept than the laboratory assistant. Secondly it consists of interpreting the observed facts. To carry out this interpretation it is not sufficient to have a trained eye and alert senses, one must know the relevant theories and how to apply them: one must be a physicist.[6]

From this it follows that:

> a physical experiment is the precise observation of a group of phenomena, combined with the INTERPRETATION of these. This interpretation replaces the concrete particulars, the results actually obtained by aid of observation, with abstract and symbolic representations which correspond to the particulars on the basis of theories which the observer accepts as valid.[7]

Two components may accordingly be distinguished within those informative statements of geoscientific texts which we assign to the empirical basis. The first component relates to facts which can be stated even at a prescientific level, such as that a rock mass fell to Earth from the heavens on a certain day and in a certain place; in particular it also contains scientific descriptions which rely on fundamental theoretical concepts, as, for instance, data on temperatures. Secondly, however, observation is always 'observation in the light of theory',[8] which can be questioned or modified. Observation and experiments are planned within the framework of theoretical preconceptions, questions, or expectations; they are carried out using methods based on theories, and their results are expressed in terminologies and nomenclatures which in turn rest on a theoretical foundation. The texts abstracted in Section 3.1 illustrate, for instance, that statements forming the empirical basis contain a core of intersubjectively testable facts in the form of fundamental data which could be expressed in pre-scientific form. These facts, however, are obtained on the basis of theory-guided questions and by means of theoretically founded methods; they are finally described using scientific terminology implying interpretations based on theories which, though not questioned at the moment, are in principle open to doubt.

Among the theoretical preconceptions entering into statements at the empirical basis of geoscientific texts, the concepts and laws of physics and chemistry play a fundamental role. Like all natural sciences, the Earth sciences build on a foundation of general laws of nature developed by physics and chemistry. This foundation is not regarded in the Earth sciences as problematic, but as a basis for observational techniques and for the interpretative formulation of observational results. Every advance and every change in physical or chemical knowledge can therefore produce basic changes in Earth science. Thus the discovery of the diffraction of x-rays by a crystal lattice and the development of physical theories on these phenomena caused basic changes in the classification of minerals, and hitherto unknown domains were discovered for geoscientific research, such as the mineralogic composition of clays and soils. In addition to preconceptions from physics and chemistry many other non-empirical elements are also included in the empirical basis. Concepts and theories from the biological sciences are found in paleontology, and observational results are described using concepts belonging to the theoretical level of Earth science. Statements on the formational provenience of rocks, the naming of rocks according to petrographic nomenclature, the description of tectonic texture, the classification of geomorphologic forms are all not only descriptions but claims implying hypotheses and presupposing geoscientific theories.

In summary, we may state the following regarding the interrelatedness of non-empirical preconceptions and statements at the empirical basis of geoscientific texts.

1 Statements about the results of scientifically relevant observations and experiments are based on a core of empirical findings which are invariant with respect to the theory under discussion and which are recognised as true no matter how they may be named or interpreted.
2 Observational and experimental results imparted as part of the empirical basis depend on non-empirical preconceptions in three ways. Firstly observations and experiments are planned in light of theory-guided questions. Secondly, the methods and instruments used to obtain results are based on theories. Thirdly, the facts obtained are communicated through scientific language in which hypotheses are invariably introduced by using names which imply interpretations.

3 Since the results and descriptions of observations and experiments depend on the methods, hypotheses, and theories recognised at the time, the possible content of the empirical basis of geoscientific texts is at any given time determined by conventions which change with the advance of science.[9]
4 Discoveries which in the context of one investigation seem indubitable statements of fact may be questioned in another context, as when checking the preconceptions underlying the observational or experimental methods or the interpretative naming of phenomena.

The empirical basis of geoscientific argumentation must be examined with regard to content and form. Its content consists of the data obtained by certain methods in nature or in the laboratory. The form in which the empirical basis appears in papers is determined by the verbal and non-verbal means available within scientific language, especially the terminologies and nomenclatures of the individual disciplines. In the following sections we will examine those two aspects, first analyzing a few examples of geoscientific terminologies and nomenclatures, and then investigating the foundations of the methods used in the Earth sciences to obtain empirical data.

A critical look at the empirical basis of an investigation must take both components, content and form, into account. First the methods used to convey facts must be examined to see whether they fulfill the demands of precision, reliability, and testability, and whether their theoretical or hypothetical preconceptions are clear. Second, the terms must be examined to see if they are being used according to their established or explicitly defined meanings.

4.2 Examples of descriptive terminologies

4.2.1 *Terminology, nomenclature, and systems of minerals*

The definitions of mineral names and the systems of minerals are of basic importance for all Earth sciences, since all geosciences consider the Earth to be a composite of minerals. Minerals are the substrate of all phenomena treated by the geosciences. The scientifically defined mineral names establish from the outset and at all times how a part of the Earth's body can be differentiated and described using mineralogical language, and how such a description can then be used in argument at a higher level of discourse. A publication by Freiesleben in the year 1817, *Abraham Gottlob Werners letztes Mineralsystem*, lists the names of 317 minerals. H. Strunz's *Mineralogische Tabellen* from 1966 contain 1630 names of well-defined minerals. Leaving aside advances in the definition of names, the very numbers alone show how significantly the descriptive possibilities have increased in 150 years.

The categorical meaning of the term 'mineral' has changed in the course of the history of science. Up into the first half of the 19th century the natural world was divided into the three 'kingdoms' of animals, plants, and rocks, corresponding to the 'natural' division of the natural sciences into zoology, botany, and mineralogy. By analogy with zoology and botany, it was believed that mineral taxonomy should be based on the concept of mineral species. The definition of species was understood to imply the identification of the shared essence of similar individuals, that is, it was assumed that individuals could be found in the world of minerals as in the world of plants and animals. How to base the definitions of mineral species and what differences or similarities there might be between species of the animate and inanimate world could not be agreed upon. The controversy between those who wanted to base the species of the mineral kingdom on chemical composition (e.g. Berzelius) and those who wanted to found them on external features of mineral species such as colour, hardness, and crystal shape (e.g. Mohs and Breithaupt) could not be resolved. Judging by the introductory chapters of more recent textbooks on mineralogy the controversy still seems undecided. Clearly there is no unified opinion on the definition of the term 'mineral', as is shown by the following examples.

> The honogeneous building blocks of the Earth's crust are called minerals. Scientific research must endeavour to group individual minerals into species.[10]
>
> Minerals are homogeneous natural bodies in the physical and chemical sense, which occur almost exclusively in the form of crystals, or at least as fine to coarsely grained crystalline aggregates.[11]
>
> Minerals are types of crystals which form or can form in nature without the influence of organisms . . . The natural world in which we live is divided up into the kingdom of minerals, the kingdom of plants, and the kingdom of animals.[12]
>
> Mineral: a naturally formed chemical element or compound having a definite chemical composition and, usually, a characteristic crystal form . . . Any naturally formed, inorganic material, i.e. a member of the mineral kingdom as opposed to the plant and animal kingdom. A naturally occurring, usually inorganic, crystalline substance with characteristic physical and chemical properties that are due to its atomic arrangement.[13]

Leaving aside these contradictory attempts of current textbooks to define explicitly the concept 'mineral', and turning instead to the actual use of mineral names in science, we obtain a clearer picture.

Mineral names are indeed always applied to singular things such as individual quartz crystals, grains of feldspar in a rock, or diamond crystals, of which the most famous, such as the 'Kohinoor', may even be designated by a proper name. Yet statements about minerals always appear in the form:

x consists of M

where x and M can be replaced by designatives for different categorical entities. x can be replaced by the name of a certain crystal or grain, i.e. the designation of a singular thing. M on the other hand stands for the name of a mineral. This mineral name does not signify a class of individuals of one species, as the name *Taraxacum officinalis* signifies the class of all individual dandelion plants. The predicate 'quartz' does not stand for all quartz individuals but rather for the substance of all things consisting of quartz, whether they be grains of sand, individual granules in a granite, crystals in a druse, or even artefacts such as the resonating crystal in a quartz watch, made from a naturally occurring quartz crystal. Mineral names thus permit the descriptive differentiation between things occurring in nature according to the mineral matter of which they consist. It should be noted that German, in contrast to English, still retains traces of the former concept of 'mineral individuals'. Mineral names are still used in the plural in German, referring to '*die Quarze*' or '*die Feldspate*' in a rock, instead of to *grains* of quartz or feldspar (*Quarz- und Feldspatkörner*), just as one speaks of 'the foxes' or 'the fir trees' in a forest. In English, mineral names (insofar as they are substance designations) cannot be used in the plural, and one must speak of 'quartz grains' or 'feldspar crystals'.[14]

Minerals may be categorically described as substances. As far as mineralogy is concerned, they are those substances of which all things dealt with in the Earth sciences consist and for which the following is true: each mineral name designates a substance having specific spatial and temporal constants and in which, within certain limits of external influences and conditions such as temperature and pressure, certain persistant characteristics can be observed in all singular entities consisting of the given mineral. Naturally occurring minerals form a discrete complex with regard to characteristic properties. As in chemistry, materials of a more or less persistent nature, forming a discrete complex are designated 'substances', and are the elements and compounds of chemistry. Each mineral name accordingly designates a mineral substance. The various mineral substances differ in many characteristics, and for each mineral name, textbooks and handbooks of descriptive mineralogy list long series of qualitative and quantitative properties, as well as internal and external spatial relationships observed on things consisting of the given mineral. Various types of properties can be differentiated in objects composed of a certain mineral, according to their significance for the definition and practical identification of the mineral.

All predicates pertaining to the composition and behaviour of a given thing consisting of mineral y we call *properties* (P).

Of particular interest are the *necessary properties* N of a mineral substance. These are those properties which must be shared by all things consisting of the mineral y; the complete set of necessary properties constitutes that which may be called the 'nature' of the mineral. The list of necessary properties of a mineral is never closed. Older handbooks list just a few properties, such as colour, lustre, hardness, specific gravity, cleavage, crystal form, chemical composition, and melting point. Newer handbooks have considerably longer lists, giving data on elastic properties, optical properties for light of various wavelengths, electrical and thermal conductivity, thermodynamic parameters, details on crystal structure, and so on. Research is constantly turning up new properties, so that the list of necessary properties of a mineral substance is endless.

From the potentially infinite set of necessary properties we must select a subset of those features which form the explicit definition of the mineral substance. *Defining properties*, D, are those necessary properties sufficient to establish that a given thing consists of the mineral substance y. The selection of the defining properties from the set of necessary properties depends on theoretical considerations establishing which of the properties of a mineral substance are to be regarded as fundamental in the sense that other necessary properties can, in fact or in principle, be deduced from them. The decisive point of view in this regard has changed during the history of mineralogy, so that the definitions of minerals have their own significant history which we cannot enter into here.

Today the definition of each mineral is based on two groups of necessary and sufficient properties:

1 chemical composition, and
2 arrangement of atoms in space.

Both groups of defining properties are physico-chemical in nature. This means that describing natural objects by using mineral names opens up the possibility of discoursing at higher levels of knowledge using chemical and physical concepts, laws, and theories. We will use the example of the mineral rutile to clarify the definition of mineral substances through chemical composition and arrangement. The mineral substance 'rutile' according to our first group of features (chemical composition), can be identified as the chemical substance 'titanium dioxide', designated symbolically by the formula TiO_2. This formula expresses the quantities of the substances titanium (Ti) and oxygen (O) found when any substance named 'titanium dioxide' is analysed. Mineralogical nomenclature subdivides all 'TiO_2' substances (which are indistinguishable at the level of chemical analysis), into three mineral substances according to their atomic arrangement. One of these is the mineral rutile.

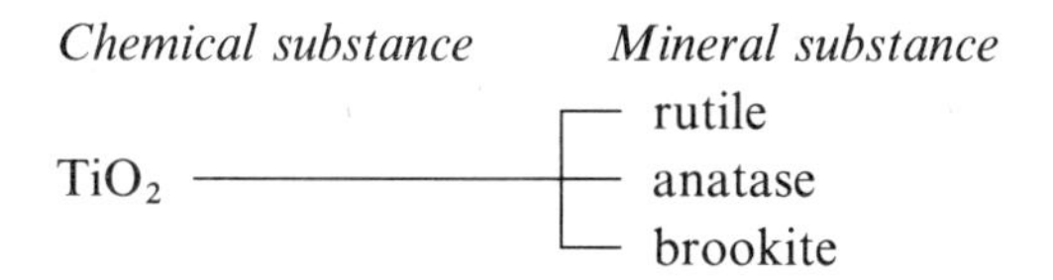

Rutile, anatase, and brookite differ in the arrangement of their atoms in the ordered crystal lattice, and, when the substance forms well-developed crystals (which is not necessarily the case), through macroscopically recognisable crystal forms.

The relationships between chemical substances and mineral substances are not always as simple as in the case of the TiO_2 minerals. Many minerals have a chemical composition which varies continuously within certain boundaries while maintaining the same arrangement of atoms (crystal structure). Such mineral substances are called solid solutions. The mineral substance olivine, for example, is defined chemically by the formula $(Mg, Fe)_2SiO_4$ indicating that the name olivine refers to all minerals of like atomic arrangement (olivine structure) composed of any mixture ranging from pure iron to pure magnesium orthosilicates (Fe_2SiO_4 to Mg_2SiO_4). For pragmatic reasons such continuous series are generally subdivided according to convention. Names of the subdivisions (e.g. forsterite for the Mg-rich, and fayalite for the Fe-rich olivine types) designate members of continuous complexes and differ in this respect from the supraordinate names of the discrete mineral substances.

There have been repeated attempts in the past to name the minerals systematically so that each name would indicate the mineral's position within the system of all minerals. This has failed above all because no system could be developed out of the older definitions of the mineral substances. Mineral names are, therefore, in no sense 'telling', they say nothing about a mineral's defining properties, nothing about its position within any system. Some names are derived from antiquity or from the Middle Ages, some from miners' jargon, some from striking characteristics, chemical composition, place names, or names of persons to be honoured.[15] There is, therefore, no nomenclature *sensu strictu* of the individual minerals, though each mineral name is an explicitly defined term. There is, however, a nomenclature of the mineral system, that is, of the concepts constituting the system. This system has the semantic character of a hyponymous arrangement of concepts into classes, divisions, groups, and species. The higher

Table 4.1 *Scheme of mineral classification, using the mineral quartz as an example*

Classes (chemical)	Divisions (chemical)	Groups (chemical)	Mineral species (crystal structure)
Elements	M_2O	TiO_2	quartz
	MO	MnO_2	high quartz
			tridymite
Sulfides	M_3O_4		high tridymite
			cristobalite
Oxides →	MO_2 →	SiO_2 →	high cristobalite
			coesite
			stishovite
etc.	etc.	etc.	

ranks, that is, the classes, divisions, and groups, are chemically defined, while the lowest level of mineral species is differentiated according to atomic arrangement (crystalline structure). Table 4.1 shows the mineral species consisting of SiO_2 as an example of this hierarchy. On each individual rank there are several cohyponyms (e.g. the oxide division contains many compounds of metals with oxygen). The supraordinate concept (here the class 'oxide') is extensionally broader and intensionally narrower than a concept at the next lowest rank (e.g. the concept MO_2, where M is a metal).

The two properties according to which minerals are defined today, chemical composition and atomic arrangement, are based on the recognition that minerals are not simply substances, but are made up in turn of lower level substances. This substance level belongs to the domain of physics and chemistry and is investigated by means of chemical and physical methods. Considerable apparatus and knowledge of physical and chemical laws are necessary to identify the defining properties by which a mineral substance can be recognised with precision. Chemical composition can be established only by chemical analysis, and atomic arrangement can only be established clearly by means of experiments involving X-ray diffraction. In practice, however, such identification techniques using the defining properties are not ordinarily applied. Usually it is sufficient to recognise a mineral by a group of necessary but insufficient properties which can be observed by those methods which are at hand. This identifies the mineral substance according to pragmatic characteristics.

Pragmatic characteristics (PC) are a subset of the necessary properties of a mineral substance y. This subset is not fixed, but is chosen according to convenience to determine whether an object consists of mineral substance

y. PC always forms a subset of N, but need neither coincide nor even overlap with D. Usually no element of PC is an element of D. Examples of pragmatic characteristics are the so-called external characteristics, the properties which can be observed by 'simple' means. Here the 'tables for the identification of minerals according to external characteristics' are generally used as a guide. The pragmatic characteristics on which such tables are based are lustre, colour, streak, hardness, tenacity (qualitative solidity behaviour), crystal form, cleavage, fracture, and aggregate shape. Since all these properties are necessary but not sufficient attributes of the mineral substance, they cannot provide a basis for unequivocable placement in the system, especially as they are ascertained only by qualitative or at most comparative data, permitting no exact demarcation. Objects may agree in pragmatic characteristics so closely with one another as to be indistinguishable, even when according to the defining properties they consist of different mineral substances. This is why identification tables always have a column of 'similar minerals' noting possible confusions, for these can always occur when limited to pragmatic characteristics and are only excluded by the set of defining properties.

Another commonly used group of pragmatic characteristics is the set of 'optical characteristics'. A mineralogical microscope and knowledge of the laws of crystalline optics are needed to discern these. Although the use of quantitative data here allow greater precision, the optical properties of a mineral substance must still be classed as necessary, but insufficient properties. If a mineral is found corresponding exactly to the mineral substance quartz in terms of light refraction and other optical qualities, though it is empirically established that no other known substance possesses these properties it may yet be another (as yet unknown) mineral whose defining properties differ from quartz. Thus, basic confusions can in principle exist even in the much more exact field of optical properties, so that any guide for the optical identification of minerals also contains a column for 'similar minerals'.

Finally the *contingent properties* (C) of the mineral substance y are also important. These are properties which can be observed in objects consisting of mineral y, but which do not belong to the necessary properties of y. They are the remnant set, therefore, once the subset of necessary properties has been sorted out of the set of all the properties of y. Since the contingent properties of mineral substance y do not necessarily occur, they will be observed only in some of the objects composed of y. Properties such as colouring, crystal development, or inclusion of foreign substances can be used to name mineral varieties which may show important clues to special formational conditions, or may be of commercial value, e.g. jewellery or

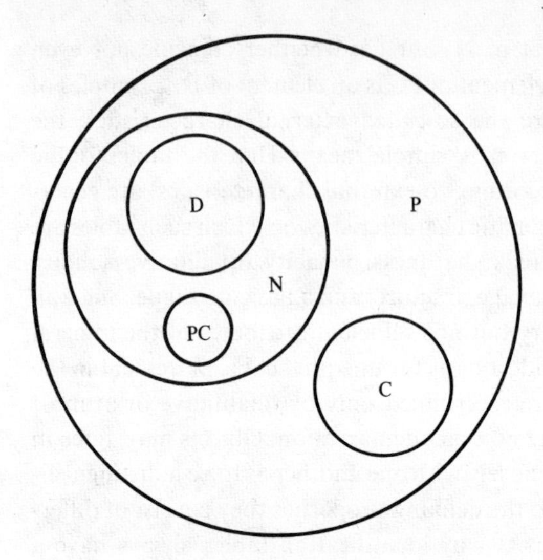

Fig. 4.1. On the definition of properties (P), necessary properties (N), defining properties (D), pragmatic characterisitics (PC), and contingent properties (C) of minerals.

precious stones. Some of the many varieties of the mineral substance quartz will serve as examples.

Rock crystal: colourless, transparent. Smoky quartz: grey. Citrine: yellow. Amethyst: violet. Rose quartz: rose colour. Chrysoprase: green, through the inclusion of hornblende needles. Chalcedony: finely-crystalline aggregate of varied colour and shape (agate, onyx, carnelian).

Fig. 4.1 shows the relationships between properties (P), necessary properties (N), defining properties (D), pragmatic characteristics (PC), and contingent properties (C). N is a subset of P, D a subset of N. PC, which is chosen by convenience, is also a subset of N. C is a subset of P, but does not belong to N.

4.2.2 *Terminology, nomenclature, and systems of rocks*

Rocks make up large units of the solid Earth and other planetary bodies. One might say that the peak of the St. Gotthard mountain in the Swiss Alps consists of granite: a word usage which shows that rock masses, like mineral names, categorically designate substances. There are, however, differences between mineral substances and rocks.

In the first place the terms 'rock' and 'mineral' are hierarchically related, corresponding to the relationship between 'mineral' and 'chemical element'. Just as chemical substances are the substance of minerals, so minerals are the substance of rocks, for rocks consist of various minerals. In the second place, unlike minerals, which are homogeneous right down to the atomic level, rocks are spatially inhomogeneous even at the macro-level. They consist of spatially bounded bodies such as grains or flakes, which consist for their part of mineral substances. Every piece of rock can be described by features belonging to the categories of substances and of configurations, that is, by its qualitative or quantitative mineral composition, and by the spatial texture of its mineral components and other structures. The description of larger bodies of rock occurring in nature (e.g. Brocken granite for the granite making up the Brocken mountain) is made difficult by the fact that both the mineral composition and the texture of such a mass usually vary from one spot to another. In the third place, unlike minerals, all compositional and structural properties in rocks form continuous complexes with indeterminate coordinate axes since there are from all aspects various ways of ordering and defining the properties of rocks. A 'natural' system of rocks is only sketchily evident, if at all, and any ordering must be founded on a consensus. For this reason both the definitions of rock names and the construction of a nomenclatorial system have always been debated in the scientific community. This must continue to be the case, for advances in mineralogy and petrology continually produce new viewpoints and new needs for appropriate subdivisions of the continuum. Improved international communication in the last few years has permitted a certain degree of (tentative) consolidation.

The following features associated with the categories of substances and configurations are now used for defining and systematising rocks.

I. Substance properties:

(a) Composition of chemical substances according to qualitative and quantitative chemical analyses.
(b) Composition of mineral substances according to qualitative and quantitative mineralogical analyses.

II. Configuration properties:

(a) Internal fabric, e.g. size, shape, and relative position of the homogeneous mineral components.
(b) External fabric, e.g. overall configuration of the rock mass due to discontinuities, surficial forms, and other structures.
(c) Associations, i.e. spatial relations to other rock masses.

For the observing Earth scientist rocks are indices for those processes and events of the past which formed the rocks or produced certain features of their mineral composition or structure. Rocks found in the field are, therefore, immediately evaluated as to their origin and assigned to one of the following three main genetic classes.

I Igneous rocks, formed by the cooling of magmas.
  (a) Plutonic rocks, formed by slow cooling at depth.
  (b) Volcanic rocks, formed by rapid cooling on or near the Earth's surface.

II Sedimentary rocks, formed at the Earth's surface through the effects of wind, water, or ice at 'normal' temperatures.

III Metamorphic rocks, formed by alteration of the fabric and/or mineral composition of rocks of classes I and II through the effects of increased pressure and/or temperature.

A given rock is assigned to one of these major classes according to features of its substance's composition or of configuration. Thus, certain mineral compositions and features of internal fabric constitute the necessary and sufficient characteristics of igneous rocks. Among these, plutonic and volcanic formations can be distinguished by special features of internal fabric and association with other rocks. Peculiarities of external structures such as stratification are among the characteristics of sedimentary rocks. Metamorphic rocks are distinguished by characteristic minerals and specific internal and external textural and structural features of igneous and sedimentary rocks. A given rock is assigned to one of the major groups on the basis of abduction, starting from the observed substance and configurational properties and abductively inferring the conditions under which these originated. The assertion that a given rock was produced through igneous, sedimentary, or metamorphic processes is nevertheless treated as a statement at the empirical basis, at least in any investigation which is not expressly concerned with showing that a rock belongs to a given class. It is assumed that there is agreement as to which features are necessary and sufficient to assign a given rock to one of the main genetic classes. A glance through the textbooks, however, reveals that this agreement is largely tacit. Instead of an explicit, systematic enumeration and discussion of the characteristics necessary and sufficient for class membership, petrological textbooks describe the physical and chemical conditions which, according to the latest research, obtain during the formation of igneous, sedimentary, and metamorphic rocks. It is left up to the researcher to infer those properties that can be used (abductively) to recognise which of the three to five processes of rock formation were responsible for forming a given rock.

The abductive basis for assigning rocks to the main genetic classes necessarily entails that fundamental problems in assigning a given rock within this scheme can in principle arise, and it may have to be reclassified in the course of research. Around 1800, many geologists of the so-called

Neptunist school believed granites and basalts to be sedimentary, until the triumph of the Plutonist view that these rocks show features of igneous origin. A significant number of problems still exist today. The texture of so-called migmatites suggests that they consist of mixtures of magmas and material altered while in a solid state. Some granites have been thought to be of metamorphic origin instead of igneous as is generally assumed. There can be doubt as to which features indicate incipient metamorphism and how to distinguish these from the less intensive alterations involved in diagenesis when a sediment is buried at shallow depth. Coarsely crystalline rocks of calcite may have been produced by metamorphism from limestone (marble) or may have solidified out of a magma (carbonatite). In recent times, rocks of so-called impact metamorphism have been discovered, formed by the extreme pressures and temperatures produced during large meteoritic impacts; under some circumstances these can be very difficult to distinguish from normal igneous and metamorphic rocks.

The definitions of rock names and the design of the nomenclatorial systems are different in the three main classes. In *igneous rocks* the defining features are based on mineral composition. The quantitative mineral composition is either determined directly (modal mineral content) or calculated according to agreed methods from the chemical composition (normative mineral content). The latter procedure has the advantage that even very fine-grained, glassy rocks can be assigned a place in the system. Of the many minerals which occur in igneous rocks, the main or rock-forming minerals are selected and grouped as follows:

Q: minerals composed of SiO_2, usually quartz;
A: alkali feldspars (potassium and sodium-rich feldspars);
P: plagioclase (calcium and sodium-rich feldspars);
F: feldspatoids, usually leucite and nepheline;
M: iron and magnesium minerals and ores.

An initial subdivision distinguishes leucocratic and melanocratic rocks according to their M mineral contents.

Leucocratic rocks: under 90% M,
Melanocratic rocks: over 90% M.

The great majority of rocks of the Earth's crust are leucocratic. These are differentiated according to their content of A, P, Q, or F minerals.[16] Due to chemical reasons, Q and F minerals are mutually exclusive, so that any leucocratic igneous rock can be represented graphically as a point in the A–Q–P or A–F–P triangle. The double triangle in Fig. 4.2 is a diagram of type Qt–Qt–Qt (see p. 66), into which every leucocratic plutonic rock can be fitted.

If the mineral compositions of the many thousands of rocks described in

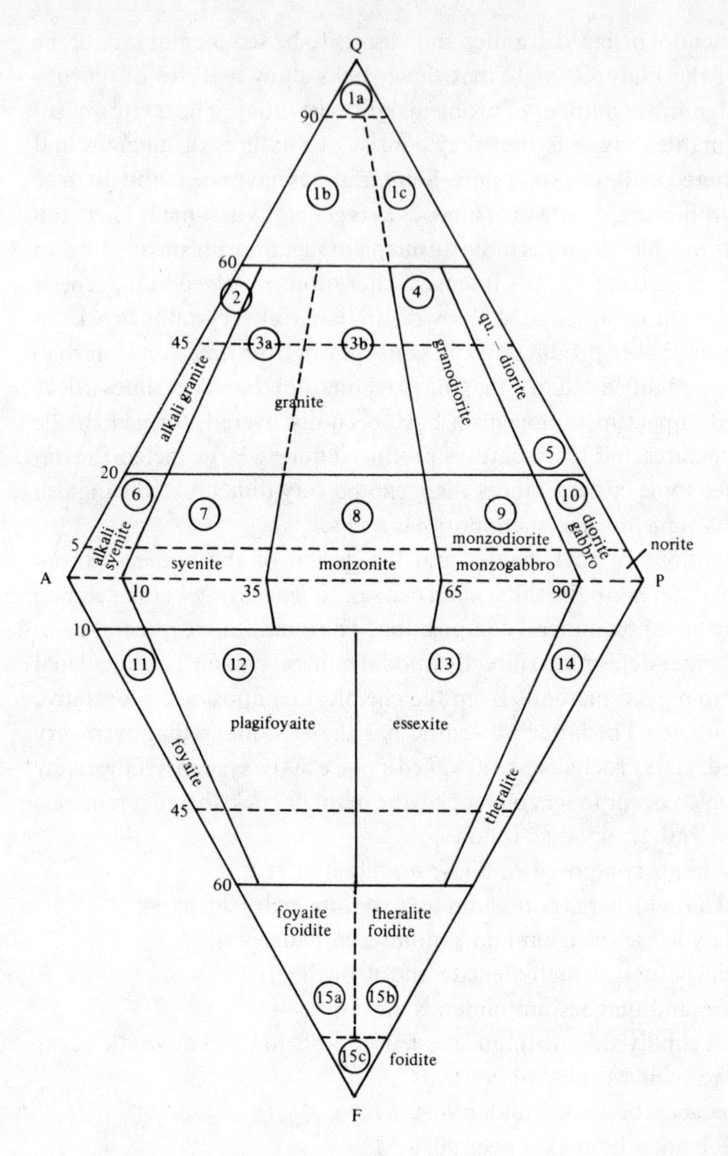

Fig. 4.2. Nomenclature and taxonomics of plutonic leucocratic rocks. Q = quartz; A = alkali feldspar; P = plagioclase; F = feldspatoids. (Streckeisen, 1967.)

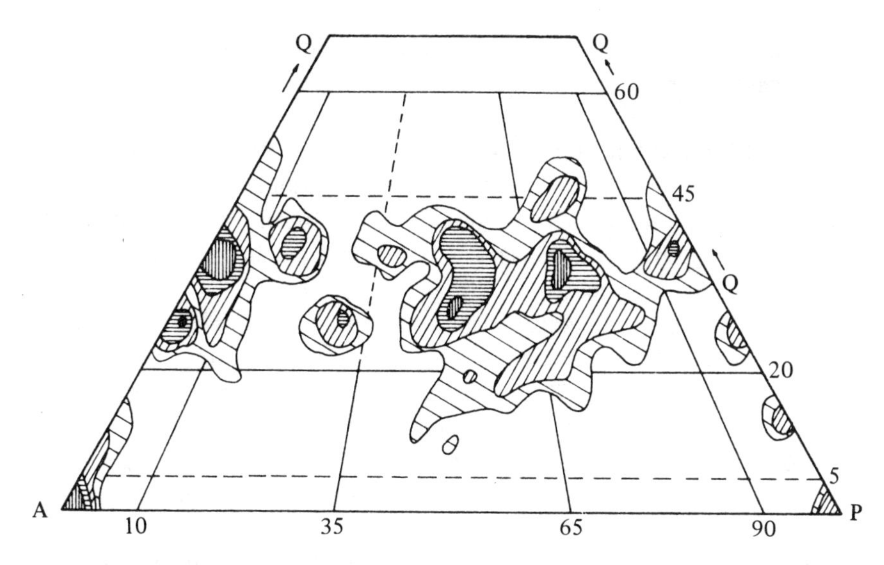

Fig. 4.3. Distribution of 319 analyses of plutonic rocks from various sites, shown in the Q–A–P triangle (see Fig. 4.2). Density of shading indicates increasing frequency (boundaries: 1.2, 20, 28, 31%). (Streckeisen, 1967.)

the literature from all quarters of the Earth are entered into these triangles it will be found, as Fig. 4.3 shows, that the points cluster in certain areas, but form an overall continuous complex which can only be subdivided by convention. Fig. 4.2 shows the boundaries of the fields and the corresponding names of the plutonic rocks as agreed on by a commission of the International Union of Geological Sciences. In establishing these boundaries and names the following points were taken into account. The system of divisions and names should:

1 agree as closely as possible with the clusters of frequently-occurring mineral compositions in nature;
2 remain as close as possible to historical tradition[17] while appealing to as many geoscientists as possible;
3 be simple and easy to use.

The same divisions are used for the volcanic leucratic rocks. The individual fields are given corresponding names of volcanic rocks, such that each plutonic rock name corresponds to the name of a volcanic rock of the same mineral composition. The systems used for the rarer melanocratic rocks of plutonic and volcanic origin are based on the proportions of various M minerals.

The divisions and nomenclatures used to order the continuum of *sedimentary rocks* are based on different parameters. Again, the genetic aspect predominates. Sedimentary rocks are classified on the basis of genetically significant characteristics of fabric and substance composition into three main groups.

I Clastic sedimentary rocks, consisting of particles with or without a matrix, thought to have been transported and deposited as a mass of separate particles by moving water, air, or ice. (An example is sandstone, which consists of quartz, feldspar, and mica particles in a calcareous matrix.)

II Chemical sedimentary rocks, consisting of more or less compacted, chemically deposited material which has been dissolved and transported in water. (Examples are gypsum, which is precipitated from sea-water, or limestone, secreted by organisms from sea-water.)

III Causto-biolitic sedimentary rocks (fossil fuels), consisting of carbon-rich substances produced by the decomposition and alteration of organic material. (Example: coal.)

As in igneous rocks, the boundaries between the three main classes here are not sharp. The features characterising these three classes of naturally-occurring sedimentary rocks from a continuum so that there are transitional and mixed forms of clastic, chemical, and causto-biolitic sedimentary rocks. Clearly, the rock complexes to be assigned to the individual classes form an even smoother continuum. In subdividing these we again use properties which are genetically relevant within the given context.

The *clastic sedimentary rocks* form a continuous complex with regard to the size of their constituent particles. Particle size is used as a quantitative parameter of internal fabric and, therefore, as a defining characteristic of the subclasses, for particle size is invariably related to the processes and conditions of transport and deposition by which the sediment was formed. Where deposition occurred under water, for instance, coarse-grained sands represent conditions of faster flow in rivers or along coasts, while fine-grained clays were formed in quieter flood waters along rivers, in lakes, on tidal flats, or far out at sea. Since clastic rocks never contain particles of one size only, but rather a spectrum of various grain sizes, the division and naming of such rocks according to grain size involves defining statistical parameters to establish the mean (average) grain size and degree of sorting (degree of distribution around the mean). Table 4.2 shows how the continuum of clastic sedimentary rocks is divided up according to the measurement of grain size.

This scheme only permits the naming of sediments consisting of one grain size group but with insignificant admixtures of larger and smaller particles. Where the grain size spectrum is too broad a given rock is described as a mixture of the three components clay, silt, and sand. Such a mixture can be

Table 4.2 *Clastic sedimentary rocks*

Grain size in mm	unconsolidated	consolidated
63–2	gravel	conglomerate (rounded components) breccia (angular components)
2–0.063	sand	sandstone
0.063–0.002	silt	siltstone
<0.002	clay	mudstone

plotted as a point on a triangular diagram whose three corners represent clay, silt, and sand. The grain size classes of clastic sedimentary rocks can be further subdivided according to the mineral composition of their particles. This leads to names such as quartz sandstone, arkose (sandstone with feldspar grains), greywacke (sandstone with rock fragments), and bentonite (clay with montmorillonite particles).

Chemical sedimentary rocks are produced by processes which are largely characterised and differentiated by chemical parameters. For this reason the main groups are defined by their chemical composition. Further subdivisions are drawn within each group according to those properties regarded as genetically the most relevant, i.e. those properties indicating formational processes and environments. The most important groups of chemical sedimentary rocks and their subdivisions are shown in Table 4.3.

Causto-biolitic sedimentary rocks are also systematised according to genetic considerations (Table 4.4). The kerogene of oil shales derives largely from lower plants, while coal originates largely from the remains of higher plants. The series from peat to anthracite, characterised by increasing carbon content, corresponds to the alteration of the plant substance during diagenesis.

The genetic aspect plays a major role in defining and subdividing *metamorphic rocks*. Abductive inference is involved here more than in the naming of any other class of rocks, for observed states of affairs of material composition and structure must be used to infer the conditions and processes that led to the formation of the rock, keeping in mind a number of theoretical assumptions. This class is defined as including all rocks produced by alteration (metamorphism) of primary rocks and there are consequently two considerations involved in their classification and naming.

I. *Naming according to rock of origin.* Here the relevant properties are those of the original rock type which were at most only partially destroyed by metamorphism. Based on the substance of relict minerals of the primary

Table 4.3 *Main types of chemical sedimentary rocks*

Chemical substance	Rock	Subdivisions
Calcium carbonate	limestone	according to components, fabric, & genesis:
		calcilutite
		granular limestone (calcarenite)
		pellet limestone
		detrital limestone
		coquina
		oolitic limestone
		limestone stromatolites
		reef limestone
Magnesium-calcium-carbonate	dolomite	
Silica (silicon dioxide)	siliceous rocks	according to organic components:
		diatomite (diatoms)
		radiolarite (radiolarians)
		spiculite (sponge needles)
Calcium sulphate	anhydrites	
	gypsum	
Water soluble salts	salt deposits	(a) according to main minerals
		(b) according to formational environment
		marine evaporites
		continental evaporites

rock which have survived metamorphism, and on preserved features of chemical composition, one can distinguish between, e.g. sedimentary or igneous origin of the rock. Relict structures may suggest that sedimentary, plutonic, or volcanic origins are recognisable. Naming by origin of a rock is done by using the prefix meta, giving rise to names such as metapelite, metagreywacke, metagranite, or metabasalt.

II. *Naming according to conditions of metamorphism.* This procedure is based on the theory of rock metamorphism and supposedly indicates which combination of minerals is produced as a stable system out of a material of given chemical composition under given physico-chemical conditions (the most important parameters being pressure and temperature). The taxonomy is, therefore, based on a knowledge of the stability states of chemically defined mixtures under given conditions. This kind of knowledge is gained primarily through experimental petrology and is being steadily deepened and expanded. A taxonomy of metamorphic rocks according to metamorphic conditions is, therefore, still not complete today and is constantly undergoing alteration and improvement.

Classification according to metamorphic conditions depends fundamentally on so-called critical minerals, mineral substances which are only stable

Table 4.4. *Main types of causto-biolites (organic fuels)*

Organic material	Rock	
Kerogene	oil shales	
Plant remains	peat	
	lignite	Increasing
	bitumin	carbon
	anthracite	content

within certain pressure and temperature ranges, and which change into other minerals or mineral combinations when conditions exceed or drop below critical pressures and temperatures. These alteration reactions, whose conditions are being experimentally established, mean that the two-dimensional continuum of pressures and temperatures occurring in metamorphism can be divided into several large fields or 'metamorphic facies'. Based on the content of critical minerals in a metamorphic rock, the rock is assigned to one of these fields, or metamorphic facies.

A complete naming of a metamorphic rock should include two identifications, one relating to the rock of origin and the other to the metamorphic conditions. An example of such a name would be metagreywacke of the amphibolite facies: a sedimentary clastic rock recrystallised under conditions present in the amphibolite facies, characterised by pressures between 2 and 10 bar and temperatures between 400 and 600 °C, and recognisable by the presence of minerals such as andalusite, sillimanite, cordierite, staurolite, and muscovite.

Besides these scientifically desirable names of metamorphic rocks, qualitatively defined names are also used, mainly for preliminary 'field designations', such as phyllite (finely-flaky rock, consisting largely of mica); schist (coarse-grained mica-rich metamorphic), gneiss (medium to coarse-grained granite-like rock with parallel banding), and amphibolite (a metamorphic consisting of hornblende and plagioclase).

It is evident that the nomenclature of rocks is based on widely divergent principles of concept-formation, so that conceptually incompatible systems are used for the three genetically defined groups of igneous, sedimentary, and metamorphic rocks. The reasons for these differences are revealed if one considers that the purpose of petrographic nomenclature is more than just descriptive. The name given to a rock observed in nature should be tailored to that observation in such a way as to serve as an empirical basis for future argumentation. Such argumentation will ultimately always be aimed at (abductively) reconstructing the conditions of formation and alteration. Thus igneous rocks are classified according to their constituent minerals and textures, as these properties may be used to develop hypotheses regarding crystallisation from magmas. Clastic sediments are understood to be products of the shaping and sorting of particles through mechanical processes and their taxonomy is, therefore, based on parameters of particle size. Chemical sediments, being products of physicochemical processes, are named according to their chemical composition. The naming of metamorphic rocks serves the hypothesis that they were formed through alteration under pressure and temperature conditions which are to be more precisely identified.

4.2.3 *Terminology, nomenclature, and taxonomy in paleontology*

The subjects of paleontological observation are structures in the rocks which are interpreted as evidence for the existence of organisms in the geologic past. *Fossils* and *trace fossils* are distinguished. Fossils are things which are explained as the remains of plants and animals preserved in their original substance by being 'fossilised', that is altered into mineral substance, or occurring as a mould or cast. Trace fossils are shapes imprinted in the rock which are understood to be the remains and representations of the activities of organisms. These may be tracks, burrows, or structures interpreted as the excretions of organisms, such as coprolites.

Looked at categorically, fossils are things and most trace fossils are configurations. Looked at semantically, fossils and trace fossils have a different meaning from the things and configurations dealt within in mineralogy and petrology. They are not studied, or at least not exclusively in terms of their material and morphological composition, but are regarded from the outset as *signs* referring to other things: to the true subjects of paleontology, the organisms of the Earth's past. Fossils are *icons* (p. 53), since by virtue of morphological likeness they represent the shapes of past living creatures; trace fossils are *indices* (p. 53), since they do not resemble the living creatures to which they refer, but were brought about by them.[18] The objects to which the terminology, nomenclature, and taxonomy of paleontology relate, namely the organisms of the past, are thus not directly present but are transmitted by icons and indices. The correct interpretation and classification of these signs is the primary task of paleontological research.

The history of paleontology shows what difficulties it has had to face and which continue to lie ahead. Late into the 18th century it was disputed whether fossils were representations of organisms or products of a *vis plastica*, 'tricks of nature', which only accidentally resembled the shapes of plants and animals. Even today 'pseudofossils' are known, structures of inorganic origin which resemble organic remains. 'Problematica' are structures and shapes in sedimentary or metamorphic rocks whose organic origin is dubious. Up till recently certain structures in Precambrian strata constituted such problematica. These were called '*Eozoon bavaricum*' in the Bavarian Forest, and '*Eozoon canadense*' in Canada, and are now attributed to marine algae. Dubious fossils are most abundant in geologically old formations. Their interpretation is difficult for two reasons. The organisms of the most ancient times were probably particularly simple in structure, did not resemble forms living today, and possessed no resistant armouring or shells. At the same time, it is the oldest rocks which on the average have most frequently undergone morphological and chemical alteration through metamorphism and diagenesis, deforming and rendering unrecognisable the fossils and trace fossils in them.

The naming of organisms of the geologic past, therefore, inevitably involves abductive reasoning, working from more or less complete fossils and trace fossils back to the life forms as they actually existed at one time. Such an 'inference' based on incomplete remains and representations is hypothetical and implies a knowledge both of organic forms and behaviours and of the rules and laws of fossilisation processes. These processes include the dismembering, decomposition, and alteration in substance and deformation of content which the organic remains underwent before and after becoming embedded in sediment. Before becoming embedded, parts belonging to organisms such as teeth and vertebrae of mammals or shells and opercula of ammonites, may have become separated from each other, making it difficult to attribute parts which originally belonged together to one and the same species. The classification of the remains of tap roots, trunks, leaves, sexual organs, and fruits in coal deposits is so difficult that separate parts have been given different names. Some remains underwent extensive changes through so-called fossil diagenesis after embedding making it difficult to identify defined remains of organisms. Particularly hypothetical in nature is the attribution of trace fossils to the organisms which produced them.

Organisms reconstructed from fossil remains are named in a manner closely related to the nomenclature and taxonomy of neontology,[19] a biological classification system based on the principle of hyponymy (p. 119). The organisation of plant and animal systems consists of a hierarchical sequence of taxonomic classes, each taxon except the last including one or more cohyponyms. The organisation used both in neontology and in paleontology today distinguishes seven taxa of decreasing generality.

Kingdom – phylum – class – order – family – genus – species.

The lowest rank, the species, is designated using the Linnéan model by a binomial or two-part name, e.g. *Cidaris coronata* (a sea urchin of the Upper Jurassic). The first part designates the genus, the second the species.

The taxonomy of neontology is based on morphological distinctions and uses the genetic principle, thus combining two of the modes of classification elaborated on p. 84. In biology, individuals of one species (the lowest taxonomic rank) form a reproductive community, that is, only individuals belonging to the same species can reproduce themselves. In principle, therefore, the definition of species is genetically based. Since, however, this criterion cannot always be demonstrated experimentally even for Recent species and since interbreeding occurs in exceptional cases, the genetic feature of common descent is supplemented by various morphological features deemed necessary and sufficient to assign an individual to a particular species. At higher taxonomic levels differentiation is based exclusively on morphological features. This classification is, therefore, less 'natural', and the history of biological science shows that there are various ways of ordering living things into a system of genera, orders, and classes. The decision to adopt one or another system cannot be based on empirical grounds alone (morphology), but must also be based on theory. Since the time of Lamarck and Darwin the theoretical foundation for the newer systems used in neontology and paleontology has been the theory of evolution. One attempts to define the taxonomic classes of living and extinct organisms and their hierarchical order in such a way as to agree with the phylogenetic descent, the evolution during the geologic past. The ideal goal of such an ordering is a system comprehending all living things which have ever lived on Earth, a system encompassing both neontology and paleontology.

Paleontology has special problems in classifying fossil organisms. Lacking the genetic criterion, the definition of a species as a reproductive community, morphological features alone can be used to determine whether a fossil individual belongs to a given species. The definition of taxonomic groups of fossil organisms is, therefore, based on morphological features at all hierarchic levels of the system. The shape of individuals of a paleontologic species, however, unlike the composition of a mineral substance occurring in crystals or grains, varies within given limits. The morphological definition of a paleontologic species is, therefore, not based on set values of morphological parameters but on a statistically obtained range of variation. As an example, Fig. 4.4 shows the morphometric statistics for individuals of an ammonite species. The morphometrically investigated feature here is the proportion of the shell diameter to umbilical width. The measured distribution can be described statistically, e.g. by the mean and standard deviation.

In place of the previously customary definition of species based on so-called types or type individuals, paleontology increasingly relies on

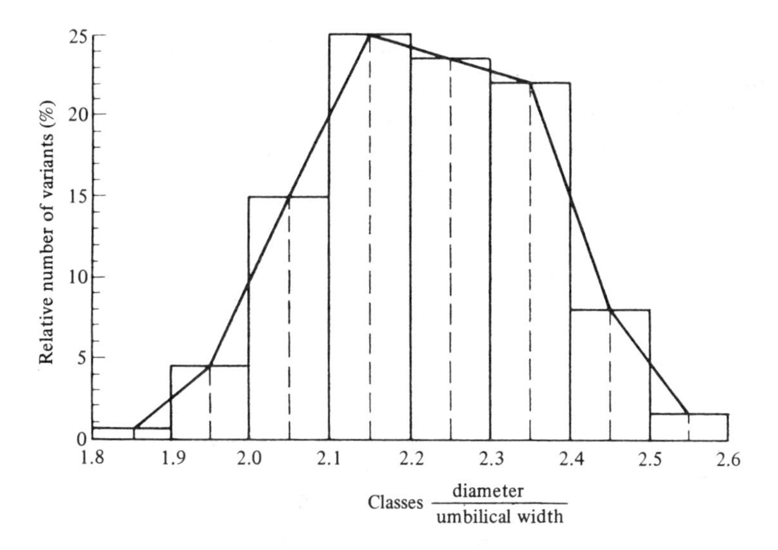

Fig. 4.4. Distribution function for the proportion of diameter to umbilical width in shells of 316 specimens of the ammonite species Dinarites arisianus *from the Triassic formation of the Alps (after Bubnoff). (Müller, 1963, vol. I, p. 149.)*

characterisations based on statistically-derived morphometric data determined on the basis of larger populations. This means that the definition of a species is increasingly becoming the result of an inductive generalisation (see p. 140).

4.2.4 *On the terminology of geomorphology*

The 'world' of geomorphology is the uppermost layer, particularly the surface of the solid Earth, or more precisely the boundary where the lithosphere contacts the atmospheric envelope. This boundary layer presents a durable structure, three-dimensionally shaped and segmented, yet at the same time is both the setting and the subject of rapid and slow changes. Textbooks describe geomorphology as the science of the 'appearance', the 'occurrence', and the 'creation of surficial forms' (Penck), the 'formational science of the Earth's surface' (Hettner), the 'explanatory description of land forms' (Davis), the 'science of the formation and form world of the solid Earth's surface'[20] the 'science dealing with the shapes of the surface of the Earth's sphere'[21] and the 'science of the physical processes which shape the Earth's surface and the forms which they create.'[22]

Phenomena of the Earth's surface are described both under the category of configurations as long-lasting shapes of material masses and under the category of processes as far as their development, alteration, and destruc-

Table 4.5 *Classification of geomorphological forms according to size.* (Cailleux, after Tricart, 1965)

Class	Order of size km^2	Examples
I	10^7	Continents, ocean basins
II	10^6	Large areas of uniform structure, such as the Scandinavian shield, the Congo Basin
III	10^4	Large structural units such as the Paris Basin, the Franco–Swiss Jura, the Central Massif
IV	10^2	Tectonic elements: mountain massifs, horsts, grabens
V	10	Anticlines, synclinal folds, and other tectonic elements; mountains, valleys
VI	10^{-2}	Relief forms such as ridges, terraces, end moraines, alluvial fans
VII	10^{-6}	Microforms such as solifluction structures, polygonally patterned ground
VIII	10^{-8}	Microscopic forms such as details of corrosion, polishing

tion are concerned. The concepts of geomorphological language therefore relate both to configurations and to processes. This, incidentally, is also true of tectonics, whose terminology refers both to configurations (deformation in the rocks) and to formative processes. The terms of geomorphology are drawn to a great extent from the vernacular but in many cases have been made more precise or redefined altogether for scientific use.

The many different forms on the Earth's surface make up continuous complexes with respect to both morphological (qualitative) and morphometrical (quantitative) parameters, and there are various ways of subdividing them. A very broad classification system uses the size of the structure formed. Table 4.5 shows eight unnamed categories, each described by an average size parameter and containing qualitatively different forms. Table 4.6 shows a qualitative classification conforming only partially and inconsistently with the hyponymous system. The complexes are further defined according to different coordinate axes, that is, according to different properties. Classes IA and B and IIA are defined morphologically, type 2 under IIB is defined according to the substrate being formed, while IIC and IID are separate altogether.

Since forms are not only described as such, but also as the results of certain processes, the qualitatively-morphologically defined classes are further subdivided genetically. Unlike rocks, where it is primarily the

Table 4.6 *Classification of geomorphologic forms* (Maull, 1958)

- I. Simple forms
  - 1. Concavities
    - (a) Valleys
      - (i) large
      - (ii) small
    - (b) Passes
    - (c) Niches
      - (i) large
      - (ii) small
    - (d) Basins
      - (i) large
      - (ii) small
    - (e) Caves
  - 2. Convex forms
    - (a) Escarpments
      - (i) large
      - (ii) small
    - (b) Mountains, hills, ridges
- II. Compound forms
  - 1. Flat and concave forms
    - (a) Plains
    - (b) Depressions
  - 2. Convex forms
    - (a) Mountains
      - (i) domed mountains
      - (ii) axially uplifted mountains
      - (iii) block-faulted mountains
      - (iv) plateau mountains
      - (v) cuestas
      - (vi) rises
    - (b) Glaciers
      - (i) land ice
      - (ii) mountain glaciers
      - (iii) continental glaciers
      - (iv) coastal glaciers
      - (v) shelf ice
  - 3. Deserts
  - 4. Coasts and islands

higher-level classes that are differentiated genetically, formative conditions are used in geomorphology to differentiate even at the lower classificatory levels. Such genetic definitions imply abductive inferences so that the attribution of a given form to a particular class must here, too, remain hypothetical. As an example we show in Table 4.7 the genetic subdivision of the morphological class 'mountains'. It should be noted that a morphologi-

Table 4.7 *Classification of 'mountains'* (Maull, 1958)

1. Mountains formed by accretion:
   (a) volcanic cones
   (b) well knolls
   (c) fluvial kames and levees
   (d) dunes
2. Mountains formed by erosion:
   (a) mountains along range edges and valleys
   (b) mountains in areas of ridge dissection
   (c) glacially formed mountains
   (d) monadnocks

cal system of subdivisions also exists, though not a systematic one, in which mountains are divided up into 'plateaus', 'ridges', 'knolls', 'cuestas', 'sugarloafs', and the like.

As in paleontology where qualitative morphological features of species are being quantitatively refined by statistically treating measurements carried out on large populations, so in geomorphology quantitative data (morphometry) are increasingly replacing qualitative descriptions of forms. Quantitative measurements relate for instance, to the angles of mountain slopes or scree slopes, river banks and river beds, or to the geometry of branching in river systems. The aim is to define 'types' of forms by quantitative measurements. One characteristic problem, however, is that while paleontology generally has large populations available from which to derive average, 'typical' values, this is often not the case for geomorphologic structures. The number of individuals in a geomorphological class is inversely related to the spatial size of the form concerned. Continents, ocean basins, sedimentary basins, and similar large structures are only present in small numbers upon Earth, so that it is impossible to determine their typical parameters by statistical means. Statistical methods can only reasonably be applied to structures whose spatial areas are no greater than about Class IV in Table 4.5.

The matter does not end with a simple morphological description of forms, however, as our divisions under the class 'mountain' show. Even simple descriptions attempt to set each form in relation to formative conditions. Morphogenesis, not morphography, is the aim. Empirical material is collected in such a way as to provide a 'form assemblage' for an area assumed to be fairly uniform with regard to formative processes and conditions, especially climatic conditions, geologic history, and lithologic substrate. The textbook presentation of geomorphology is, therefore, not based on forms but rather on complex morphogenetic scenarios, each characterised by a typical 'form assemblage' such as fluviatile forms, large complex forms (flatlands, cuesta landscapes, mountain belts), forms of the humid, semi-humid, and arid climates, coastal forms, sea-floor forms, and glacial forms.[23]

Since forms are analysed according to their genesis, geomorphological observation also focusses on processes which produce and alter forms. Descriptive terms for processes are therefore defined either according to temporal structure or physical, chemical, or geological nature.

Following temporal structure, three main classes of processes are distinguished.[24]

1 *Catastrophic processes* are divided into a preparatory period of slow, often hardly visible change, a paroxysmal event of short duration and high intensity, and a relaxational phase usually involving slow modifications.
2 *Periodic processes* are discontinuous processes which over the course of time repeat themselves more or less similarly and rhythmically.
3 *Continuous processes* continue over very long periods of time at more or less constant speed and intensity.

Based on the morphogenetic causes or nature of the processes various classifications can be set up, e.g. the following.

1 Endogenously caused processes:
   (a) volcanically caused
   (b) tectonically caused.
2 Processes caused by exogenous terrestrial forces:
   (a) climatically caused weathering and soil formation
   (b) caused by gravity
   (c) caused by flowing water
   (d) caused by flowing air, and
   (e) caused by flowing ice.
3 Processes caused by extraterrestrial forces (impacts of large meteors or comets).

The examples show that two conflicting goals underlie the geomorphological vocabulary. One aims at as unbiassed a description as possible, the other attempts an explanatory naming of phenomena. In the case of landscape forms, unbiassed, descriptive nomenclature is based on qualitative (or in the case of morphometry, quantitative) morphological features. Thus valleys named according to their shape are 'gorges', 'V-shaped valleys', 'box-shaped valleys', or 'U-shaped valleys'. The simple description of formative processes can lead to classification according to temporal structure, as in the above-described divisions into catastrophic, periodic, and continuous processes.

Morphological classification of configurations on the one hand, and temporal classification of processes on the other, are often unproductive for purposes of further discussion. What is important is to name moraines and dunes, e.g. as such and not simply to designate them according to morphological parameters. It is for this reason that a nomenclature of genetically defined terms has developed based on interpretations drawn from abductive hypotheses. Thus one distinguishes and designates river valleys of various forms, for instance, as belonging to the humid regions, the seasonal tropics, glacial valleys, etc., and names the formative processes according to the forces which caused them, as listed above. Which genetically defined terms in geomorphology may be attributed to the empirical basis and which belong to the theoretical level of knowledge depends entirely on the state of research and the consensus of those involved.

4.3 Observation

The procedures used to acquire the knowledge of the facts discussed in the statements of the empirical basis are called observation. Scientific observation is not passive perception, but a planned and methodologically standardised activity. This is expressed even in the vernacular: one 'has' perceptual experiences, but one 'makes' observations.[25] Everything a researcher observes in nature or in his laboratory is the result of active, theory-guided involvement with reality.

There are two kinds of scientific observation, depending on the researcher's intent. Francis Bacon differentiated between the observation of 'free nature' as manifested in the heavenly spheres, plants, animals, and celestial mechanics, and methods leading to deeper knowledge in which nature is 'tried and tortured' by artificial means.[26] We may accordingly distinguish phenomenological and experimental observation as the extreme forms of observational activity through which the researcher acquires the facts of the empirical basis. Under *phenomenological observation*[27] we include all methods aimed at summing up nature in its 'natural state' uninfluenced by the observer. Under *experimental observation* we include all procedures which investigate the behaviour of objects and systems under 'artifical', intentionally set up and controlled conditions.

4.3.1 *Phenomenological observation*

Phenomenological observations are intended to lead to a description of those phenomena occurring in nature which are relevant to the given discipline or to the given problem. They are most closely related to passive perception since they presuppose an open-mindedness, i.e. a passive attitude. At the same time phenomenological observation also contains an active component. Out of the infinite number of phenomena observable in even the minutest segment of reality, only those are chosen for attention which are or may be significant for the specific problem of a special discipline, or for a particular theory or hypothesis. Phenomenological observation moreover implies action in that the observer must so select his methods and instruments that the conditions, processes, or relationships observed will be affected by the act of observation as little as possible, and in such a way that reconstruction of the unobserved state seems feasible.

Various procedures are used for phenomenological observation in the geoscientific disciplines: these may be categorised by the amount of instrumentation they involve and by the extent to which they interfere with natural relationships. Purely observant procedures, which do not 'affect' the object correspond most closely to pure phenomenological observation. Phenomenological observations may include many other procedures as well, however, in which experimental techniques influence the object to some extent, not with the intention of studying its behaviour under various conditions but to identify it and determine its composition. This is phenomenological observation with experimental interference, which in technique, though not in intention, grades into experimentation proper.

In general we may overlook the experimental interference involved in observing geomorphological configurations, large and small structures in tectonically deformed rocks, trace fossils, and organic remains, where the only instruments used are those like microscopes, electron microscopes, and cameras. Such instruments only serve to reveal a microworld to the optical senses and refine and objectify optical perceptions without essentially changing the object itself in the process.

Experimental interference is also fairly minimal in the identification of minerals according to pragmatic characteristics, e.g. according to optical properties. Usually part of the object must be crushed in order to determine under a microscope how a mineral grain alters a given light ray. Optical properties such as light refraction or double refraction inferred from this treatment serve to place a mineral taken from a natural context within the mineralogical system. The calculation of the grain size distribution of a sediment needed to name the rock using the nomenclature of sedimentary petrography also represents an interference with a natural structure.

Experimental interference is more marked in observational procedures aimed at identifying the chemical substance of objects such as minerals or rocks. We still consider such procedures to be phenomenological observations, however, since they are intended to identify the given composition of natural objects even when these objects must be reduced to their basic chemical elements in the course of routine chemical analysis.

In geophysical research various instrumental techniques from physics are used for phenomenological observation. Determining the gravity, magnetic, or electric field of the Earth's surface requires no experimental interference since the instruments simply record physical parameters. In other procedures experimental interference is needed to phenomenologically observe the existing composition and depositional environment of rocks underground. Seismic blasting, for instance, creates elastic waves whose velocity and propogational directions yield the desired data.

Geological maps constitute particularly important documents of phenomenological observations in geoscientific research as they incorporate the results of different specialised investigations into morphology, petrography, mineralogy, paleontology, and geophysics. The problems involved in geological mapping will be treated separately in a later section as an example of collecting and processing phenomenological observations.

4.3.2 *Experimental observation*

The experimenter does not contemplate *natura naturata* as does the phenomenological observer, but rather assumes the role of *natura naturans* of the Creator or of the creative natural forces. The material with which the experimenter operates derives from that selfsame Nature which the observer examines phenomenologically. The conditions to which the experimenter 'subjects' this material are also 'natural' in that Earth scientists, unlike atomic physicists, experiment only with conditions assumed to exist in the realms of nature investigated by geoscience. For the purpose of the experiment, however,[28] substances and things are isolated from their 'natural' context and subjected to artificially created, controlled influences. This triggers processes and creates conditions which are the subject of the observations.

The results which experiments are intended to produce should be of general validity and should have the character of inductive regularities or laws needed to explain scientifically the facts gathered by phenomenological observation. Two types of experimental function can be distinguished here:[29] experiments can be *methods of testing* where the results are intended to confirm or refute previously conceived regularities or laws. Experiments can also be *methods of discovery*, when the results are intended to uncover or refine new, conjectured or only vaguely understood regularities or laws. In the case of experimentation as a method of testing, the argumentation follows deductively; where experimentation is conducted as a method of discovery, the argumentation is inductive.

The experimental method used in Earth science functions largely as a method of discovery. The experimenting geoscientist behaves quite like a physicist or a chemist. The laboratories, apparatus, and experimental methods used in Earth science are in principle like those of physical and chemical research and geoscientific experiments also look for regularities and laws of a chemical and physical type. The Earth scientist cannot simply leave it all up to physicists and chemists, however, as the laws which he needs to explain the phenomena of interest to him are not always of the kind investigated in physics and chemistry. Earth science is mainly interested in the behaviour of special substances and systems, in the effects of very special conditions, and above all in what happens inside structurally and materially very complex 'impure' assemblages. The changes appearing in the investigated object while certain conditions are being changed (e.g. the elastic changes in length of a rock prism being subjected to tensile forces) are observed during the experiment. Three types of experimental conditions should be distinguished. First, certain conditions whose influence is not to be measured are held constant (e.g. temperature in the above example). Second, those conditions whose influence is to be measured are varied (e.g. tensile forces in the example). Third, some conditions are ignored as it is assumed that they have little or no influence on the outcome (in the above example, the geographic position or time of day of the experiment). From this it follows that each experiment must be preceded by a well-considered plan, based on theoretical expectations and insights (in our example, e.g. the expectation that each rock will undergo a change in length when subjected to tensile forces, the knowledge that a change in temperature would also involve a change in length, and the conviction that time of day can have no influence on the outcome).

The experimentally obtained findings concern previously unknown relationships, such as whether a certain functional relationship exists (e.g. whether the melting point of olivine is influenced by pressure), how a functional relationship is to be formulated (e.g. by what function the solubility of quartz in water increases with rising temperature), or what parameters or constants obtain (e.g. the permeability of a given rock, defined as the relationship of the pressure gradient to the amount transmitted).

Experimentation in geoscience is confronted with some specific problems. Some natural conditions can be reproduced fairly easily in the laboratory, such as the temperatures and pressures normal to the Earth's surface or to the upper regions of the crust (temperatures of up to circa 200 °C, and pressures of up to around 10 kbar), time periods which are neither too long nor too short, and spatial dimensions which can easily be handled in a laboratory. The experimental difficulties grow, however, when investigating regions of very high temperatures and pressures, processes which are very fast or slow, or events which involve very large or very small spatial areas. There are conditions and processes that cannot at present be simulated, and some will probably never be reproducible in the laboratory, such as the conditions of substances in the deep mantle or the Earth's core, very slow plastic deformations (folding) of larger rock units, alterations and reactions of minerals during weathering, diagenesis, and metamorphism, and processes occurring during the impact of large bodies from outer space upon the planetary surface, during which tremendous energies are released and enormous temperatures and dynamic pressures are momentarily produced.

In some cases such difficulties can be overcome by experimenting with models. A model must be scaled to laboratory conditions in terms of spatial size, mass, and temporal dimensions, yet at the same time it must so 'resemble' natural configurations that the properties and behaviours of natural structures can warrantably be inferred from observations on the model. To guarantee this 'resemblance' or applicability of the experimental results to nature, the model must be designed according to theoretical considerations. The relevance of the model experiment rests upon the validity of these.[30] Thus, to simulate deformation produced over geologic time-spans in many cubic kilometres of tectonically strained rock using a model with small test bodies and short periods of time, physical analysis indicates that to maintain resemblance the reductions in time and space necessitate a corresponding reduction in the solidity and density of the material used in the experiment. This is the justification for basing conclusions concerning the deformation of large masses of solid rock on deformational experiments involving small test bodies of clay.

Major difficulties are involved in experimentally simulating impacts, whose significance for earliest Earth history first became clear during the exploration of the Moon, Mars, and Mercury. Experiments on the effects of rapid projectiles on minerals and rocks can only be carried out with small projectiles and velocities of at most 10 km/s, while large bodies from space commonly have velocities of around 25 km/s or even much more. In order to at least approximate the energies released during large natural collisions, results of nuclear tests and large chemical explosions have been used as models for the deformation of rocks and the craters produced by impact. The validity of the resemblance is particularly open to question in such cases.

Finally, where real experiments simply cannot be carried out on very complex structures and processes, they may sometimes be simulated in computer programs. Such computer simulations have of late come into use in various fields.

4.4 Geological maps as components of the empirical basis

As we showed elsewhere (p. 68), maps, as non-verbal presentations of facts, play a significant role in geoscientific texts. The geological map is the most widespread form of disseminating phenomenological observations. Geological mapping is fundamental to geoscientific research and at any given time the majority of geologists are probably concerned with surveying, drawing, and interpreting geological maps.[31]

The geological map, usually drawn at a scale of 1:25 000, uses topographical and geological signs to convey various types of information. The topographic basis of the map rests on facts gathered by photogrammetric or other means; these are highly objective and leave little room for personal interpretation on the part of the observer. Topographic maps, therefore, have a high degree of iconicity. Geological maps created by entering geological signs onto the topographic map are also based on phenomenological observations which, as such, are highly objective. The points, lines, and areas representing geological states of affairs, however, state at once 'more' and 'less' than the sum of all the individual observations: 'more', as the presentation does not result automatically from recording all the facts, but is the result of a hypothesis as well; 'less' because in view of this hypothesis, only the important things are represented and the unimportant left out.

> At first the geologist seeks out all the exposures in or near the area to be mapped and works up their petrography, stratigraphy, and tectonics. With the results he then sets up a hypothesis about the rock distribution in the areas which lack exposures.[32]

That which the mapping geologist seeks to capture in his observatons and to represent on the map is the uppermost layer of the Earth's crust, the so-called lithosphere, consisting of solid rock or as yet unconsolidated formations of the geologic past. Except where freed of loose material in high mountains, steep coasts, banks, and other slopes, the surface of the lithosphere is covered with vegetation, recent soils, or weathering debris, and is only visible in places. Every geological map, even if it does not expressly state so, is an 'uncovered' map, showing the composition of the largely invisible surface of the lithosphere. Some special maps 'uncover' not only Recent, but also older unconsolidated masses. In the areas of the Pleistocene glaciations in North America and Europe there are geological maps showing the solid bedrock beneath Ice Age moraines which are sometimes of very great thickness.

Since cores are generally not available for geological mapping, the most important observations are made in 'exposures', that is, in places where human activity (as in stone, clay, or sand quarries, roadcuts, or foundation diggings) or natural erosion has laid bare the bedrock otherwise buried under vegetation, soil, or weathering debris. In particularly interesting spots the mapping geologist will also carry out small diggings. The mapping geologist is largely dependent on the contingencies of available exposures in an area. The poorer the existing exposures, i.e. the more bedrock buried under vegetation, soil, and debris, the more hypothetical is the resulting map. Moreover the exposures available in an area do not stay fixed. Artificial exposures such as road cuts and quarries are good for a while, then disappear or decay through disuse or even filling in. Other artificial exposures are only briefly accessible, as during the digging of foundations, allowing glimpses of the bedrock. Natural exposures along river banks, sea coasts, and mountain slopes are also not permanent but undergo a constant process of change.

If we accept the observations of the mapping geologist in exposures as empirical, i.e. as open to intersubjective review, we can only do this with reservations. Directly after surveying a map its author will be able to show the exposures which he has investigated to all interested parties and thus render the observations made there more objective. In a short time he will no longer be able to do this. Old exposures disappear and new ones may replace them, and these may sometimes reveal states of affairs which could not have been known to the author of the map. In Europe exposure conditions have deteriorated in the last decades. Many more stone quarries, sand, and clay pits were formerly needed to obtain material for road and building construction, but as transport facilities have improved local materials have increasingly been replaced by those from further away. The old stone quarries and pits have been filled in for safety reasons, or used as refuse dumps.

To read and use a geological map correctly, more is needed than just an appreciation of its semiotic character (see p. 68), that is, the ability to understand the meaning of the signs explained in the key. The critical reader must also know that certain 'non-empirical' assumptions underlie the map and which ones these might be. He must know that every geological map says more than what the author actually observed. This is due to a peculiar and irremediable malproportion between the semiotic potential of the map and the nature of the certifiably valid subject matter to be represented in the map. Between the level of the subject matter and the level of its semiotic representation intervenes the interpretation of the subject matter, interpretation which allows greater or lesser leeway for different possible states of affairs whose many ambiguities cannot semiotically be represented. That which the map shows is invariably an interpretation of the basic state of affairs, or only those states of affairs which 'agree' with the interpretation chosen by the author. In this manner plausible alternatives fall by the wayside and can at most be discussed in the accompanying text. The mapping geologist is thus constrained by the semiotic instrument of the map and must accustom himself to supporting a firm and unambiguous opinion – the one recorded in the map. The apodictic certainty with which geologists sometimes speak of 'facts' in geological maps and the trust which the user places in the apparent factuality of a geological map, are consequences of this habituation.[33] New exposures which appear or are deliberately created in an area by excavations or cores may so change the empirical basis that older geological maps must be viewed as outdated and be revised or remapped entirely.

The interpretive achievement of the mapping geologist consists firstly in identifying and classifying the phenomena observed in the exposures and secondly in constructing a model of the entire underlying bedrock.

That which is observed in a specific exposure must be worked up in such a way as to produce statements on the minerals and rocks occurring there, on the fossils found in the rocks, on the internal spatial texture of the rocks, on the formational age of the rocks and on the spatial relationships of the rock units to one another, to the Earth's surface, and to the compass directions. The discussions of the previous chapters regarding terminology, classification, and nomenclature in each of these fields should be kept in mind throughout. The non-empirical and hypothetical elements of the entire descriptive repertoire of the Earth science are incorporated into the descriptions which the mapping geologist makes of his observations in the exposures. We stress a few aspects here.

Identifying and classifying the rocks is basic and requires decisions which rely on inferences as to formational conditions. This is true even for the distinctions between the major classes of igneous, sedimentary, and metamorphic rocks. There has long ceased to be any doubt as to some genetically relevant features, e.g. those by which sedimentary rocks are recognised. The distinctions between igneous and metamorphic rocks, on the other hand, have been a subject of very recent research, so that masses shown as igneous on older maps may today be interpreted as metamorphic complexes. As we have seen, further subdivision of the major classes also depends on the state of research. The mapping geologist can, therefore, do no more in describing the rocks than what is possible given the systematics and nomenclature accepted by his day or by his school of thought.

Further problems confront the mapping geologist even in the exposures themselves. The exact classification of metamorphic and igneous rocks which is important for any further discussion in the framework of modern research can only be carried out today in the laboratory through microscopic and chemical analysis. When mapping, therefore, the geologist must often be content with imprecise 'field designations' (see p. 44 and p. 117). Yet rocks form continuous complexes which at times show significant variations, even within small areas. Small samples must be taken for identification in the laboratory, and these must be so selected as to be representative while taking into account the variability of mineralogic composition. Which rock types are distinguished upon the map as being metamorphic and which igneous depends largely on the personal decision of the mapping geologist, which again is determined by the theoretical considerations which guide him in mapping.

In the case of sedimentary rocks it is extremely important to find, identify, and interpret fossils in order to determine the formation to which the given stratum belongs. Here again there is room for interpretation, as fossil faunas cannot always provide the exact placement of a stratum within the geologic time scale. Geohistoric interpretation of fossil-free strata is even more difficult: it can only be based on lithologic features, and is often highly questionable.

Not only must the rock types and fossils be observed, but also morphologic features must be recognised and their shape and position measured. This involves noting sharp boundaries between different rock bodies, and interpreting these genetically. They may have been produced while the rock was forming (e.g. the boundary between an intruding magma and the neighbouring rock or the stratigraphic boundaries in sedimentary rocks, produced by sudden changes in the supply of material) or they may be faults, tectonically-caused juxtapositions of masses which were originally located far apart. The spatial positions (strike and dip) of sedimentary boundaries and faulting surfaces must be measured. Since it can be assumed that bedding surfaces were roughly horizontal during formation any deviation from the horizontal indicates later tectonic movement. Along with faulting or tectonically-caused fracture in the rock assemblage, plastic deformation appearing as warping or folding of the primary structural elements must also be measured.

The second task of the mapping geologist consists of using the states of affairs recorded in the individual exposures to create a structural model of the upper lithosphere that can be presented in two-dimensional projection on the geological map. Such a model unites into a coherent picture the states of affairs observed and interpreted at various points in the map, so that it may be said that the model presented in the map explains the field observations. Since only a tiny portion of the substructure shown in the map is visible in nature the model cannot be constructed without reasonable assumptions as to the composition of the hidden parts of the substructure, that is, it cannot be constructed without hypothetical assumptions. There will always be a certain amount of room for alternative hypotheses and models. A given combination of states of affairs obtained in widely-separated exposures can be explained by several models differing to a greater or lesser degree, all based on differing hypothetical assumptions about the structure of the hidden bedrock. This leeway varies with the complexity of the underlying structure and with the type of exposure. Large exposures are common at very different heights in high mountains, and the composition of the upper lithosphere is revealed more completely here than anywhere else. Nevertheless the interpretation of the observations in a spatial model may be difficult and ambiguous, since tectonic deformation through folding and faulting is generally very intense and complex in high mountains. The opposite extreme of a nearly completely covered substructure is found primarily in the tropics with their thick soils and lush vegetation. Vegetation and soils also hamper the investigation of the lithosphere's surface in many other areas of the temperate climatic zones.

When choosing between alternative possible structures of the bedrock buried between exposures the geologist depends on geomorphologic clues to indicate something about the deeper structures. Where these are lacking the principle of greatest possible 'simplicity' is followed. In a tectonically very disturbed area the preferred model is the one which interprets the evidence from the exposures in terms of the fewest and geometrically simplest deformations and faults. This principle is only a heuristic aid having no objective significance: there is no law of nature according to which the structure of the lithosphere must be 'simple'.[34] The objectively most enlightening concepts of the structural form to be expected are therefore those inductively known to occur in certain areas of the lithosphere or in the neighbourhood of the area being mapped.

In conclusion we will present two examples of the hypothetical character of geological maps.

Example 1. Mapping in the Rhine Plateau in the years 1883 and 1928. Influence of tectonic hypotheses on the map.

Fig. 4.5 shows details from geological maps made of the same area in 1883 and 1928. The exposures contain sedimentary shales interlayered with greywackes and volcanic rocks (diabase). The compiler of the map in 1883 chose to construe the unexposed boundaries between greywacke and shales as rounded, elongated contours. This would mean that each greywacke

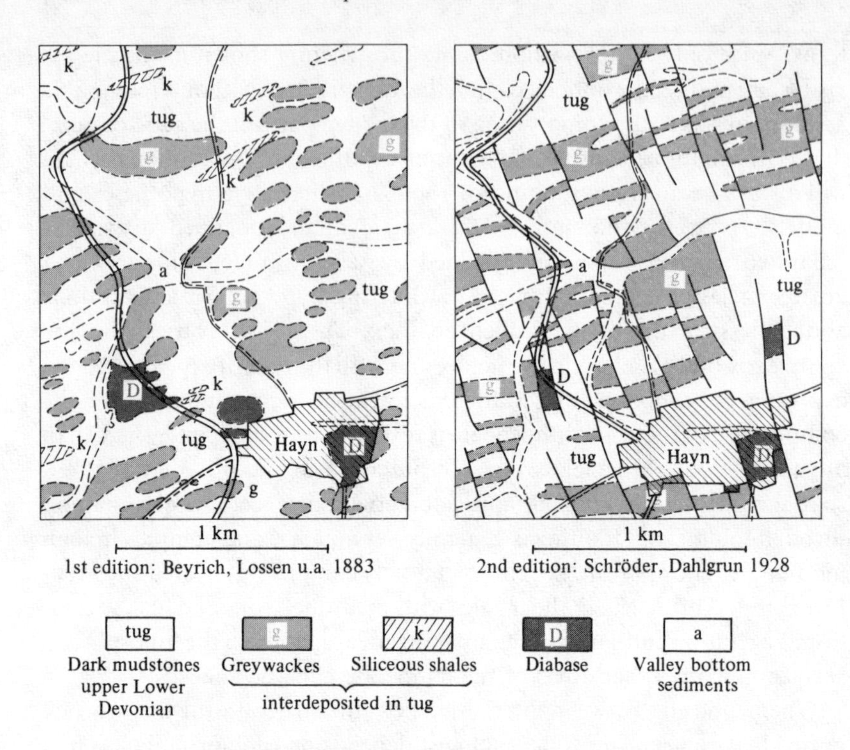

Fig. 4.5. Section from the geological map (1:25 000) of the area around Schwenda on the Rhine Plateau, according to the first edition (1883) and the second edition (1927). (Wagenbreth, 1958, p. 23.)

deposit would consist of either a hollow or a hummock, assuming that the greywacke was deposited as thin sheets between the shales. The later compiler in 1928 assumed that a more or less steeply-dipping series of shales with thin interbedded greywackes occurred in this area. This was presumed to have been broken up by a series of uniformly oriented cross-faults along which horizontal displacement then occurred. According to this view he has shown the greywacke occurrences as parallelipeds, each bounded by parallel faults and sedimentary boundaries. The hypothetical construction of this author is based on the fact that appropriately oriented jointing planes were observed in exposures of the surrounding area.

The user of this map must know that the 'true' shape of the boundaries of the greywacke occurring in the field cannot be seen. A map which shows only that which is observable could not show any boundaries at all, a technical impossibility. The geologist must, therefore, decide which hypothesis he regards as the most likely. This is then presented to the reader as fact; it is accepted by his colleagues if it can be integrated without contradiction into the other states of affairs known about the structure of the area concerned.

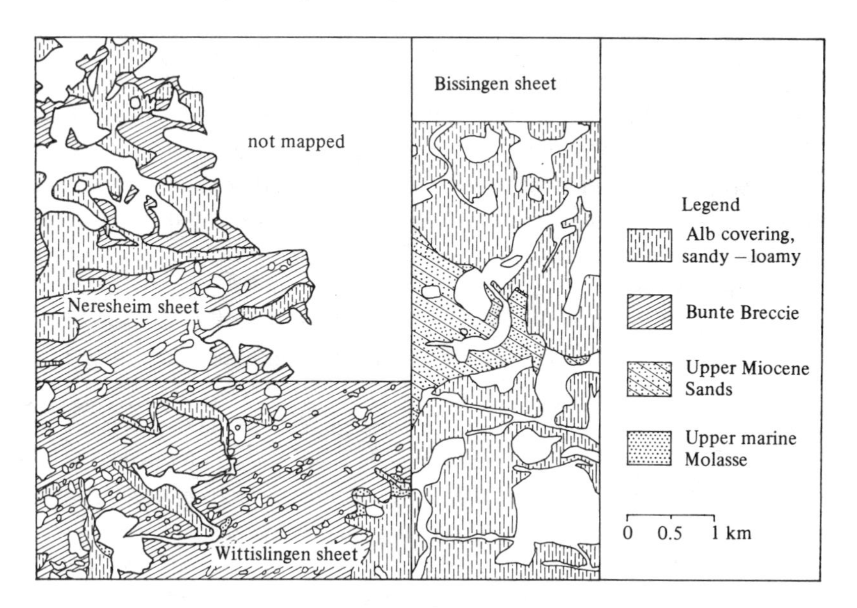

Fig. 4.6. Section from the geological map of the area SW of the Ries Crater, 1:25 000. The three sheets of Neresheim, Bissingen, and Wittislingen were mapped by different authors. (Hüttner, 1956, Schalk, 1957, Gall, 1971.)

Example 2. Mapping in the region of the Nördlinger Ries. Influence of hypotheses about the geologic history of a region on mapping technique.

Fig. 4.6 shows simplified details from three surveys made at a scale of 1:25 000 in the area of the southwestern Ries Basin whose origins were much debated at the time these maps were made. The entire area is covered with forests and fields and contains few exposures. The rocks visible at the surface are not shown in the figures and consist of sporadic occurrences of Upper Jurassic limestones and Tertiary sediments (upper marine Molasse, Upper Miocene sands). Besides these, fragmented rocks also occur in this area, known as *Bunte Breccie*, whose origin is connected with the formation of the great Ries Basin. Since this breccia is a soft, easily weathered rock it can only be recognised in occasional evanescent exposures created by road building, drainage work, or house construction. Otherwise it lies buried under the vegetation and soils of forests and fields and is scarcely recognisable even when weathered out.

The geologist mapping the Neresheim sheet in 1958 assumed that the Bunte Breccie had been formed by an explosive process which created the Ries Basin and spread the breccia masses around the Ries. According to this hypothesis an areal distribution of the breccia should be expected. To test this he used a coring device in his investigations, with which he made cores through the upper soil layers. He did in fact find the expected areal distribution of Bunte Breccie under the surficial covering of forest and field soils, and showed it as such on the map.

The compiler of the Wittislingen sheet adjacent to the south went on the hypothesis published in 1968 that the Ries Basin had been formed by the impact of a large meteor. This hypothesis also suggests an areal distribution of Bunte Breccie round about the Ries. Hüttner's method of mapping with a coring device was, therefore, used and a nearly uninterrupted distribution of breccia was shown to exist under the soil blanket. The Wittislingen sheet, therefore, forms a continuation of the Neresheim sheet and does not contradict it.

In 1957 the compiler of the Bissingen sheet adjacent to the east worked on the older hypothesis current at that time, according to which the masses of Bunte Breccie occurring outside the Ries Basin were not ejected from a central basin, but were produced by many small, local volcanic eruptions. According to this view the Bunte Breccie should not be areally distributed but occur only locally, a view which was not contradicted by the few scattered exposures. The idea of using a coring device, therefore, occurred to no one at the time, and the Bunte Breccie was only mapped where it was found accidentally exposed. This did not occur on the Bissingen sheet excerpt shown in Fig. 4.6. Where no exposures were present, the surface was simply mapped as 'loamy Alb covering'. The result is a geological map contradicting the sheets of the neighbouring areas which were mapped in the light of quite different theories on the formation of the Ries.

5

Problems of inductive organisation of empirical data

In most of the geoscientific disciplines one has to deal with large quantities of empirical data derived from phenomenological and experimental observation. These findings on individual facts are not generally suited as such for further discourse in the form recorded in the observational reports. The raw empirical data must be organised and ordered in a certain way before it can be used either for practical purposes or for argumentation at the theoretical level of research. We therefore find that works based on observation rarely list the individual data; usually they contain statements derived from them, describing the composition, relationships, or behaviours of a greater or lesser number of individual cases by means of graphic presentations, formulae, averages, or verbal statements.

These summarising descriptions occupy an intermediate position between the singular statements constituting the empirical basis *sensu strictu*, and the universal laws of the theoretical level. Due to their basic dependence on the individual observational data, they cannot on the one side be sharply differentiated from the statements at the empirical basis. On the other hand they grade into laws in the form of summaries of empirical data to which a general validity exceeding the individual case can temporarily and reservedly be attributed; they are therefore often found fulfilling the function of laws. It is thus impossible to distinguish this domain of summarising descriptions sharply from the empirical basis on the one side and from the theoretical level on the other. We will nevertheless examine it in isolation, since in the geoscientific disciplines in particular one must continually handle large amounts of data using systematisations and generalisations in order to summarise and work with them before any ordering or explanation by means of hypotheses, laws, or theories is possible.

In ordering the empirical data at this intermediate level of research, the operations of systematising and of inductive generalising may be distinguished. *Systematising* unites several singular statements into one finding which should have no other function than to explain clearly what was observed in many individual cases. The systematisation 'simplifies' the underlying data to some extent, but this data could in theory be systematised in other ways. There is therefore a certain amount of leeway in choosing the representational system. A *generalisation* follows the pattern of an inductive 'inference' (p. 81), and is present when a systematising statement is assumed to be valid for all representatives of a given class, i.e. including those which were not observed.

Systematisations and generalisations can be verbally articulated or expressed through mathematical formulae and graphic presentations. In the following sections we will start by treating some methods and problems of systematisation and generalisation using examples in which the observational results consist of quantitative data. In these cases mathematical and graphic procedures and representational forms are preferred.

5.1 Systematising and generalising

Systematisations become necessary in Earth science when the value of a certain property (a one-place relation) is measured on many objects in order to characterise either the substance of which the objects consist or the objects themselves. We can differentiate here between two cases. In the *first case* it is assumed on theoretical grounds (for reasons which are not empirically derived) that there is one 'true' value for this property, a 'theoretical fact' attributable to all the objects of the class under investigation, while the individual observations yield 'practical facts' which vary from one case to another. This variation is attributed to unavoidable inadequacies of measurement or to impurities in the objects.[1]

In this case the divergent values must be systematised, i.e. reduced to a chief or mean value. Given the theoretical expectation that the property concerned has a value which applies to all objects in the class, the mean is regarded as an approximation of this 'true' value. With this we have already gone beyond the stage of systematising to the stage of generalising by maintaining that a value has been obtained which applies to all unobserved members of the class as well. Where sufficient theoretical grounds are present such a generalisation of individual observed values can take on the character of a law.[2]

Thus it may be expected, based on the theory of crystalline structure, that the refractive indices of certain mineral substances will have definitive and characteristic values, since these belong to the necessary properties of a mineral. Repeated measurements of the light refraction of various particles of one mineral will in fact yield, as a 'practical fact', values varying within certain boundaries. The observational result is not, however, presented as a list of all the measurements, but rather as a systematisation containing two components: a *mean*, which is assumed to approximate the 'theoretical fact' as closely as the methods permit, and a number expressing the *scattering* of the measured values around this mean. Out of n measurements $x_1, x_2, \ldots x_n$, the mean $\bar{x}$ can be calculated according to the formula:

$$\bar{x} = \frac{x_1 + x_2 + \ldots x_n}{n}$$

The most common measurement presently used for the scattering is the standard deviation, defined as follows:

$$s = \sqrt{\frac{\Sigma (\bar{x} - x)^2}{n - 1}}$$

The calculation of the mean value according to the above formula (arithmetic mean) is based on the assumption that the larger the number n of measurements, the closer the mean will approach the true value. Correspondingly the value of the standard deviation approaches a limiting value of deviation as the number of measurements increases. The light refraction value 'obtained' this way is generalised and applied to all objects consisting of the mineral, i.e., a law is indeed derived from the measurements, since the theory dictates that a mineral defined by its chemistry and atomic arrangement must have definitive, non-varying optic properties.

In the *second case* it is assumed that the distribution of measured values of a property is not due to interferences or to difficulties of precise measurement, but to the nature of the object itself. The goal of observation is therefore not to obtain a single value which approaches the 'true' value of the 'theoretical fact' but rather to describe the *variation* of the property within a given population of investigated objects. Here the frequency with which certain values of the property occur within the population is looked for with the aim of deriving a *frequency distribution*. In all the geosciences the systematisation of data by deriving frequency distributions is very important and with the aid of mathematical statistics and electronic data processing, large quantities of data can be worked through within a short time.

The data to be processed consist of the values obtained by observational measurement. These are characterised by a dimensional *parameter* and lumped into suitable classes each identified by an interval of the parameter. The systematisation process assigns a frequency parameter to each class expressing what percentage of the cases investigated falls into the given class. The grain size distribution of an unconsolidated sediment, for instance, can best be shown using data on the diameter of the individual grains. Optic methods can be used to determine the diameter of an isolated grain, so that the resulting statistics are based directly on the individual case. More often, however, individual cases are lumped into classes through the very methods of observation. Sieve or elutriation analyses are used to determine the quantity of grains whose diameters lie within given intervals. The frequency parameter is thus the weight percentage of each class out of the total sample.

The *scale* of the numerical parameters and the selection of class intervals are decided by the observer. It is usually convenient, though not necessary, to choose intervals of equal size. Of the various possible scales, the simplest to use is a linear scale, especially when dealing with numbers of similar magnitude. In other cases, and this is true of most sediment analyses, the linear scale is cumbersome, as the sizes of grains occurring together within one sample may range over several orders of magnitude. The grain size distribution of sediments is therefore generally plotted logarithmically. When the class intervals are defined by equal-sized logarithms a geometric series of grain size diameters is produced, as in the interval series 2–4, 4–8, 8–16, etc.

Graphically the frequency distribution of classes can be represented as a so-called *histogram*,[3] and the discontinuous histogram can be replaced by a continuous *distribution curve*, as shown in Fig. 5.1. This construction takes us beyond systematising observations to a generalisation, for the distribution curve asserts that the grain size distribution is unbroken and continuous down to the smallest of intervals, which is more than was observed. The conclusions which may be deduced from the continuous distribution curve go beyond what was empirically ascertained. It is possible, for instance, to predict the quantity of grains contained within a diameter interval smaller than the intervals set by the sieve sizes upon which the initial grain size classification was based.

Graphic representation of the frequency distribution is often considered insufficient: for easier comparison of different distributions the frequency distribution is characterised by certain numbers. This must be done by further systematisation through which the empirical data are ordered mathematically. In the simplest case it is assumed that there is a single midpoint or main point in the distribution. There are various ways to define such a central value, of which we may mention the following: (1) the arithmetic mean of all the diameters, also called an average; (2) the median of the distribution, that is, that value of diameter which divides the distribu-

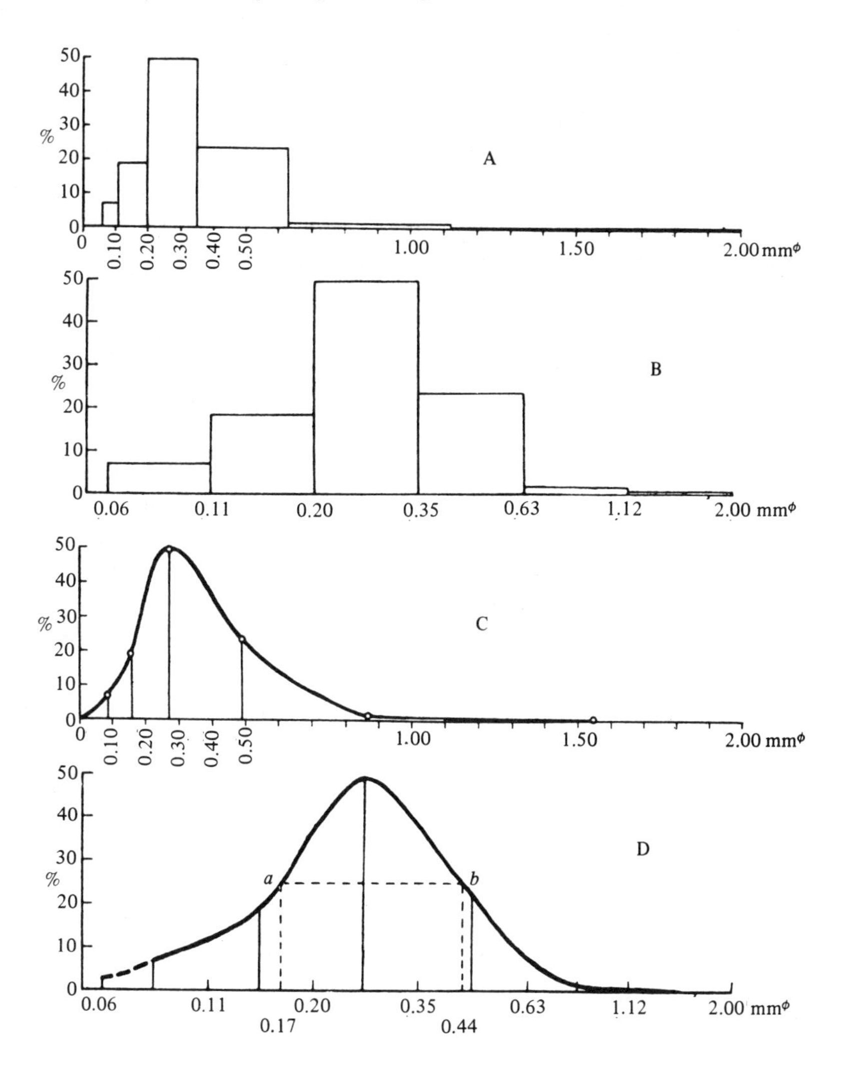

Fig. 5.1. Grain size distribution of a beach sand, shown in various ways.
A: *Histogram using a linear scale.*
B: *Histogram using a logarithmic scale for the grain sizes.*
C: *Distribution curve using a linear scale.*
D: *Distribution curve using a logarithmic scale for the grain sizes.*
(Köster, 1964, p. 264 ff.)

tion into two halves, 50% of the grains being smaller and 50% larger than the central value; (3) the mode of the distribution, namely that diameter which forms the peak of the distribution curve, or that class which contains the largest number of grains. These three central values coincide when the distribution has a single maximum (single-mode distribution), and is symmetrical with respect to this maximum. The central values incidentally depend on the scale chosen, that is they are determined by the mathematical scheme by which the data are organised and are generally different on a logarithmic scale than on a linear one. The empirical data on their own give no indication of which mean will best characterise their distribution.

A more precise mathematical treatment of the natural frequency distributions is based on the ideal shape of the distribution, shown graphically as a single-peaked, symmetrical bell curve. Distributions of this type are called *normal distributions*. Where the number of variants or classes is finite, as in the sieve analyses, the mathematical shape of a normal distribution fulfills the *binomial function*. In continuous distribution curves, where there is an infinite number of classes or possible variants, the mathematical shape of the normal distribution fulfills the Gaussian function. The transition from the histogram to the continuous distribution corresponds to the transition from the discontinuous binomial function to the continuous Gaussian function.

It is of great advantage when an empirically-determined distribution of some property can be shown as a Gaussian function without too many discrepancies between the measured data and the mathematical function, for the entire complex of data may then be reduced to two numbers: the *mean of the distribution* (in a Gaussian distribution the arithmetic mean, median, and mode all coincide), and the *standard deviation*, which measures the width of the symmetrical distribution around the mean. Empirically determined frequency distributions are therefore, wherever possible, described by the parameters of the mean and standard deviation derived from the Gaussian function. It has become evident that this is sometimes only possible when certain conditions are met, and it is for this reason, for instance, that grain size distributions of unconsolidated sediments are conventionally measured on a logarithmic instead of a linear scale. Linear scales often give skewed distributions, falling off steeply and asymmetrically towards the smaller grain sizes. Symmetrical functions are frequently obtained more easily using a logarithmic scale, as shown in Fig. 5.1. If this empirical finding holds in general, there should be some theoretical reason why empirical distributions, particularly grain size distributions, tend to yield better symmetrical functions when the measured properties are shown on a logarithmic scale, yet as far as we can see, no theoretical basis has been

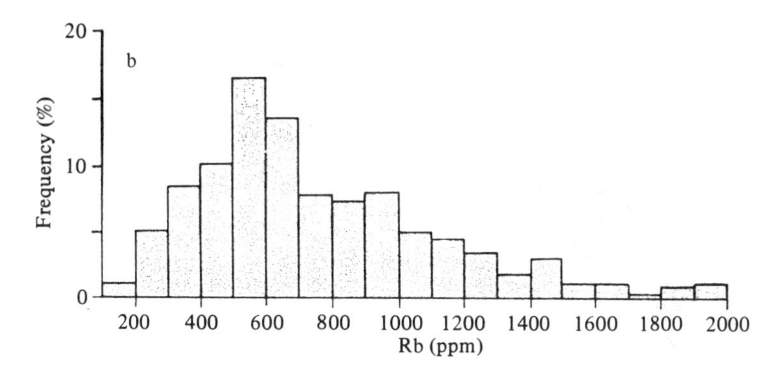

Fig. 5.2. Histogram of rubidium contents of 261 biotite analyses from various provenances. (Heier, 1972, p. 37, D-2.)

found for this state of affairs. We can only note that asymmetrical distributions are most likely to be found where, as in the case of the grain sizes, the curve must end at the origin, and cannot proceed beyond 0 into the negative numbers.

If an empirical distribution does not fulfill the Gaussian function, as in the case of the grain size distribution of many sediments, this ideal is still used in describing the deviation from the normal distribution by a certain value. Thus asymmetrical distributions can be characterised by specifically defined parameters of 'skewedness', and multi-modal or otherwise irregular distributions can be interpreted as several overlapping Gaussian functions with differing means and standard deviations.

Fig. 5.2 shows a further example of a frequency distribution. 261 analyses of biotites from various rocks showed that these had rubidium contents varying from less than 200 to 2000 p.p.m. (mg/kg). Within this range as the histogram shows, there is a clear frequency maximum between 500 and 600 p.p.m., around 17% of all investigated biotites having these amounts of rubidium. In this case the scale used is linear.

The systematisations and generalisations we have looked at so far concerned only single properties of objects. The treatment of *multi-place relationships*, i.e. of relationships involving several qualitative or quantitive parameters observed in things or processes, is very important. It is from these above all that we arrive at empirical generalisations which may then yield the laws of the theoretical level. Relationships between two quantitative parameters can be shown particularly clearly in the right-angle coordinate system of the plane. Hence out of any given observational material the relationships usually isolated are those which can be understood as functions of two variables only. In geoscientific literature, therefore, systematis-

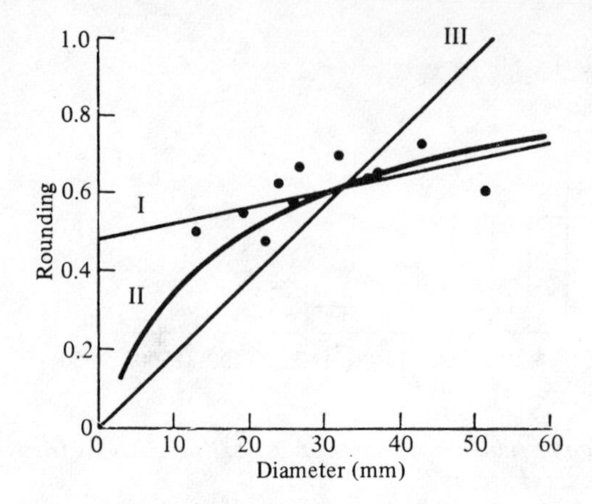

Fig. 5.3. Relationship between rounding and diameter in beach clasts. (Krumbein & Graybill, 1965, p. 238.)

ations consisting of two-place relationships and the consequent rules of the relationship between two parameters are particularly common.

Thus of the many types of data which could be measured and observed on the transported pebbles of a certain beach, the average diameter (D) and the rounding (R) (determined by a process we will not describe here) were chosen in the expectation that these values would yield a simple relationship which could then be applied to other beaches. The empirical material here consists of a set of paired numbers relating to the diameter and rounding of the measured clasts. These data can be represented graphically as in Fig. 5.3 with the rounding R shown on the y-axis and the diameter D on the x-axis. Each clast, represented by its R and D values, is entered as a point in the xy plane. According to our previous discussion (p. 66) on the use of graphic representations, we are dealing here with a *diagram* which is being used for *information processing*. The diagram presents a summarising systematisation of the measured data which can be shown either as a drawn curve or as a mathematical function.

It might be supposed that a graph such as that shown in Fig. 5.3 must necessarily yield a certain curve as the systematisation of the individual values and would then serve as the basis for an inductive generalisation on the 'true' general relationship between the measured parameters, in our case between rounding and diameter. Such a hope is fed by a widespread overestimation of the powers of inductive reasoning, and is not justified, for the researcher is free within certain limits imposed by the data to choose between various forms of systematisation. The figure shows three possible curves, each of which could satisfy the observed values of diameter and rounding, given a certain amount of statistical scatter. An indefinite number of further curves would do the same. The question of which of these curves or their corresponding mathematical functions represents the 'true' relationship between R and D cannot be answered using empirical material alone. A decision must be made which is not simply empirically determined and which must take into account two criteria. Firstly, the chosen curve or function must represent a good systematisation of the observations, fitting the measured values as well as possible *(criterion of optimal systematisation)*. Secondly, the shape of the curve should agree with theoretical expectations or postulates which justify regarding it as a graphic representation of a generalising statement which is true not only for the small number of measured cases, but can also be applied to analogous objects and relationships *(criterion of plausible generalisation)*. The criterion of plausible generalisation is particularly important. Greater scattering of the individual values around the generalising curve will be accepted more readily than a contradiction of well-founded theoretical expectations. For this reason, curve I is rejected, as it runs into the y-axis above the origin. This would mean that the smallest clasts would have a degree of rounding which would be the minimum degree present. There is no apparent reason why this should be the case, for, based in theory on qualitative reflection on the action of waves upon a beach, one would expect the curve to go through the origin, for the smallest particles are not rolled and therefore do not become rounded. This expectation is satisfied both by line III and the parabolic curve II. Of these two, the authors prefer curve II, as they believe that they can show a better fit with a plausible model of the processes occurring within the zone of breakers.

In this case the distribution of the data in the diagram led along with theoretical considerations, to a generalising curve expressed mathematically, e.g. by a parabolic function. In general research practice, however, generalisations of two-place relationships shown graphically as lines and mathematically as linear functions are preferred. This is for two main reasons. Firstly, lines and linear functions are easy to deal with. Secondly, only in the case of linear functions is it possible to use correlation statistics to calculate exactly the so-called regression, that linear function satisfying the criterion of optimal systematisation which best fits the observed data. The criterion of plausible generalisation may demand that one function be preferred even if it does not fit the observed data as well, but where no such theoretical grounds exist it is the criterion of optimal systematisation which determines the choice of function. Statistical methods are used to calculate the correlation line from the measured data. The measurement of the average deviation of the individual data from this line is calculated as the

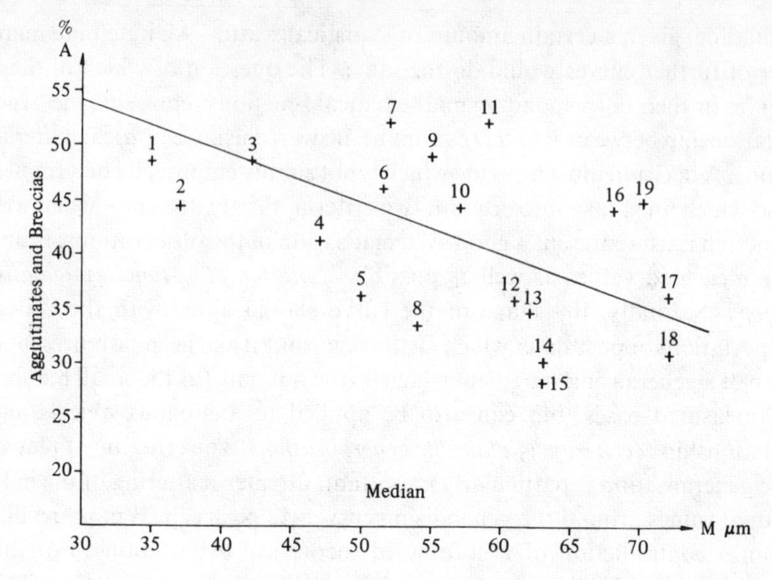

Fig. 5.4. Lunar soils from the landing sites of Apollo Missions 12, 14, 15, 16, and 17. Relationship between content in the fines (<0.5 mm) of agglutinates and breccias (A) *and the median of the grain size distribution* (M). *(von Engelhardt* et al., *1976, p. 373.)*

correlation coefficient. Where there is no relationship, this coefficient (r) has the value 0. Where there is a clear and constant relationship between the two values, $r = 1$. r may be either positive or negative, according to whether the values for y rise or fall with x. For the calcualtion of correlations and correlation coefficients, we refer the reader to statistical textbooks.[4]

Fig. 5.4 shows the graph of the results of an investigation of 19 samples of lunar soil collected during the Apollo Mission at various sites on the Moon's surface. The relationship between two variables is shown: the median (M) of the grain size distribution of the soil sample, and the soil's content (A) of breccia and agglutinate fragments (rock and mineral fragments in a glass matrix). It is assumed that both these components were produced through bombardment of the Moon's surface by small meteors. Each cross in the graph corresponds to the paired value M and A obtained on a given soil sample. The cloud of points is systematised by a line calculated by correlation statistics. The initial decision to use this inductive summary of the observed data was made entirely on the basis of the criterion of optimal systematisation. The equation of the line is

$$A = 68.72 - 0.49\ M,$$

where A is the content of agglutinates and breccia fragments in per cent weight and M is the median in millimetres. The calculated correlation coefficient has the value

$$r = -0.70.$$

This is not a very high correlation and is expressed graphically in a considerable scatter of the data around the regression line. Possibly a non-linear curve would achieve a better fit of function to the observed data. In that case, however, there would be no simple way of finding the function that would best fit the data and of calculating a correlation coefficient. Moreover the authors found no theoretical reasons for preferring a non-linear function. According to the present state of knowledge all that can be said is that with increasing meteoric bombardment of the Moon's surface, A should increase and B should decrease, so that the negative correlation between A and M in the function obtained corresponds to the criterion of plausible generalisation.

Since relationships expressed as curves when shown in a linear scale often flatten out and approach lines when shown in logarithmic scale, a favourite method of obtaining linear regressions is to *introduce logarithmic scales* for one or both variables. In discussing grain size distributions we found that the logarithmic scale is preferable when dealing with numbers of greatly differing magnitudes. In systematising multi-place relationships where no theoretical assumptions exist as to their mathematical structure, the following are further reasons for preferring the logarithmic scale.

1 The difference between two variables is more evident in a linear scale than on a logarithmic one (e.g. the logarithms of 11 and 15 equal 1.04 and 1.18). Scattering around a mean or around a regression are therefore more obvious on a linear scale than on a logarithmic one.

2 Observations on relationships between two or more variables which can only be expressed as curved lines when using linear scales for the variables can often be systematised as straight lines when one or more of the variables is presented on a logarithmic scale. This has been described as a 'fact born of experience' known to every natural scientist. 'It is a hard empirical fact found in many laboratories all over the world that observational results when plotted in log : log scale do give straight lines.'[5]

3 Linear functions which in many cases can be obtained through transforming the observed data onto logarithmic scales often permit the calculation of regressions and correlation coefficients using statistical methods, especially with the aid of computers.

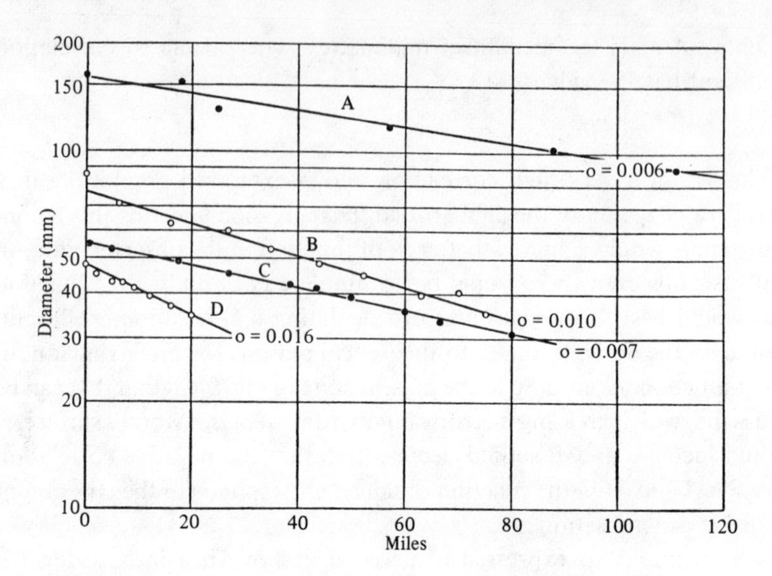

Fig. 5.5. Relationship between the clast size in a river (in mm, shown logarithmically) and distance transported (in miles). A: Rhine; B: Mur; C and D: experimental results on the abrasion of limestone clasts in a rotating drum. (Pettijohn, 1949, p. 529.)

Fig. 5.5 shows as an example the semi-logarithmic graph of a two-place relation known as 'Sternberg's Law', a relationship derived from observations in various rivers between the reduction of diameter in bedload of a river and the distance transported. A straight regression is obtained when the logarithm of the bedload diameter is plotted against the distance transported. This corresponds to the following function:

$$\text{Log } D_0 - \log D_s = ks;$$

or

$$D_s = D_0 e^{-ks},$$

where D_0 and D_s are the diameters of the clasts at the beginning and end of the transportation distance s, and k is a constant that varies from one river to another.

Fig. 3.1 showed a relationship which yielded a straight line only when plotted in log: log form. A relationship believed to exist between the suspended content C_s of a river and its average velocity U can be shown as a regression line when both variables are entered in log form. The regression corresponds to the function

$$\log C_s = m \log a + b \log U$$

or

$$C = a' U^b \text{ (with } a' = a^m)$$

where a and b are constants.

To sum up we list below the most important kinds of function which, taking into account the criterion for optimal systematisation, are generally used to represent two-place relationships and for which it is possible to provide the best fit to the data by calculating the regression and the correlation coefficient.

1 Linear functions with a linear scale:
$y = mx + c$.

2 Linear functions in a simple logarithmic scale:
$y = m \log x + c$
(corresponds to the exponential function
$x = ae^{by}$,
where $a = e^{-c/m}$ and $b = 1/m$).

3 Linear functions in a log : log scale:
$\log y = \log c + m \log x$
(corresponds to the parabolic function of the form $y = cx^m$).

In many cases it is impossible to reduce observed relationships to an interplay between two parameters. Multi-place relationships must be taken into account and the mathematical formulation of the systematised observations involves the difficult task of setting up functions with several variables and choosing from several possible graphic types that one which best fits the measured data and which least contravenes theoretical expectations. We cannot here explore the available methods of mathematical statistics, particularly the so-called *factor analysis*. We will limit ourselves to a single example of the systematisation of interdependencies observed between more than two variables.

A relationship observed or conjectured in New Guinea between the age of an erosional valley and the parameters describing the valley shape was expressed by setting up the following function:[6]

$$V = aL^x t^y \sin^z \alpha;$$

where V is the talweg cross section (that is, the area of a longitudinal section of the valley bounded by the valley floor and the rim of the older surrounding countryside), L is the length of the valley, α is the angle of inclination of the countryside into which the valley is cut, and t is the age of the valley. α, x, y, and z are constants which must be chosen such that the function correctly reproduces the measured morphological parameters and the age of the valley.

In the attempt to organise morphological data obtained from valleys of various ages, the author started off with the hypothetical expectation that for every valley the size of V would depend on the variables L and t and only on these. This expectation rests on theoretical preconceptions regarding the mechanisms by which erosional valleys are formed. The equation chosen to systematise the observations, however, was not derived from theoretical preconceptions, but was chosen according to the criterion of optimal systematisation because it can be transformed logarithmically into the linear form:

$$\log V = \log a + x \log L + z \log \sin \alpha.$$

Correlation statistics can then be applied to this linear equation which yields those values for the constants best corresponding to the observed values of V, L, t, and α.

Not only quantitative data, but also the results of particular qualitative and comparative observations can be summarised, systematised, and expanded into inductive generalisations going beyond the realm of empirically proven facts. These generalisations include statements of qualitative and comparative features which, though not part of the class-definitions, can be ascribed to all examples of a descriptive class. Such properties include qualitative identifications such as colour, lustre, or transparency as well as comparative qualities such as scratch hardness and elasticity of mineral substances, some textural features in certain rock types, or morphological characteristics of paleontological specimens. Such qualities are initially ascertained by observing many individual cases; by an act of generalisation they are then attributed to all specimens of the given class, where this seems plausible based on theoretical expectations.

Along with generalisations of qualitative and comparative properties, all Earth sciences contain many inductive generalisations in the form of qualitative multi-place relationships and lawlike statements derived from individual observations. Their validity is generally acknowledged if the number of individual observations on which they are based is large, if they seem plausible in the light of theoretical expectations, and if they suggest possible theoretical explanations of relationships. Observations on the occurrances of minerals and rocks, for instance, gave rise to the following inductive generalisations before any theoretical explanation existed for these relationships:

Pentlandite deposits are always found in association with SiO_2-poor rocks.

Muscovite in igneous rocks occurs solely in granites.

Primary tin deposits are found only in or near biotite-free granites.

Complexes of igneously-formed carbonatites are usually found in tectonic grabens.

Paleontology contains many inductive generalisations derived as regularities from observations on fossil animal species occurring in strata of various ages. These include such rules on 'key fossils' as the following:

Strata containing the remains of the ammonite species *Parkinsonia parkinsoni* are invariably older than those with the remains of *Oppelia aspioides*.

Less certain are the following more general rules,[7] once regarded by some researchers as 'laws', applying to the evolution of animal species in the geologic past.

In the course of evolution of an animal group the individuals become increasingly large (so-called Cope's Law).

When mammals adapt to a colder climate their legs become shorter (Allen's Law).

Geomorphological observations led to the following generalisation about landscape development whose general validity seems plausible on physical grounds:

On slopes formed by erosion, the angle of slope increases as the divide between watersheds is approached.

We have shown through examples how in all areas of geoscience, verbally, mathematically, or graphically expressed systematisations of qualitative, comparative, and quantitative observational results are at first used merely to fulfill the task of summarising the bewildering multiplicity of observations. They can then be expanded inductively into generalisations which also embrace unobserved states of affairs. This act of inductive reasoning opens the path from the empirical basis to the theoretical level of research. As the examples show, two functions of the inductively obtained generalisations may be discerned.

Inductive generalisations can *firstly* serve to support conclusions drawn from hypotheses or theories and to confirm these qualitatively or quantitatively in detail. This occurs when, for example, given that crystallographic theory initially merely states that certain properties of defined mineral substances should be characteristic constants, data on these properties are actually obtained. These are the necessary properties of minerals, including quantitative parameters (e.g. refractive indices), comparative data (e.g. scratch hardness), and qualitative descriptions (e.g. cleavage). Inductive generalisations of this kind impart new insights whose validity for an entire class of objects (e.g. for all things consisting of a given mineral substance) is underwritten by theory or hypothesis. They fulfill a *confirmative function* in showing that a theoretical assertion is corroborated. Statements derived like this through systematisation of empirical data, therefore, belong to the laws of the theoretical level.

Inductive generalisations are established *secondly* where there is no hypothesis or theory to order or explain the given phenomena. Out of the multiplicity of data, induction then serves to develop statements which are at best plausible, but having no theoretical basis can only cautiously be accepted as generally valid. Such statements we distinguish from the laws of the theoretical level by calling them regularities. Generalisations of this kind fulfill *heuristic function* since they may temporarily, tentatively, and for practical purposes be used in the manner of theoretically-founded laws.

Examples of regularities serving heuristic purposes may be found among the generalisations of quantitative relations treated above. Other examples include statements on the properties of paleontologic species which cannot be derived from the definitive features of the species but which are assumed to be true of all its members, and assertions on the properties of mineral substances which, like some qualities of colour, seem to be typical although their general validity cannot (as yet, anyway) be derived from crystal structure or chemical composition.

The examples show that there is no sharp boundary between the confirmative and heuristic functions of inductive generalisations. In most cases such generalisations should fulfill both purposes and one can really only speak of the dominance of one or the other function. Thus, implicit in all generalisations is the vague suspicion that out of a mass of observable qualities the ones chosen were those which will yield some generally valid regularity. Where such an expectation is not supported by any established theory, this may be termed a heuristic procedure. Where such observation, however, confirms a conjecture which may be regarded as the forerunner of a hypothesis, then a confirmative function is at least indicated. Where qualitative theoretical preconceptions are involved in choosing the mathematical form of the regularity, there can scarcely be any clear boundary between heuristic and confirmative functions: consider, for example, the regularity found for the relationship between rounding and diameter of bedload.

Heuristic generalisations of empirical data play an especially large role in all the Earth sciences. This is firstly because, in contrast to physics and chemistry, very large quantities of contingent facts, both qualitative and quantitative, are dealt with and must be rationally organised. Secondly, it is necessary both for research and for applied Earth science to derive statements from these empirical facts which can be used, at least on a preliminary basis, as if they were laws, even before any scientifically satisfying explanations for the phenomena are available. Frequent usage in research and application has caused these regularities to take on the character of laws in the mind of researchers so that it is not always easy to recognise whether given arguments are based on real laws or on empirical regularities. Nevertheless it should be clear that empirical regularities rank lower in cognitive terms than do the constructs of the theoretical level. Every regularity established through heuristic induction nurtures at least the promise of an ordering based on law of nature. Each such regularity poses a question which can only be answered at the theoretical level: the question of how the inductively-found regularity can be explained through universal laws and theoretical interconnections.

In order for research to proceed in this direction the given generalisation must satisfy the criterion of plausibility. Inductively obtained regularities which seem plausible from the standpoint of other knowledge, particularly of theoretically sound knowledge, can become the first step towards lawlike statements if they are used as the basis for investigations which seek to explain empirical relationships in the light of universal laws of nature. In the examples above, the heuristically found relationships between size and rounding of clasts, and between median and content of breccias and agglutinates in lunar soils are relationships which, one might suspect, could at some stage be explained through physical laws, in these cases, the physical laws governing the processes of erosion of clasts in the wave action area on the one hand, and the stressing of the lunar soil through meteoritic bombardment on the other.

This is not the case when mathematical functions are chosen to describe empirical relationships solely on the criterion of optimal systematisation, that is, to provide the best correlation regardless of theoretical plausibility. The best fit of the descriptive function to the empirical data may be of great value for heuristic and practical purposes but there is no adequate reason why that should also provide the 'true', theoretically founded explanation attributable to universal law. Often, in fact, the theoretically founded law is not distinguished by any particularly good correlation between the measured and the expected values, as other influences must also be taken into account. On the other hand, there are functions distinguished by relatively good correlations which cannot correspond to any 'true' law. Examples of the latter are the convenient parabolic functions which have become so beloved since they so usefully provide straight lines when plotted in log : log. It can be shown that many of these are not candidates for real laws, being physically senseless although they may provide good correlations. We give two examples.

As described above, investigations on the suspended load of rivers show that the paired values of suspended load C and velocity U of a river yield, when graphed, a cloud of points, which, when plotted on a log : log scale can be represented by the line whose equation is

$$C = a\,U^b;$$

or

$$\log C = \log a + b \log U$$

where a and b are constants having special values for each river. When the physical content of this function is examined, it turns out that the equation for the dependence of suspended load on velocity can only make sense physically when the right and left sides are of the same dimensions: C is measured in g/cm^3, and U in cm/s; b as an exponent has no dimension, a on the other hand must have a dimension such that equivalence is maintained on both sides of the equation. The value of the exponent b determines the dimension of a. Where $b = 1$, a must be of the order of $cm^4/g\,s$; where $b = 2$, however, a must be on the order of $cm^6/g\,s^2$, etc. Where b is not a whole number, the dimension of a must be expressed in fractional exponents. Since b should have a different value for every river, a would have a different physical dimension and meaning for each river as well. Such uncertainty in constant a, however, contradicts every conceivable physical model of suspended load transport. Regardless of how the model looks and what laws it is based on, the constant a must have the same physical meaning for all rivers: it may very well take on different values, but must always have the same dimension.

Other inductively obtained parabolic functions also contain exponents which, though constant, are not whole numbers and whose physical meaning is often very dubious. Thus, the following relationship has been described between diameter D (in km) of a meteoritic crater on Earth in crystalline rocks and the kinetic energy E (in joules) of the meteor:

$$D = 1.96(10^{-5})E^{1/3.4}$$

where $D > 2.4$km. It is very unlikely that there is a physical model of the impact process explaining that crater size is determined by an energy power having the exponent $\frac{1}{3.4}$, (0.296), especially as other authors give exponents which are different, though still not whole numbers.[8]

Modern methods of instrumental observation (e.g. automatic procedures in chemical analysis) have made it possible in the last decades to obtain a much greater body of empirical data in far less time than was previously possible. The rapid development of computer technology makes it possible today to process vast amounts of data quickly and painlessly, and in particular to try out various possible relationships between measured quantities and search out the regressions with optimal correlation coefficients. All the many publications which describe empirical functions using such aids alone without any guiding thread of a hypothesis should be regarded critically as heuristic models posing questions for reserach. The theoretical plausibility of such functions should first be tested and/or improved as a starting point in striving toward the desired incorporation of such inductive regularities into the concepts of the theoretical level.

5.2 The concept of probability

The postulate that inductively found regularities should be theoretically plausible takes into account that in inductive inferences (as in abductive ones) the truth of the premises does not warrant the truth of the conclusion so that inductive generalisations are not considered true but merely more or less probable. If it can be shown that an inductive regularity is plausible this then confirms and increases its degree of probability.

This brings us to the concept of *probability* whose basic significance must be discussed in order to understand the epistemological status of inductive generalisations, and to evaluate their role in geoscientific research.[9] Since the time of Carnap's trail-blazing investigations,[10] two different concepts of probability have been differentiated; subjective or personal probability (also called inductive probability) and the statistical or objective probability. *Statistical probability* is attributed as an objective property to types of things or events.[11] It does not deal with isolated or unique events, but with the observable relative frequencies of massive or repeated occurrences of phenomena, events, or processes. Thus, the inductively established frequency distribution of the rubidium content of biotites shown in Fig. 5.2 implies the statement that there is a statistical probability of 0.30 that biotite crystals from various rocks will have a rubidum content between 0.03 and 0.07%. It can, therefore, be predicted that out of a sufficiently large quantity of biotite to be analysed, 30% will have the property of containing 0.03–0.07% rubidium. Similarly the grain size distribution of a sediment or the mineral composition of a sand obtained by counting the grains of a sand sample imply statements on statistical probability. When analyses of a measured sample of sand grains from a sand stratum (i.e. from a limited number of grains) are used to derive a grain size distribution which states, for instance, that 15% of the grains have a diameter between 0.06 and 0.12 mm, this can then be generalised to the entire sand stratum to state as follows: 'should individual grains be removed from the stratum, the probability that a grain's diameter falls within 0.06 and 0.12 mm is 0.15; of a greater number of grains, 15% by weight will have this size.' What holds for these generalisations of the frequency distributions of certain parameters in complexes existing at one time also holds for generalisations about chronological sequences of events. From the frequency of the formation of meteoritic craters in the geologic past, for instance, it has been inductively established that the statistical probability of a large meteor striking the Earth is 1:1 000 000, that is, that for a time period of n years, where n is a very large number, $n \times 10^{-6}$ impacts of large meteors may be expected.

In these cases the relative frequency, f_i, with which a given phenomenon in a limited complex (sand grains of a given size or mineral composition, impact of a large meteor, etc.) is observed to coexist or to occur through time can be used to infer the statistical probability with which the same phenomenon will occur in a complex that is numerically, spatially, or temporally unlimited. This will be equal to the empirically established relative frequency f_i. To be able to argue in this fashion the given complex must be so constructed that in repeated observations on increasingly large sets of data the empirically found relative frequences $f_1, f_2, f_3 \ldots$ will form a series whose members will differ less and less from each other until the final value f_i is attained where the difference between succeeding members of the series is no greater than the small number ε. This limiting value f_i of relative frequency serves as a measure for the statistical probability of the occurrence or presence of the given phenomenon. Applied to our example of a sand composed of grains of various sizes or minerals, this means that the sample used for the grain size analysis must contain enough grains that any increase in their number would bring about only a negligably small change in the relative frequency of the individual grain size classes. Given the task (e.g. in the heavy mineral analyses frequently used in sedimentary petrography) of counting out prepared grains under a microscope in order to determine the relative frequency of different minerals, the results of counting out 100, 200, 300 . . . grains will be compared until the relative frequencies cease to change visibly with continued increase in grain counts. Experience shows that such a stabilisation occurs at counts of around 800 to 1000 grains.

Where a single parameter is obtained from many individual observations, as in measuring the value of light refraction or the density of a mineral substance, n observations will yield a frequency distribution of the parameter's measured value; this yields the statistical probability with which future measurements will find a value of the parameter within a given interval. Thus, for instance, the standard deviation s, according to Gaussian theory, is equivalent to the prediction that from a large number of measured values, 68% will lie within the interval $M \pm s$, where M is the mean of the given variable. To determine such a parameter exactly, the individual measurements must be repeated until s ceases to vary noticeably.

The statistical probabilities of inductive generalisations in more complex cases involving multi-place relationships are defined in the same way. In graphically or mathematically represented functions the statistically probable statement is represented by that form of the regression line or function (with its scatter or correlation coefficients) showing itself to be the limiting value with increasing numbers of the individual things, events, or whatever. Even in the case of regularities involving only comparative or qualitative concepts one can still speak of their statistical probability. The relative frequency of some observation out of a large number of cases forms the measure of the statistical probability for its occurrence in all further similar cases. The statistical probability of some of the examples of verbal regularities mentioned above is assumed to be 1 (statistical certainty), as none of the investigated cases have turned up evidence refuting the regularity.

The confidence which can be placed in the validity of the inductively obtained generalisation, that is, in the statistical probability derived from the frequencies depends on the certainty with which it can be assumed that the relative frequencies and the regression function curve will indeed converge with the number of observed cases to limiting values and limiting functions, and furthermore that these convergencies will continue to hold true even into the realm of the unobserved. This certainty is not guaranteed by simple induction but must be supported by other grounds. Thus, where determining the values of physical mineral properties, we must have confidence in the validity of crystallographic laws; in the case of analyses of grain sizes and mineral compositions of sands, our confidence is based on the hypothesis of the homogeneity of the sediment body from which the sample was taken; in predicting future impacts of large meteors, we have confidence that the distribution of these bodies in planetary space has not and will not change. In the case of relationships between several parameters, we are supported by the expectation that regular relationships exist between the variables. Usually we are led at least by an idea which may be called the principle of sufficient reason: we assume that the relative frequencies of certain phenomena derive from many cases which as far as we can judge are 'similar' and will continue to hold true for all further cases which are 'similar' in the same sense, since there is not sufficient reason for thinking that the relative frequencies will change as long as the 'similar' cases are simply repeated. Thus, there are grounds (grounds in fact, which differ from case to case, which remain tacit, and of which we may be unconscious) for justifying our confidence in the convergencies upon which the inductively inferred statistical probabilities are based. These grounds are open to debate and statistical probabilities consequently cannot be taken for granted; their reliability can be questioned and must therefore be evaluated for each case.

There is such a thing as a 'probability of statistical probabilities', a probability of a higher and more decisive level. This is so-called personal or *subjective probability*, a concept more closely related than that of statistical probability to the pre-scientific and colloquial use of the word 'probability'. The concept of subjective probability can be regarded as a rational formulation of the intuitive probability used in the vernacular. Subjective probability is to be understood as the degree to which a person 'believes' or doubts something. The person meant here is not any arbitrary individual, but a researcher belonging to the scientific community in general and the community of geoscientists in particular, who is familiar with or has access to the scientific, empirical, and theoretical knowledge of his time. The rather inaccurate word 'believe' here is supposed to refer to a rationally justifiable judgement of 'taking-to-be-true', or various degrees of such conviction, determined by a familiarity with the empirical data and theoretical insights, and by a certain weighing-up of these components. Since this familiarity varies from one individual to another and changes during time with the growth of scientific knowledge, we speak of a personal or 'subjective' probability.

Subjective probability is, therefore, not an objective variable attached to things, events, or processes as a property ascertainable by observation. Subjective probability is rather the reproducible and communicable result of evaluative reasoning on the part of the rational and well-informed individual in scientific research. It is subjective probability which above all, in contrast to statistical probability, can be attributed not only to the generalisations and regularities which interest us here but also to statements on isolated events. The scale ranges from values such as 'practically certain' to 'practically impossible'. Furthermore, the concept of subjective probability can and is widely used comparatively: we speak of a regularity or an event being 'more probable' than, or 'equally probable' as another.

The criterion of plausibility introduced above can now be explicated by the concepts of statistical and subjective probability. The generalisations and regularities established by induction from observations are statements about statistical probabilities. Further research can only build on these to the extent that they seem plausible in the light of currently accepted empirical and theoretical findings, that is, to the extent that a non-vanishing degree of subjective probability can be accorded them.

The two-fold meaning of the term 'probability' is a case of polysemia (p. 45). As with all ambiguities, this can easily lead to confusion, a danger to which Earth science is particularly prone, as probabilities are often dealt with there. We will close with an illustration of this.

The points on the Earth's surface which have been hit during the geologic past by large extraterrestrial bodies must form a random pattern unaffected by any constructions of the Earth's crust. At best they should be characterisable by statistical measurements derived as inductive generalisations from the frequencies of impact craters observed on the surface of the Moon: the number of impact craters, for instance, having diameters over 10 km, formed over the last 10^6 years in an area of 10 000 km^2. This number designates a statistical probability and says nothing about how many impacts would actually be expected in a particular area of given size, or even in an area of the Earth's surface characterized by special geologic structures.[12] In debates about whether certain structures were caused by endogenous forms or by an extra-terrestrial impact, the defenders of the endogenous origin have frequently used arguments of probability. The Vredefort Dome in South Africa, for instance, a possible impact structure, lies on a line of large magmatic intrusions into the Earth's crust. It has been argued that it is 'highly improbable' that a collision would have occurred at precisely this spot. Here we are obviously dealing with subjective probability or improbability, which has nothing to do with the statistical probability of an impact event on the Earth's surface. The defender of the endogenous origin will not be able to call up statistical generalisations on the frequency of impacts on underground bodies of magma, but the 'improbability' will encourage him to search for further arguments on which to base the plausibility of his hypothesis. The supporter of the impact hypothesis finds the argument of 'improbability' insignificant since an impact can take place anywhere at all. Nevertheless, the probability argument will also spur him on to find further support for his hypothesis. Similar arguments were used in the debate on the endogenous v. impact origin of the Ries near Nördlingen. Here again it was called 'improbable' that a meteor would have landed on a spot distinguished by the protrusion of the crystalline basement, the crossing of various fault lines, etc. Such arguments appealing in this way to subjective probability incidentally bring up the subject of the so-important process of argumentative scrutiny of hypotheses and theories which we will treat in detail in Chapter 6.2.

6

Problems of theoretical knowledge

LAWS

6.1 Criteria of lawlikeness and the problem of counterfactual conditions

It is only possible to carry out basic operations at a theoretical level of knowledge on the basis of universal propositions, which when the theoretical level is involved, we call lawlike statements. Just which lawlike statements may be used in abductive and deductive reasoning is apparently as clear to any initiate in the geoscientific discipline as it is to researchers in other areas of the natural sciences. The tie which unites individual researchers into a scientific community consists above all of the shared understanding (partly tacit, partly codified in textbooks) of a canon of lawlike statements admissable at the theoretical level of research. Anyone engaged in research supported by the international community will therefore, as Stegmüller says,[1] 'probably assume that it should not be difficult to find a suitable criterion' for distinguishing lawlike statements from statements which are not truly universal, and he will suppose 'that any further discussion on this is just philosophical sophistry'. A closer look shows however, that this impression is misleading. 'All the likely methods which at first suggest themselves for distinguishing between lawlike and accidental statements prove upon closer look to be inadequate'. Contemporary philosophers of science therefore appear to agree that even today despite many efforts 'the problem of lawlikeness is one of the most basic and difficult problems of empirical knowledge'.

Without claiming to solve the problem of lawlikeness within a general theory of science, we must nevertheless attempt despite these unresolved questions to characterise and systematise the lawlike statements involved in geoscientific research. The lawlike statements upon which all scientific inferences are founded are so central that it is impossible critically to understand geoscientific argumentation and theorising without knowing how to distinguish them from other general assertions, and how to justify the universal validity ascribed to them.

We will structure our investigations according to three aspects. First we will attempt to capture descriptively the characteristic features of those statements which are recognised as laws in geoscientific research and which are admissible in deductive and abductive argumentation. Secondly we will ask how confidence in the universal validity of established laws is justified. Finally we will attempt to classify the lawlike statements used in geoscientific research.[2]

Every lawlike proposition can be put into the form of a *universal conditional* which, according to the pattern 'whenever A is the case, then B is also the case' asserts that every thing, substance, process, etc. which fulfills condition A also satisfies condition B. Such a proposition should moreover be synthetic, that is, it should not be true merely on logical grounds, but should have an empirical content, which is confirmed only by assessing observational results. We will give two examples for established lawlike statements in conditional form.

(1) If a thing consists of fluorite (according to the definition of this mineral substance), it has a light refraction index of $n = 1.434$.

(2) If a mudstone (of corresponding chemical composition) is exposed to temperatures of between 500 and 600 °C, then at pressures of less than around 500 kbar, andalusite will be created, while at higher pressures, kyanite will be produced.

Some laws can also be expressed in *biconditional* form, according to the pattern 'B is the case if and only if A is the case.' If statement (1) were true in biconditional form then it could with certainty be inferred that any substance having a light refractive index of 1.434 would be fluorite, which, however, is not the case. Sentence (1) is a law only in conditional form. Sentence (2) is generally held to be true in biconditional form. It is believed that the presence of andalusite or kyanite in a metamorphic rock with certainty indicates that the pressure during metamorphism was less than or more than 500 kbar, respectively. Biconditional lawlike statements are particularly popular because of their argumentative strength in abductive reasoning. This can tempt one to assume biconditional validity where only conditional validity has been established.

It is a necessary, but not all all a sufficient condition for the lawlikeness of a proposition that it can be rendered in conditional form. There are many conditional statements to which we do not ascribe lawlikeness. Thus the analysis of many water samples taken during deep coring for oil deposits in various sedimentary horizons of the Illinois Basin, USA, gave rise to the following generalised finding.[3]

(3) If a water sample is taken from the Genevieve Limestone in the Illinois Basin at a depth of 3000 feet, it will contain 130 ± 15 g salt per litre.

This result is an inductively obtained regularity relating to an unlimited number of water samples taken from a limited and locally fixed volume of rock. The difference between a lawlike statement and a regularity may easily be recognised as follows: the said regularity initially describes a particular state of affairs, namely the empirically confirmed fact that a large number of water analyses from the Genevieve Limestone at 3000 feet show an average salt content of 130 g/l. This result is then extended as a regularity to include all future water samples obtained from this horizon. This extension beyond the actual observation is based on a certain degree of subjective probability (see p. 159).

While such regularities only note that something *is* such and such or behaves thus and thus, laws such as propositions (1) and (2) do not simply describe certain facts or responses, but assert that something *must* be such and such or respond thus and thus. It is because of the powerful connection expressed by the word 'must' between the antecedent and the consequent that laws, unlike regularities of type (3) can function fully as universal statements at the theoretical level of argumentation. In the course of abductive or deductive reasoning, only lawlike statements have the power to join known and unknown (to be inferred) states of affairs in such a manner that they strictly fulfill the requirements of a deductive nomological explanation.

Following Rescher (1970) we designate as nomological necessity that feature which all universal statements must possess in order to fulfill their function as lawlike propositions in scientific discourse. Nomological necessity is manifested by the power to generate hypotheses which are transfactual (go beyond the facts) or counterfactual (are contrary to the facts).

It is evident that laws (1) and (2) can easily serve as a foundation for *transfactual hypotheses*. For instance, (2) can be used to derive abductively a hypothesis on the maximal burial of a particular complex of metamorphic rocks, or deductively to predict the presence of andalusite or kyanite-bearing rocks in mountain units of known geologic history. Little, however, can be deduced from (3). The most that can be predicted from it is the salt content of further water samples from the limestone.

The power of lawlike statements to generate hypotheses becomes particularly apparent in the production of *counterfactual hypotheses*, i.e. hypotheses in the form of *irreal conditionals*.[4] Thus law (2) may be used to derive the following counterfactuals.

(4) If the metamorphic rocks of the Bavarian Forest contained kyanite (as is not the case), then metamorphism there would have taken place at more than 10 km depth (which was not the case).

The if-clause sets up a counterfactual hypothesis, for the rocks of the Bavarian Forest contain no kyanite. For this hypothesis an equally counterfactual conclusion is drawn in the then-clause, for the metamorphism in the rocks of the Bavarian Forest took place at shallow depth. Counterfactuals are common in geoscientific discourse. Argumentation of type (4) is used repeatedly to elucidate and test the stringency of explanations of given phenomena on the basis of hypothetical conditions and in accordance with certain laws. Counterfactuals act as thought experiments and pose problems for logical analysis which have been much discussed but which we will not go into here.[5] For our purposes it is merely important that the truth of counterfactuals, unlike normal conditionals, cannot be checked logically. According to the truth-functional interpretation, an if–then connection is always true when the if-clause is false, irrespective of whether the following then-clause is true or false. Logically, therefore, the following counterfactuals would be as true as statement (4):

(4*) If the metamorphic rocks of the Bavarian Forest contained kyanite (which is not the case), then the metamorphism would have taken place at less than 10 km depth (which was the case).

We hold statement (4) to be true and statement (4*) to be false, although there are no logical grounds for preferring (4). The connection between the if- and the then-clause is therefore conceived not as a truth-functional connective, but as a natural law, and we choose (4) because we recognise assertion (2) as a law. Based on this connection we can simultaneously test the counterfactual power and lawlikeness of propositions 'experimentally' by constructing counterfactuals. General non-lawlike statements can be recognised because they are not suited for setting up counterfactuals. It is evident, for instance, that a regularity such as sentence (3) has no counterfactual power, for no sensible counterfactual can be derived from it, as the following attempt shows.

(5) If a water sample with 5 g salt per litre were drawn from the Genevieve Limestone, it would contain 130 g salt per litre.

Lawlike statements can therefore be distinguished from regularities of the type of sentence (3) by testing for transfactual and counterfactual powers. A closer look shows, however, that certain difficulties arise in drawing a sharp distinction between statements to which we unhesitatingly accord the character of lawlikeness, and certain inductive generalisations, for some inductively inferred regularities are intermediate between the two. On the one hand we do not hesitate to recognise propositions on the physical properties of mineral substances, on phase transitions in metamorphism, etc. as laws, due to their transfactual and counterfactual powers. On the other hand we are certain that empirical generalisations on the salt content of the ocean waters of all seas or on the geographic distribution of laterite weathering on the South American continent have no more lawlike character than example (3) above. Then again, inductively inferred regularities, such as our earlier examples of the relationships between rounding and cobble size, between transport distance and cobble diameter, or between the various components of the lunar soil (p. 148), while not laws, are still possible candidates for laws since it seems reasonable to use them at times in the context of deductive or abductive argument in support of hypothetical statements, and since to some extent they stand the test of irreal conditionals.

If we ask for the reasons which justify according laws in the narrower sense a validity reaching beyond the realm of observable empirical facts, and which justify viewing certain inductive regularities as possible candidates for laws, we must take up where our discussions on the justification of inductive generalisations left off, (p. 155). As in the case of inductive regularities, laws can be justified in two ways: by the direct evidence to be gained from empirical facts and by the non-empirical, indirect evidence. Together these build the foundation for recognising the universal validity of laws.[6]

The *direct evidence* includes all empirical facts which support a given lawlike statement. No amount of empirical support, however, can justify its claim to be a law, that is, to have nomological necessity and the power to produce hypotheses. This justification can only be accomplished by the non-empirical factors which collectively make up the indirect evidence. The role of indirect evidence in support of laws corresponds to the role of plausibility or subjective probability in support of inductive generalisations,[7] the difference being simply that the items of the indirect evidence for laws can be expressed more fully and incisively than the often only vaguely discernable plausibility of inductive regularities.

Indirect evidence is provided when the lawlike statement involved does not contradict other recognised laws, especially the natural laws of physics and chemistry, and when it furthermore belongs to a theory as a complex of laws which in varying degrees of generality are supported by, related to, and derivative from each other. The strength of indirect evidence grows as the connections and interdependencies become broader and involve more disciplines. Indirect evidence is particularly important when it can be shown that the lawlike statement concerned is a consequence of particularly fundamental and therefore universal natural laws such as the two main laws of thermodynamics.

Since the crucial elements of indirect evidence cannot be observed but only developed by contemplation and interpretation within a wide domain of theoretical knowledge, the act called 'discovery of law' resembles a subjective *imputation*,[8] but is nevertheless well-founded, and as such it is accepted and recognised by the scientific community. The lawlike statements which in a given situation are recognised and used by Earth science have the character of *conventions*[9] which at the time optimally fulfill the demands of the direct and indirect evidence. All lawlike statements must remain basically hypothetical, however, since it can never be ruled out that new facts will at some time be discovered which will contradict the recognised laws, or since it may turn out that there are natural laws of greater general validity which contradict the hitherto accepted lawlike statements. There are no 'eternal' laws proof against all revision, and historical changes in the canon of lawlike statements used abductively or deductively to create hypotheses and to constitute theories is one of the features of progress in geoscientific inquiry which will be discussed in more detail in a later chapter.

6.2 Classification of laws

The laws set up and used in the various Earth sciences cover a wide and diverse field of subject matter. In the preceding chapter we ignored their differences in content, and attempted to reduce all laws and lawlike propositions to that shared logical form which makes their use in scientific discourse possible.

Looking now at the differing subject matter expressed by laws, we find that the main categories underlying the descriptive language of the geoscientific disciplines (see p. 83) provide a way of classifying laws according to content.

As we have found in a previous chapter, the universe of geoscientific research contains things, substances, and configurations, which change with time. We have, therefore, laws governing the composition of things, substances, and configurations: these are the *laws of determinate properties*. Then we have laws governing the changes through time to which things, substances, and configurations are subject: these are the *laws of process*, which we may separate into causal laws elaborating on the origins of processes, and developmental laws which conceive of change in terms of unidirectional processes. Finally there are the *laws of states*, relating to the conditions under which things, substances, and configurations represent stable, metastable, or unstable states; so long as these conditions remain constant, these states are not subject to processes of change. In the following sections we will be looking at the peculiarities of each of these types of law insofar as they pertain to the Earth sciences.[10]

6.2.1 *Laws of determinate properties*

Things and substances can be defined, classified, and identified according to properties (see p. 83). We have described the various kinds of properties using the example of mineral substances (p. 101); of these properties, it is the difference between defining characteristics and the remaining necessary properties which is significant for laws.

The defining characteristics are already imputed as properties when a certain thing, substance, or configuration is named by a descriptive term of a scientific nomenclature. When an individual crystal is designated as 'calcite', this means it has the defining characteristics of a particular crystal structure and chemical composition. The identification of a rock as 'granite' means that this rock possesses the properties of mineralogical composition and texture which define granite. Defining properties are not the subject of laws of determinate property. That is, statements such as

> the mineral substance of calcium consists of calcium carbonate;

or

> the grains of every sand have a diameter between 0.063 and 2.00 mm;

are definitions of the mineral substance 'calcite' and of the sediment 'sand', but not laws of determinate property.

This restriction does not apply to the necessary properties, which are not contained in the definitions of the names. Laws of determinate property are universal statements which attribute to all objects of a set certain predicates which do not belong to the set definition. Quantitative laws of determinate property include optical data on mineral substances, statements on other physical parameters (density, elasticity constants, thermal expansion, thermal conductivity, etc.), or dispositional identifications (melting point, phase transition temperature, solubility, etc.) of minerals. Comparative laws of determinate property include the so-called scratch hardness laws, while qualitative laws of determinate property include mineral cleavage. Qualitative laws of determinate property are wide-spread in all the geoscientific disciplines. It is frequently impossible to transform them into quantitative form, though this is a common research goal.

In order that laws of determinate property really have the character of laws and represent more than mere inductive generalisations, their nomological necessity must be ensured by indirect evidence. This is the case, for instance, for the necessary physical properties of minerals: from the theory of crystal structure it follows that a defined mineral substance must have definite and constant values of density, light refraction, and so on, all of which constitute the corresponding laws of determinate property. This does not hold for contingent properties. The brittleness of a mineral substance does not depend on defining characteristics, but on 'accidental' flaws in crystal structure, and can therefore only be derived as an inductive generalisation for certain sets of items of the given substance (e.g. the mean brittleness of sodium chloride crystals growing from solutions under specified conditions).

Even when biconditionality (p. 163), so popular because so useful for argumentation, is assumed for laws of determinate property, it must also be supported theoretically by indirect evidence if it is to be relied upon. Biconditionality is often employed as though it were assured, even when there is no theoretical foundation for it, but only a greater or lesser degree of (subjective) probability that it obtains. Thus its probability is relied upon when using pragmatic characteristics to identify mineral substances (p. 106).

It has been established as law that fluorite has a light refractive index of $n = 1.434$. While not certain, it is merely highly probable that there is no other optically isotropic substance having exactly the same light refractive index. Pragmatically, however, this law is employed as though it were biconditional, and the property of light refraction used to identify the mineral fluorite.

It is common for new discoveries made in the course of research to refute the supposed biconditionality of laws. Thus, until recently, the following law was held to be true in biconditional form:

> If and only if rock is produced by rapid cooling from a melt, its structure will be glassy.

On the basis of biconditionality, it was believed that the igneous origin of every rock having glassy structure could with certainty be abductively inferred. Today we know that glassy structures are formed not only by rapid cooling from a melt, but that diaplectic glasses can be produced in a solid state without melting by the effect of shock waves on crystallised substances. Glassy structure is therefore a necessary property both of quickly cooled igneous rocks and of rocks exposed to certain pressures under the impact of shock waves.

Pragmatic reasons often occasion the attempt, particularly in abductive inferences, to assume the biconditionality of laws of determinate property. This source of error may arise much more often in the Earth sciences, where abductive argument plays such a major role, than in other natural sciences. The division of rocks and geomorphologic forms into genetic classes, for instance, postulates biconditional laws, that is, it is assumed that certain properties indicate one and only one type of formation. We will give a few examples here of such laws of determinate property; their indirect evidence is based on the validity of such laws as the laws of process, which we will discuss later.

The presence of rounded hollows in a rock is viewed as a feature of igneous origin, on the basis of the law that hollows bounded only by rounded surfaces without corners or angles can only be formed by the separation of a gaseous or immiscible fluid phase out of a fluid. The biconditionality of this lawlike statement seems to be warranted by the physical theory of surface tension between immiscible phases (indirect evidence).

Stratification is usually seen as a clear characteristic of sedimentary rocks, based on the biconditionality of the lawlike statement that the qualitative property of stratified structure can only be produced by the deposition of separate particles in air or water. In fact, stratification cannot solely be relied upon in identifying sedimentary rocks. Stratified structures can also appear in a cooling magma reservoir or by sorting during metamorphism.

The configurational property of surface polishing by fine parallel lines on bedrock or on separate clasts, called 'glacial striation', is a characteristic of rocks and geomorphologic forms produced by the action of glaciers. These morphologic characteristics were therefore used to classify certain rock masses (tillites or moraines) and landscape forms (e.g. cirques) as glacial forms. The breccias around the Nördlinger Ries were once explained as moraines for this reason, and striations on the bedrock along the Ries edge were interpreted as glacial striations, the products of a 'Ries glacier'. Today we know that both striated bedrock and tillite-like breccias can be produced by mass movements resulting from an impact. As a consequence of this discovery, the explanatory classification of forms and rocks, based, among other things, upon the supposed biconditionality of the lawlike statement mentioned above, must be reviewed and in some cases revised. There may be other configurations and rocks, hitherto interpreted as glacially formed, which, like the Ries, may have been brought about by impact processes.

6.2.2 *Laws of processes*

These are laws which establish an invariable chronological sequence of events or properties, and may be broken down into two subtypes: causal laws and developmental laws.

Causal laws

It has been established[11] that the terms 'cause' and 'effect' have nearly disappeared in the 'exact' sciences. Physical laws of nature are formulated as mathematical functions, without the words 'cause' and 'effect' ever appearing. In other natural sciences, however, and in the Earth sciences, it is quite legitimate to designate a certain temporal sequence of cause and effect and to view their succession as produced by causal necessity. It is, for instance, quite reasonable to state that a wind of given velocity causes the formation of ripples on a surface covered with loose sand, and a causal law may be formulated that explains this relationship by establishing generally that given the fulfillment of certain conditions, streaming within a fluid medium will cause asymmetric ripples to appear on a surface consisting of loose grains. At an early stage such a law can be stated qualitatively. At levels of higher precision, cause and effect are described quantitatively, by relating the density, viscosity, and velocity of the streaming medium to the size and density of the sand grains and the geometric measurements of the resulting ripples.

Direct evidence of causal factors is based on experimental observations and on the results of thought experiments. A causal law describes what the outcome of conditions set up in an experiment must be. As von Wright suggests, therefore, causal laws may be distinguished operationally from inductive generalisations. A relationship of causal law is present 'when, by performing p, we bring about q, or by leaving out p, we eliminate q or prevent it from coming into existence. In the first case the cause is a sufficient, and in the second case a necessary condition of the effective factor.'[12] In the case of causal laws then, a lawlike statement is recognised in so far as one can assume the basic possibility of experimentally manipulating the causal factors. This relationship of causal law to action is what Kant had in mind when he characterised the development of the natural sciences since Galilei in the following way:

> They comprehended that reason has insight into that only, which she herself produces on her own plan, and that she must move forward with the principles of her own judgements, according to fixed law, and compel nature to answer her questions, but not let herself be led by nature, as it were in leading strings.[13]

The experimental method used to find causal laws showed

> that all natural science proceeding by this method is aimed at bringing about through human labour either in reality or in thought, those configurations and structures to which the individual sciences are dedicated.[14]

Since the geosciences usually deal either on the one hand with objects which are too large or distant for actual experiments, or whose natural conditions cannot be manipulated, or on the other hand with processes which are too fast or slow or which cannot be reproduced as a whole, causal laws and causal explanations of phenomena in the Earth sciences are often tested by thought experiments instead of by the actual experiments which predominate in physics and chemistry. The use of causal lawlike statements in the Earth sciences shows that the phenomena and processes being dealt with are far more complex than in physics and chemistry, since the natural world to be investigated must be taken as it is, and its conditions cannot be manipulated at will in the laboratory. Thus the impact of a mass from planetary space upon the Earth's surface creates a shock wave which spreads through the various rocks composing the surrounding area. The high pressures and temperatures along the front of the shock wave cause changes in the rocks and their mineral components; nearest the point of impact they are vaporised and melted, further away they are altered to glasses and new crystal types, and they are plastically deformed. Still further away they are subjected only to elastic stress. A causal event thus induces a series of events linked by causal laws, each event being both a cause and an effect. In order to explain the entire sequence, various causal laws must be employed.[15]

According to scientific parlance, there are several conditions which clearly must always be fulfilled if one wants to speak of causal relationships and causal laws. Following Nagel,[16] we name the following.

(A). The *causes* of an event are taken to be all the conditions which must be fulfilled in order for this event to take place, that is, all the necessary and sufficient conditions for the occurrence of the event. In actual fact, however, explanations of an event never include all the conditions and causal relationships; rather a single event or a limited complex of events are selected as the 'causes'. We will illustrate this with an example.

Where a porous sand of quartz grains is tectonically buried at depths where it becomes subjected to increased pressure, the small surfaces of the quartz grains lying roughly parallel to the Earth's surface (i.e. at right angles to the direction of pressure) will become subjected to so-called pressure solution where they touch one another, that is, their quartz substance will be dissolved. This is because the solubility of crystals is greater at places where they are under pressure. The event of pressure solution, through which a loose sand is altered into a compacted sandstone, depends on the following conditions.

1 The pressure at the points of contact must exceed a certain minimum;
2 The pore space of the sand must be filled with water in which the quartz can dissolve.
3 The contact surfaces of the grains must be covered with a thin film of clay, through which the components dissolved under pressure can diffuse into the pore spaces.

These three conditions correspond, according to situation, to three different causal relationships and causal laws.

1 If sands at various depths are compared, marked pressure solution in accordance with the dependency of solubility on pressure will be found only in those sands buried more than 1000 m deep. The cause of the pressure solution is therefore burial at depths of over 1000 m, and a causal lawlike statement may be formulated as follows: 'When quartz sands are buried at depths of over 1000 m, compaction of the grains will invariably set in through pressure solution.'
2 Comparing sands buried at similar depths, but containing differing substances in their pore spaces, it will be found that sands containing oil or gas show no evidence of pressure solution. It is the water filling the pore spaces which causes the solution, and a causal lawlike statement would run: 'If and only if sands are permeated with water, will pressure solution take place at the appropriate depths.'
3 If, finally, water-permeated sands buried at similar depths but varying in their clay content are compared, pressure solution will be found only in those sands whose grains are surrounded by a clay film. The cause of the pressure solution is the diffusion of the dissolved ions through the clay film at the points of contact. A corresponding causal lawlike statement may be formulated, setting the process of solution in relation to the speed of diffusion, which is dependent on the thickness of the clay film, the diffusion coefficient of the clay, the concentration of the pore solution, and the pressure at the contact points.

(B). Causal laws *sensu strictu* relate to *nearby effects*, that is, cause and effect are spatially contiguous and take place in the same spatial region. An event A can cause a spatially remote event B, if A and B are the first and last members of a cause and effect chain of events bridging the distance between A and B. Thus a fault event taking place deep in the Earth's crust can be cited as the cause of earthquake damage at the Earth's surface, since both events are connected by a chain of elastic oscillations, a seismic wave stretching from the earthquake epicentre to the surface.

However, in the case of *action at a distance* one also speaks of a causal connection, as when the combined positions of the Sun and Moon are cited as the cause of the tides. Here the concept of a field of gravity is the mediator between the cause (position of Sun and Moon) and the distant effect (tides). Changes in the relative positions of the Sun, Moon, and Earth cause changes in the field of gravity stretched between them, whose energy vectors directly affect the oceanic water masses at a certain spot on the Earth's surface.

(C). The causal relationship always has a *temporal character*, in that the effect appears after the cause which it 'touches' in time. When an event A which happened long ago is understood to be the cause of facts existing today, this means either that event A long ago is related to the currently-existing fact through a chain of temporally adjacent and causally-linked events, or that state of affairs B came into being immediately after cause A, and has undergone no major changes in the time since. The latter case is illustrated when we say that the impact 14.8 million years ago of a gigantic meteor was the cause of the large, allochthonous rock masses which may be observed today lying in the region of the Nördlinger Ries. What this means is that a causal lawlike statement exists according to which the impact is linked to the displacement of these masses, and that their position has not changed significantly in 14.8 million years. An example of our first case is the discovery that by force of certain laws, the meteoritic impact 14.8 million years ago created a circular basin whose shape has changed through later processes in such a way as to produce the present morphology of the Ries Basin.

(D). The causal relationship between two events is *asymmetrical* in the sense that it is not reversible. According to causal law, the wind is the cause which forms the ripples on the sand surface, but the formation of the ripples cannot be the cause for the movement of the wind.

In addition to these four conditions which Nagel (1961) has named as requirements for causal laws, other authors also cite further constraints said to apply to causal laws. According to Stegmüller,[17] causality cannot be expressed by probabilistic laws based on statistical data, and causal laws should all be quantitative. The first constraint can be ignored here, as Earth science always deals with processes of the world of macro-phenomena, and we can therefore overlook problems of the non-causal, statistical laws of the world of micro-phenomena. The second restriction does concern Earth science, but requries some specification here. It is frequently the case in Earth science that certain states of affairs can only be represented in a qualitative manner, or where complexes are concerned, in a statistical fashion. The qualitative or statistical form used to characterise entities thus set in relation to each other does not, however, affect the causal nature of that relationship, since it is assumed in such cases that the underlying elementary processes are governed by causal laws which could be formulated quantitatively and deterministically were it possible to examine each individual process in isolation. Since, however, it is the integrated process which is observed, determining the conditions and effects resulting from the summation of many diverse elementary conditions and correspondingly varied effects, the parameters set in relation in the integrating laws can be described only in a statistical or even qualitative manner.

The nomological necessity of all the above examples of causal laws derives from physical and chemical laws of nature. This is invariably true for all causal lawlike statements dealt with in Earth sciences, and it expresses the relationship of the Earth sciences to the fundamental natural sciences. A law can postulate a relationship of cause and effect only where that relationship can be traced back to laws of nature. All causal laws in geoscientific research are laws produced by the specialised application of general laws of nature to states of affairs relevant to geoscience.

Developmental laws

In all geoscientific disciplines, long term, irreversible changes of that which exists at any given time play a very large role compared with other natural sciences. In physics and chemistry when investigating types of movements, alterations, and reactions which are valid now and forever for materials and their components or phases (such as the way a body falls in a gravity field or the solution of zinc in sulphuric acid), the processes are experimentally reproducible and do not take up long periods of time. In the biological sciences as well, the processes usually dealt with may be more complex, but their courses can nevertheless be observed and reproduced experimentally, e.g. the typical course of physiological reactions, the behaviour of living organisms in their environment, the ontogenesis and subsequent life cycle of organisms, ending with the death of the individual. Research into heredity and the supra-individual changes in genotypes, too, deals with sequences of events which can be reproduced and triggered experimentally.

Most of the temporal changes investigated in the Earth sciences are different in kind. True, oft-repeated processes are studied here as well: cooling of magmas, weathering of minerals at the Earth's surface, transport of material through wind and water currents. These processes are not different in principle from those studied in physics and chemistry. With regard to the substances and structures involved, their external conditions and the forces impelling them, however, they are incomparably more complex, and take place over far greater periods of time than the reactions investigated in pure substances in a chemical laboratory or in experiments with known materials under controlled conditions in physics. Apart from such processes, which, though complex and therefore variable in individual aspects, can at least be observed frequently in nature, geoscience, like astronomy, must also deal with secular, apparently unique developments. Like the stars in astronomy, the subjects of Earth science, minerals and rocks, plant and animal species, volcanoes and mountains, entire continents and oceans, the interior and the surface of the Earth and planetary bodies, are all entities in constant change which presumably take part in the unique, contingent, and irreversible history of the universe. A major task of research in the geoscientific disciplines consists of reconstructing partial sections and local areas of Earth and life history, and finally, of reconstructing geohistory and planetary history as a whole.

This is the reason for the oft-stressed relationship between the Earth sciences and the study of human history.[18] Earth science investigates and explains the developments in the natural world, much as historians of human history investigate and explain the development of mankind. These historical enterprises are analogous in that both deal with the abductive reconstruction of unique, non-reproducible, and contingent processes of the past. They differ largely in that the explanation of human history is hermeneutic, i.e. consists of an interpretive 'understanding' of the human worlds of the past, while the explanation of geohistoric processes must have recourse solely to laws of nature.[19]

One initial problem is that many processes in nature are so complex that it is impossible to break them down exhaustively into such elementary processes as can be studied as reactions in a laboratory and explained with reference to physical and chemical causal laws. In order to explain geohistoric processes, therefore, laws of a particular kind are often used, which we will call *developmental laws*. These can be preliminarily characterised as discoveries which relate to the temporal changes in individual entities or complex systems, which establish what changes will regularly take place under given conditions, and which determine how these changes will take place in a given period of time. Since developmental laws link together successive states of things, they are related to causal laws. They differ from causal laws in two points however, which justify treating them as a special class.

Firstly, developmental laws lay down initial conditions which must be met in order for a development to take place in a certain fashion. In contrast to causal laws, however, they do not cite grounds sufficient for the initiation of the events; rather they designate the *direction* or *tendency* which the development follows. Secondly, developmental laws establish a temporal order of events separated by time intervals of arbitrary length. They describe *continuous or discontinuous changes of things or systems* in the course of time. Causation, however, can only be said to obtain where there is an immediate temporal succession of cause and effect.

Since developmental laws as a rule relate to systems whose structure and component processes are so complex that the elementary components cannot be delineated, it often proves difficult or as yet entirely impossible to base them upon general laws of nature. Many developmental laws are therefore but a step away from inductive generalisations.

Developmental laws which approach the character of general laws of nature presuppose sequences of events which have taken place repeatedly in the same or similar fashion. In particular this includes processes which can be observed at present in nature, such as the mechanical and chemical disintegration of primary rocks and minerals on the Earth's surface, the formation of soils, the movement of dissolved and comminuted material in rivers, the deposition of sediments in lakes and oceans, and the activities of volcanoes. We will look more closely at some developmental laws relating to such oft-repeated processes.

V. M. Goldschmidt's 'geochemical distribution laws of the elements' may be regarded as developmental laws. They may also formally be called laws of determinate property as they establish the average content of chemical elements in the main minerals and rock types. Nevertheless they are intended as developmental laws, since the content of an element in a given rock type is seen as the result of that element's behaviour during the long process of differentiation, a process which has led repeatedly during Earth's history to the formation of specimens of the major rock types. Thus in the three main types of sedimentary rocks, shales, sandstones, and limestones, the element barium is almost invariably accumulated in the shales. This enrichment is the result of a regular development during the complex and interrelated processes of weathering, transport of the weathered material, and deposition. During these processes, barium ions freed by weathering of the initial material are bound in and on the clay minerals by adsorptive and chemical forces, and are deposited with them. Equivalent developmental laws apply to the paths of all chemical elements during the processes typical of the destruction and alteration of rocks. It is the task of geochemistry to study these laws, and with their aid to explain the sorting of chemical substances in the various rock complexes of the Earth's crust, soils, and waters, as well as to explain how certain elements become concentrated in so-called deposits. It is assumed that these briefly summarised developments led from a hypothesised, initially homogeneous distribution (in a homogeneous original planet or in the solar nebula from which the planets condensed) by means of physico-chemical laws to certain distribution patterns in the minerals and rocks. These laws determine how the atomic building blocks of the elements become distributed between a magma and the crystals segregating out of it, or between the mineral particles and solutions to which they are exposed, such as soil solutions or river water. Developmental laws, however, do not concern themselves with details of the causal links uniting the events or stages of the entire process. The nomological necessity of the developmental laws illustrated above is implied by the realisation that the given development runs in accordance with causal laws of nature. There is, for instance, a geochemical developmental law by which the alkali metal rubidium becomes enriched in the melt during the crystallisation of a silicate magma, so that the last rock type (pegmatite) produced from the remnant magma is the richest in rubidium. This developmental law may be regarded as founded upon a causal law, since rubidium ions have such a large diameter that they fit poorly into the crystal lattices of the minerals produced one after another in the course of crystallisation, and must collect in the last remaining portion of the magma.

Developmental laws of this kind, relating to processes based on many separate causally connected steps, none of which can be isolated but all of which can be explained by laws of nature, are very common in mineralogy and petrology. Another example is provided by the developmental law of progressive metamorphism, a process which rocks undergo when exposed for sufficiently long periods to increased pressure and/or temperature. These metamorphic laws, which we will discuss in more detail below, lead to the so-called metamorphic facies, or final stages of the development. These facies are associations of certain minerals, worked towards and realised to a greater or lesser extent during metamorphic development, that is, during recrystallisation of the primary material in certain pressure and temperature regions.

The concept of developmental laws also underlies the usual classification of sandstones according to their degree of 'maturity' in sedimentary petrography. 'Mature' sandstones are those which consist of quartz grains only. A very 'immature' sandstone would be, for instance, a greywacke, which in addition to quartz grains contains fragments of feldspar, other minerals, and unweathered rock. 'Mature' and 'immature' relate to early and late stages in the development of sandy weathered debris in which lengthy and recurrent processes of chemical and mechanical weathering and transport lead to the steady enrichment of mechanically and chemically highly resistant quartz grains. The soil sciences, too, use physico-chemically governed developmental laws. These establish the typical initial, middle, and end phases of the soil's gradual alteration by weathering agents for each climatic zone and each parent rock type.

These examples involve oft-repeated processes. It is also desirable, however, to understand the large, unique developments and to recognise the laws operating in them. Clearly a law can only be spoken of where several identical, analagous, or similar cases can be subsumed under that law. The unique course of the Earth's development from a hypothetical original condition to its present state does not in itself represent a law as such, and can at best be recounted as a story. But the Earth's history can be understood to be governed by law as soon as its overall course can be reduced to some physically governed process reproducible actually or experimentally in a model or in a thought experiment. Thus in the nineteenth century and into our time, it was thought that the course of Earth's history could be explained as the cooling of an originally hot body. The governing laws for such a hypothesis are the physical laws determining the gradual radiative heat loss of a hot body in a vacuum, the concomitant processes of shrinking, and the resulting tensions and deformations. With the discovery of heat production by the radioactive decay of various types of atoms, the concept of a gradually cooling Earth had to be abandoned and replaced by other models.

Recently Earth's history has lost its unique character through the 'geological' investigations of the Moon, Mars, and Mercury, and has become one among other cases in planetary history. Viewed from the planetological aspect there is a typical, or law-governed sequence of phases in a development which all the inner planets have undergone.

Focussing on the history of Earth in particular, periods, episodes, and special processes can be discovered and distinguished in which, since they have recurred in characteristic fashion, lawlike statements can be discerned. These lawlike regularities, however, have the character of inductive generalisations, though it is established or assumed that they are physically plausible. Regularities of geotectonic development fall into this category, such as the 'law' of the sequence of events leading to mountain formation (orogenesis), confirmed by research on many mountain ranges. The sequence begins with the formation of elongated subsidence zones, geosynclines, in which vast quantities of sediment are deposited from the adjacent continents. These sediments become shot through with outflows typical of submarine basaltic vulcanism. In a later phase the masses are compressed and folded at depth, and finally slowly raised up again, erosion and weathering in the rising crustal portions creating the constantly changing mountain and valley forms of young high mountain ranges.

The difficulty in setting up laws or regularities of this kind is that they must be drawn from a relatively limited number of cases which vary considerably in detail. Despite this, the cases must all be recognised as analogous manifestations of the same type of phenomenon. To set up laws on the formation of folded mountain ranges, such mountains as the Alps, the Himalayas, and other already partially destroyed mountain structures from earlier geologic epochs must be presented as analogous cases of the same phenomenon of folded mountains. The cogency of laws derived in this way depends on how precisely the analogy can be defined and on the number of event sequences to which it can be applied. As long as only one Ice Age was known it made no sense to speak of a lawful progression of glacial advances and retreats in such a period. Now that glaciations are known from periods prior to the Pleistocene, it has become reasonable to search for regularities characteristic for all analagous glacial periods.

Besides the impossibility of reducing many real geohistoric processes to elementary events which are physically and chemically wholly transparent, and in addition to the difficulty in recognising laws in events for which only a small number of instances exist, we must draw attention to a third problem in the context of developmental laws. While the laws of physics and chemistry are founded on facts which can be experimentally reproduced at any time, the developmental laws of geoscience are almost invariably based on hypothesised states of affairs. Laws on the geochemical behaviour of elements during the formation and differentiation of the Earth's crust, or regularities pertaining to the evolution of the plant and animal phyla are derived from hypothesised states of affairs of the geologic past, which in turn must be inferred abductively. It must therefore be recognised that geoscientific developmental laws are only as certain as the hypothesised states of affairs on which they are based.

The problem of teleological explanation of geoscientific processes

From the examples which we have looked at for the development of inorganic systems it is evident that 'development' does not simply refer indiscriminately to arbitrary change of given entities through time. We do not call it a 'development' when tiny particles in a suspension constantly change their configurations within a small area through Brownian molecular movements, or when on a certain spot of the Earth's surface the mean temperature of a given month undergoes irregular fluctuations over a long period of time. A development or evolution can only be spoken of when a process of change is observed in the course of which a progression is evident. Something must increase or decrease through time in size, intensity, complexity, completeness, or in some other aspect. Moreover, development always involves a coming into being or a passing away, i.e. the changes cannot simply be measured by quantitative parameters, but must be qualitatively described: homogeneous mixed weathering debris becomes qualitatively different sedimentary rock; marine troughs of geosynclines become folded mountain ranges.

Processes which are called developments have a directional character; a beginning, middle, and end can be distinguished in them. Each stage has its place in the development, and is defined not only with respect to the origin from which the development began, but is primarily also orientated towards the end to which the development progresses, whether this is envisioned as an actually attainable condition or an asymptotically approached goal. Any investigation which perceives developments in natural processes therefore appears related to the teleological or finalistic view point, according to which natural processes are regarded not only from the causal point of view, but also with respect to final causes or purposes. The question must be raised whether developmental laws of markedly teleological character are set up and applied in the geoscientific disciplines.

It is well known that the teleological view of nature was widely accepted and scientifically legitimate from the time of Aristotle to the Renaissance. That is all long past now, and there seems to be a general consensus today that scientific explanations must be based on universal laws of a non-teleological kind. Kant undertook to investigate 'whither this stranger in natural science, viz. the concept of natural purpose, would lead us' and established in his Critique:

> for since we do not, properly speaking, observe the purposes in nature as designed, but only in our reflection upon its products think this concept as a guiding thread for our judgement, they are not given to us through the object.[20]

This is clearly also the official standpoint in natural science today. The notions of goals and purposes are not admissible in the framework of scientific argument.

Nevertheless, laws with a decidedly teleological character have been formulated and used to advantage, not only in the Earth sciences, but in various other natural sciences as well. We will demonstrate this using an example provided by physics. Alongside the physics based on causation, whose origin can be roughly described by the names of Newton and Huyghens, attempts have been made since the seventeenth century to explain physical processes in nature through general principles of a teleological kind. The first of these goes back to a proposition by Fermat and Leibniz, according to which a light ray penetrating a medium having a locally variable light refractive index (n), will take such a path (s) between any two points that the product of path distance and refractive index formed over all the path elements (mathematically formulated, the integral $\int n \cdot ds$) will have a minimal value in comparison to all other possible paths. Leibniz believed he could use this law to show that not only efficient causes *(causae efficientes)*, to which the post-Renaissance scientists had restricted themselves, but also final causes *(causae finales)* could be used to explain nature. He believed that final causes not only provided the most beautiful of opportunities to marvel at the wisdom of God, but also served to elucidate processes whose inner nature is not known well enough to apply efficient causes, and to explain the mechanisms used by the Creator to produce certain effects. Since Leibniz's time, very general laws of teleological type have repeatedly been proposed to explain the direction of physical processes without considering their causal determination. These are the so-called *extreme principles*, dealt with mathematically by variational calculus. Each proposition establishes that the given process is distinguished from all other possible ones in that a suitably chosen variable will take on an extreme, i.e. maximal or minimal value.

Maupertius in the year 1747 was the first to set up an extreme principle of the most general kind, the 'principle of least action' (*principe de la moindre action*), according to which a body moves in such a fashion that the product of its mass, velocity, and distance is always a minimum. He, too, believed that such a law was particularly suited to the wisdom of the Creator. In later times, other and more practical expressions were found, e.g. Euler's proposition, which also explicitly maintained that natural causes could be understood both in terms of their effective and their final causes. According to Euler's proposition, a body will move under the influence of forces in such a way that its actual path, expressed in the form $\int v \cdot ds$, will approach a minimum, that is, it will be smaller than that of any other infinitely close adjacent path having the same starting and end points. ds refers here to any infinitely small segment of the distance, and v to the velocity belonging to this distance segment. This proposition can in fact be used to solve many varied problems of mechanics without referring to any causal laws at all.[21]

Of all other teleological type laws in the physical and chemical sciences, we stress the second law of thermodynamics, so important for all natural processes. This establishes generally the direction of all processes in nature without touching upon the causal laws which govern these processes individually. The second law asserts that all processes in nature proceed such that the energy available to the system will partly be bound in a non-recoverable fashion, in that the entropy, a thermodynamically defined magnitude of a physical state, will increase or at most remain the same. The second law states: all processes taking place in nature occur in such a fashion that the entropy of all bodies taking part in the process is not reduced. The total entropy is usually increased and remains constant only in reversible processes.

A system which is open or partially open to its surroundings can be supplied with energy from outside. It can then appear as though entropy is being reduced locally, as when plants use water and carbon dioxide to create carbohydrates, highly organised molecules of low entropy. This is only possible, however, because solar energy is being fed to them from outside. According to the second law, the ultimate fate of substances built up by living creatures in the closed system of the Earth is the irreversible breakdown of the highly ordered molecules and molecule aggregates into elementary building blocks of higher entropy.

If propositions of a teleological kind are to be found in physics and chemistry, it should hardly cause surprise to find the geoscientific disciplines teeming with teleologically formulated laws and explanations. Before we turn to these laws, however, we will look generally at the relationship between causal and teleological explanations, following closely on Rescher and Stegmüller.[22]

Teleology and a 'mechanistic' view of causality are usually juxtaposed as exemplifying mutually exclusive points of view. In fact, however, there is no simple contradiction between the two, for several relationships are possible. This can be clarified by a simple formal characterisation of teleology and the mechanistic ideal of causality. In a *teleological explanation*, current events are explained with reference to future conditions and processes, while in a *causal explanation*, the present is explained through the past. Such a reduction to temporal relationships makes no mention of goals, intentions, purposes, necessities, etc. The causal explanation can simply be described as an explanation '*a tergo*' the teleological explanation as an explanation '*a fronte*'. Now there is not just one single causal-mechanistic standpoint, but several. The main ones are the following:

M_1: Every event can *only* be explained *a tergo*.

M_2: Every event can be explained *a tergo*.

M_3: Every event can only be explained when some *a tergo* data are available.

Accordingly there are also several teleological positions:

T_1: Every event can *only* be explained *a fronte*.

T_2: Every event can be explained *a fronte*.

T_3: Every event can only be explained when some *a fronte* data are known.

This listing makes it possible to test the compatability of teleological and causal-mechanistic standpoints. Only the extreme position T_1 is incompatible with all the causal standpoints M_1 to M_3. The same is true for the extreme causal position M_1 relative to all the teleological positions T_1 to T_3. Further mutually incompatible pairs are M_2 with T_3 and T_2 with M_3. What is more interesting are compatabilities. When Leibniz and Euler believed that all events can be explained with respect both to effective *and* final causes, they were expressing the formal compatibility of positions T_2 and M_2. Positions T_3 and M_3 are also logically compatible with one another.

From this formal survey we may infer that there are two ways in which causal and teleological explanations can be compatible. One may either believe that a given event can be explained just as well by exclusively causal as by exclusively teleological law, or one may believe that to explain an event both some causal and some teleological laws are required. Causal and teleological laws relating to the same physical domain need not be mutually exclusive.

If we now look at the content of teleological explanations and laws[23] we note first of all that teleological explanations answer 'why' questions in a particular way. Teleological explanations do not begin with the word 'because', but rather with phrases such as 'so that' or 'in order to . . .'. Explanations begun in this fashion indicate a purpose, goal, or function to be fulfilled. It is said teleologically: the fish group of the crossopterygians developed nasal passages, lung sacks, and a bony framework for fins so that they could give rise to the first land dwellers with lungs and legs. The causal explanation, on the other hand, runs: the first amphibians to live on land (the ichthyostegians) were able to develop out of the crossopterygians because these fish had peculiar body structures which could be altered for the functions necessary for life on land.

Behind the apparent similarity of all teleological 'so that' and 'in order to' explanations hides a basic difference. We should distinguish between *goal-directed* and intentionally *goal-pursuing* processes. In the natural sciences today, goal-directed teleology alone is 'officially' permitted. To define this, we will look first at goal-pursuing teleology. The deliberate behaviour of a man is best explained as a goal-pursuing process, in that he elaborates his wishes and motives, his set goals and resolves.[24] Only such cases of goal-pursuing events can be explained in terms of purposes.

Applied to the realm of physical events, this means that natural objects and configurations can only be spoken of as purpose-fulfilling in a real and explanatory sense if the presence of a 'spirit in nature'[25] or a creative God is assumed. It is therefore reasonable for Leibniz and Maupertius to refer to the wisdom of the Creator when they explain the natural realm as purposeful and perfect in the sense of having an intended goal. The physico-teleological proof of God so highly esteemed in Antiquity and the Middle Ages was based on the discovery of purposeful configurations and purpose-fulfilling events in nature, and on the inference that a purposeful nature can only be the work of an ordering, goal-setting, godly Being.

Over long periods not confined to the nineteenth century, paleontologists regarded the phylogenetic evolution of living creatures in the geologic past as directed towards an intended goal in which the directing and ordering influence of a Creator could be recognized. Even into the most recent of times, true, goal-pursuing teleology could be postulated in paleontology: we refer particularly to the work of Pierre Teilhard de Chardin.[26] Many other paleontologists who were not theologians, however, have held similar views, as evidenced by a statement of the Tübingen paleontologists, F. von Huene, a respected authority on dinosaur evolution who died in 1969.

> It seems to me particularly important that belief in a personal Creator of Heaven and of Earth with its inhabitants is often stressed, particularly by natural scientists. All historians of nature whether they wish to or not, help to proclaim the glory of God, the Creator and the Lord of Earth and of Man, for they are the ones who make his work known.[27]

For the majority of researchers active in the Earth sciences today, however, the goal-pursuing explanation is ruled out, as they adhere to Kant's assertion:

> In order, therefore, to remove the suspicion of the slightest assumption – as if we wished to mix with our grounds of cognition something not belonging to physics at all, viz. a super-natural cause – we speak, indeed, of teleology in nature as if the purposiveness in it were designed, but in such a way that this design is ascribed to nature, i.e. to matter. Now in this way there can be no misunderstanding, because no design in the proper meaning of the word can possibly be ascribed to inanimate matter; we thus give notice that this word here only expresses a principle of the reflec-

> tive, not of the determinate judgement, and so is to introduce no particular ground of causality . . . Hence we speak quite correctly in teleology, so far as it is referred to physics, of the wisdom, the economy, the forethought, the beneficence of nature, without either making an intelligent being of it, for that would be preposterous, or even without presuming to place another intelligent Being above it as its Architect, for that would be presumptuous.[28]

When, therefore, purposes and purposefulness in nature are spoken of, this can only be understood by science as a '*facon de parler*'[29] an incorrect use of the word 'purpose', since there is no reference to a purpose-setting Will. The organisation of a living creature is said to fulfill a purpose well, meaning only that the organism resembles a machine so constructed by an intelligent technician as optimally to fulfill the purpose of surviving and functioning in a certain environment. Purposefulness in the context of the natural sciences is not an explanation, but a hidden description of functional relationships. This is evident when seemingly teleological explanations can be transformed into non-teleological descriptions simply by transposing the elements.[30] We will illustrate this by an example.

The following statement on the shell construction of a fossil animal group of marine ammonoids is a pseudo-explanation in teleological disguise. Ammonoids possess a snail-shaped shell of many chambers arranged serially; these chambers can be filled with air or water by means of a tube, the siphuncle, which runs through them. The purpose of this apparatus is to permit the animals to rise or descend quickly through the sea, thus tapping vital hunting grounds at different depths in the water. The state of affairs expressed in these sentences can be rearranged in order to express the same thing without reference to a purpose, i.e. non-teleologically. The ammonoids were able to rise or descend rapidly through the seas and feed at different water depths. An animal floating in the water can only perform rapid vertical movements by changing its buoyancy, so the ammonoids must have possessed some apparatus which brought this about. On the basis of physical laws, the siphuncle of the ammonoid shell must have been such an apparatus.

The teleological formulation gives the appearance of being an explanation. It could only be one, however, if a planning Creator or Spirit of Nature were present which would be interested in creating favourable living conditions for the ammonoids, and which therefore equipped them with a siphuncle. Since this cannot be intended in the scientific context, the word 'purpose' only describes a functional relationship of great biological importance for this group of animals.

After this excursus on the scientific use of the word 'purpose-fulfilling', we turn again to the question of teleological laws of development. Based on the above, we have established that the developmental laws used today in the Earth sciences express only goal-directed, not goal-pursuing teleology.

In the inorganic branches of geoscience, developmental laws in which the final condition of a system appears to govern the unfolding of a process are all in the final analysis forms or special cases of the second law of thermodynamics. In these, entropy is replaced by other functions derived from the concept of entropy and determined by the physical properties of the substances involved. It is the extreme values of these functions which govern the direction of spontaneously unfolding processes. Such laws are used to a large extent in petrology, the science which seeks to investigate the development and decay of mineral associations occurring in nature, and to explain their structure on a physico-chemical basis. The concept of *thermodynamic equilibrium* plays a dominant role in the laws of petrology. If a system is in equilibrium, it will undergo no change. Only when the conditions of equilibrium are overstepped do processes set in motion, whose direction is governed by the law of entropy: in other words the system 'seeks' to attain a new state of equilibrium. A system consisting of several mineral substances is in equilibrium as long as the measure of its energy content, the so-called free enthalpy, calculated from the physical properties of the components, maintains a minimal value. Equilibrium is disturbed once any change occurs in conditions (usually pressure and temperature) such that another mineral combination or structure of the same chemical composition would have a lower enthalpy. In the new situation the system will 'strive' towards the new mineral combination or structure in order to come to a new state of equilibrium.

It is on laws derived from this principle that all explanations of metamorphic processes are based, that is, all the alterations described previously which rocks and minerals undergo due to changes in pressure or temperature, following, for instance, burial at great depth. Such processes are involved, e.g. in altering a shale into a coarsely crystalline gneiss, or a limestone into marble. Metamorphic processes are not primarily explained by causal analysis, but on the basis of the general teleological law that all metamorphic processes steer toward the condition of lowest free enthalpy.

As our first example we will look at the alteration of graphite into diamond and vice versa. Both minerals consist of carbon. At certain pressures and temperatures the two crystal types, diamond and graphite, are in equilibrium, that is, they co-exist without any tendency for the diamond to change into graphite or vice versa. In equilibrium, the pressures and temperatures are such that both these substances possess the same free enthalpy, so that neither alteration would bring about any reduction in the

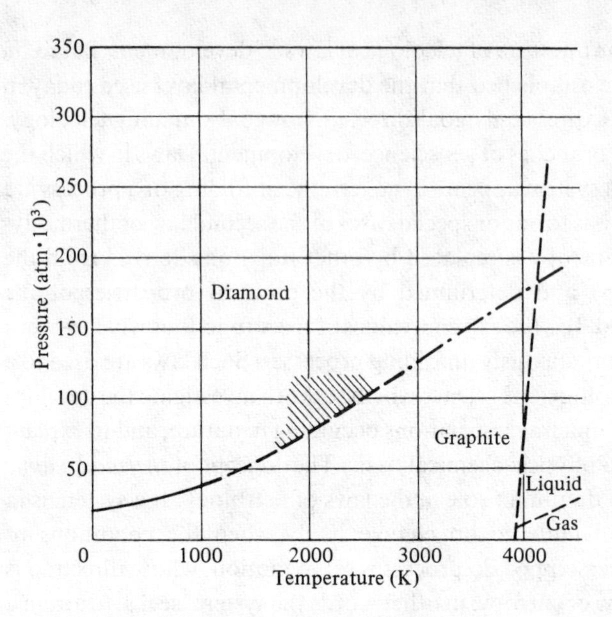

Fig. 6.1. Phase diagram of carbon in a pressure–temperature field. The hatched area shows the conditions under which diamond was synthetically produced. (Correns, 1968.)

free enthalpy of the system. The pressures and temperatures at which the enthalpy of both substances is the same can be calculated from the physical properties of diamond and graphite in a manner which we will not explain here. These conditions lie along the line or equilibrium curve shown in the pressure–temperature graph in Fig. 6.1. Under the conditions of pressure and temperature shown to the right of the curve, a change from diamond into graphite would reduce free enthalpy, while for the conditions corresponding to the realm shown to the left of the curve, free enthalpy would be reduced by a change from graphite into diamond. The diagram can therefore be taken as representing the teleological lawlike statement that at pressures and temperatures belonging left of the curve, graphite tends to change into diamond, while at pressures and temperatures corresponding to the right of the curve, diamond seeks to change into graphite. How these changes come about, at what speed the crystal lattices become rearranged, and whether this occurs within observable periods of time depend on many conditions and on causal laws. The thermodynamic law of equilibrium states only the conditions under which an alteration will be striven towards.

Teleological laws or inorganic processes which can be regarded as modifications of the second law of thermodynamics are used in great numbers for explanations and predictions in all disciplines of Earth science. Even today one can agree with Leibniz that for physico-chemical explanations of geologic natural processes there are two options: to discover the causally-determined mechanisms of a process, that is, to explain by means of causal laws of nature, or to explain teleologically using either the law of entropy and its modifications or another extreme principle of physics.

Since the law of entropy is generally true of all natural processes, teleological methods often provide an easier explanation, sometimes the only available one. The forms of the entropy law employed make up an extended series, running from strict laws based on thermodynamics to qualitative lawlike statements made plausible by the law of entropy. It is a qualitative lawlike statement, for example, that a rock consisting of many small grains tends to change to a coarse-grained rock, for in this way the amount of intergranular surface area per unit volume is reduced, equivalent to a reduction in free enthalpy. Other examples are the most general teleological lawlike statements of geomorphology, which assert that a volcanically or tectonically created relief on the Earth's surface will tend to flatten out with time. This wearing down corresponds to the tendency to reduce the amount of potential energy stored in the relief. The manner and speed at which levelling will occur must be explained by causal laws and depends on governing states of affairs such as gravity, kind of weathering, and transport processes with or without mediums such as wind or flowing water.

Teleological laws of a special kind, derived neither from the law of entropy nor from other physical extreme principles were formerly proposed in paleontology. We mentioned the problems of paleontological research when we analysed the concept of '*the purposefulness of organic forms*,' and we saw that in the context of today's science, statements on the purpose-fulfilling organisation of living creatures are nothing more than descriptions of functional relationships. But this reduction only side-steps the original intention of the researcher. The structures, properties, or behavioural traits of recent or extinct creatures formerly termed 'purpose-fulfilling' with an eye to a Creator or a Natural Spirit, are nowadays simply explained as necessary to living functions. This explanation shows causally how a given species may continue to exist despite surrounding influences which would otherwise destroy it; but what it does not answer, what it even overlooks, is the great problem of paleontological research: if 'purpose-fulfilling' cannot refer to the intentions of a purpose-setting intelligence, how did functioning life forms come to exist in the first place?

Paleontology takes up this question within the framework of the theory of evolution. This is based on the hypothesis, now generally accepted, that the chronological succession of plant and animal forms shown in the fossil record is to be explained as a real hereditary sequence. Research in the course of this and the previous century has produced an ever more complex picture of *phylogenesis*, or evolution of life, through which primitively organised forms of life have given rise over the billions of years of Earth's history to more intricately constructed and specialised ones. Ancestral relationships are shown by a branching evolutionary tree, though it might almost be called an evolutionary bush, so many of its branches break off in dead ends. This is so arranged as to display how all the species evolved from each other and from common ancestors during the course of Earth's history.

Phylogenesis is an excellent example of a development; this is a directed sequence of changes, or bundles of such sequences, during which qualitatively new forms were repeatedly produced. That qualitatively new forms resulted is directly evident in the differing structural types which, according to the theory of evolution, evolved out of one another: amphibians out of fishes, mammals out of reptiles, or birds out of dinosaurs. It is more difficult, however, to ascribe a single direction to the whole of phylogenesis. Schindewolf has emphasised that protozoans, snails, and fish were and are just as well adapted to their environment as the 'higher' vertebrates, so that evolution cannot be progressive in that sense. 'But nonetheless this does not mean,' Schindewolf continues,

'that we cannot ascribe to the various organisms in the plant and animal kingdoms quite objective differences in level of organisation or complexity of structure, equipping these organisms for very differing degrees of achievements. Thus there can be no doubt that the far-reaching division of labour among the cells of multi-celled animals gives greater functional benefit, and that they should be regarded as more highly organised than one-celled animals. Amphibians have a body structure which enables them to live on land as well as in water, and have therefore conquered a wider environment than the fish, so that they are superior to them in this sense. The birth of living young and the care of the young in mammals have ensured them greater safety for the offspring, while the improvement of blood circulation and metabolism, the division of labour in the dentition, etc., mean that they have a higher organisation, they represent a significant morphological and physiological improvement over the reptiles'.[31]

Explanations of phylogenesis have been approached from either side by *a tergo* and *a fronte* explanations. The theories of Lamarck and Darwin (the selection theory) operate *a tergo*, that is, with causal laws. According to Lamarck, physical and chemical influences of the environment cause the changes in form. Alterations in the hereditary material are brought about by environmental stimuli, or are produced by the environmentally-caused use or non-use of organs. The theory of selection, going back to Darwin and generally accepted today, maintains that phylogenesis was brought about by 'natural mating choice and selection.' 'Accidental' and inherited mutations of properties and complexes of properties appear in every population. When competing for reproduction, the individuals with the advantage are those who carry mutations which have shown themselves to be particularly favourable or 'purpose-fulfilling' for survival under certain environmental conditions. Evolution is thus causally explained through the selective action of environmental factors on changes in type produced spontaneously and directionlessly (i.e. not lawfully).[32]

In addition to causal explanations using the theory of selection, paleontologists have long chosen or even preferred to use *a fronte* explanations through teleological developmental laws. We will look more closely at some of the recently used laws. They are all more or less markedly closer to empirical regularities than to strict laws, their validity tends to rest on empirical corroboration (direct evidence) rather than on theoretical context (indirect evidence, see p. 166). For this reason, paleontologists do not agree even today about many of these laws. In our context we will only present some of these laws, without discussing the conflicting views about them. In general, we follow the presentations of Schindewolf (1950) and Erben.[33]

Teleological laws in phylogenesis relate to typical courses recurring in the evolution of different taxa of the plant and animal kingdom, in the same or similar fashion. A comprehensive law of this type is the phased evolution suggested, among others, by Schindewolf.[34] According to this, evolution within a phylum does not follow a steady course of gradual change, but is punctuated or phased. It starts with typogenesis. 'A number of different organisational structures or types are laid down in explosive upheavals and major alterational leaps.' After this follows a phase of typostasis, 'in which continued elaboration, multiplication, and differentiation take place within the established framework, leaving the basic structure itself unchanged. Evolution proceeds slowly, gradually, flowingly, in tiny little steps.' Typostasis generally lasts much longer than either the first or the third, last phase of typolysis. This is characterised 'by many features of decay, degeneration, and loosening of the general type. Overspecialisations and excessive gigantism typify the doomed lines in this period.'

Many examples have been presented to show the validity of phased evolution. Fig. 6.2, taken from Schindewolf, shows phases of typogenesis and typostasis among amphibians and reptilians, and typolysis in some extinct groups of these orders. The evolution of corals, cephalopods (nautiloids and ammonoids), placental mammals, and other taxa shows similarities, so that it is agreed, at least as regards the first two phases, that 'typogenetic and typostatic events occurred with regular frequency in the evolutionary sequence.'[35]

Ernest Haeckel suggested the so-called basic biogenetic law, according to which ontogeny and phylogeny are related such that the early ontogeny or development of the germ presents a condensed repeat of phylogeny: each living creature runs through the stages of its ancestors arriving only in the last phases of ontogenesis at the full organisational level of its own species. According to a more recent view, this palingenetic ontogenetic development is overlain by the opposite tendency towards 'proterogenesis'. The law of this development is that

'a new, characteristic complex of features is laid down in more or less early ontogenetic stages, but initially is not retained through to the final stages, but is replaced by a reversion to the ancestral structures. In subsequent

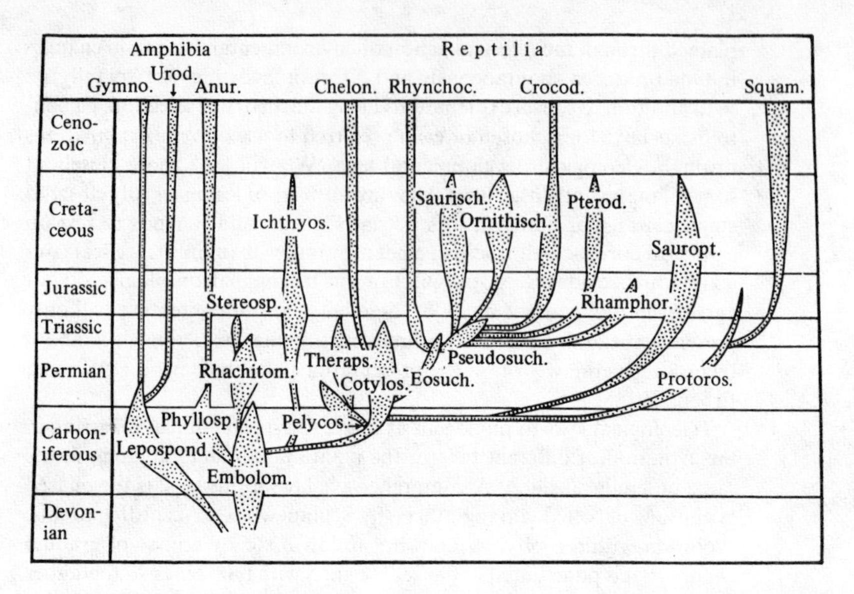

Fig. 6.2. Evolution of the amphibious and reptilian orders. (Schindewolf, 1950.)

members of such an evolutionary line, the new type complex gradually extends from the juvenile stage into maturity and old age, slowly narrowing down the period when the ancestral structures are visible, and finally causing it to disappear altogether.'[36]

This law appears to contradict the biogenetic basic law. According to Schindewolf, however, the two regularities relate to different phases of ontogenesis. The recapitulation of the ancestral condition occurs in the earliest stages of germ development. This earliest stage of the individual's development is followed by a phase of alteration and reorientation into a new path, evident in the juvenile stage, which, however, is replaced at maturity by a reversion to the specialised structures of the immediate ancestors. These are best recognised in the forms of old age, not in the juvenile developmental forms. Thus the law of 'proterogenesis' states that in some species the juvenile forms 'are an anticipation of the future'.

Further developmental laws relate to phylogenetic details. Cope's Law, or the law of non-specialised derivation, states that

'it is not the advanced, adaptively highly specialised lines of ancestry which permit the development of radically new evolutionary lines, but rather, as a rule, the conservative, less specialised lines.'[37]

The law of irreversibility, or Dollo's Law, expresses the directed, non-cyclical nature of phylogenetic change.

'Individual features, form complexes, or entire phenotypes which have once been given up in the course of evolution can never appear again in identical (homologous) fashion. Descendants can never return to the original ancestral condition. Evolution is irreversible.'[38]

Rosa's 'law of progressively decreasing variation' is better known today as the 'law of decreasing evolutionary potential.' It states that 'during the course of evolution of a phylogenetic line, there occurs a general, step by step narrowing of evolutionary possibilities and directions.'[39]

Finally we may mention the law of phylogenetic increase in size. This maintains that

'at the start of a lineage we find small forms, and that a gradual increase in body size takes place in the course of further evolution; as in the orthogenetic development of the individual organs, we find here that the increase in size eventually far exceeds any physiologically and ecologically sustainable limits; it loses all proportion and goal.'[40]

These and other developmental laws explain the individual phases and overall course of phylogenesis as a bundle of directed processes, or as a single directed process. The notion that this evolution is steered by and can be explained using teleological lawlike statements without reference to exclusively causal mechanisms, is generally termed 'orthogenesis,' or better, 'orthoevolution.' Paleontologists are still not agreed whether and how orthoevolutionary explanations of phylogenesis, with their teleological laws of development, are compatible with explanations according to the theory of selection, based on causal laws. The possible viewpoints may be examined using the scheme shown on p. 183. The most extreme viewpoint, T_1, according to which teleological laws alone are admissable, is not supported in science today. On the other hand, there are still defenders of the viewpoint M_3–T_3, according to which neither teleological nor causal laws can be dispensed with in explaining phylogenetic evolution, as some processes can only be understood through their goal, others only through their origin. As an example of such a view, we cite Schindewolf.

'The phylogenetic machinery appears to us therefore not an exogenesis, that is, a process impelled and governed purely accidentally by environmental conditions, but rather primarily an autogenesis. According to this conception, the organism is not simply the plaything of external factors: since its ability to react to environmental influences and the resultant forms of reaction lie within it, the organism is largely the unconscious director of its own evolutionary destiny . . . We must provisionally reckon with elementary life-phenomena in biology, even when we cannot for the present more closely determine or mechanically explain their nature.'[41]

Most paleontologists today would probably not support this view. The current style of research prescribes that teleological laws be interpreted as manifestations of causal laws. To this end the researcher attempts to use the results of experimental genetics to find causal mechanisms which would produce the directed processes evident phenomenologically. Such attempts might be interpreted as a modern realisation of Leibniz's viewpoint M_2–T_2, according to which each process can be explained equally well *a fronte* as *a tergo*: if this is the case, then for every law *a fronte* there must exist a law *a tergo* which would perform the same function. Modern research, however, is presumably not as conciliatory as Leibniz. Basically it overwhelmingly adopts viewpoint M_1 according to which causal laws alone count in scientific argument, and teleological laws are at best allowed as heuristic instruments. As an example of such a position we will conclude by citing from a criticism of the views of Schindewolf. The author not only attempts to find causal justification for the teleological laws stressed by Schindewolf (the organism as 'director of its own evolutionary destiny'), but goes beyond this to doubt the 'phased evolution' which Schindewolf found in some lines. He doubts that this can be considered a general law, as there are not a sufficient number of cases to support it. This is the typical objection to inductively found developmental laws relating to a limited number of natural processes whose similarity can be doubted in view of differing details.

'Schindewolf's typostrophic doctrine was internally a logical, impressive construct . . . All his theses were based on a rich observational material, all were in accord with what was known at the time of biology and genetics. Schindewolf in Germany and George Gaylord Simpson in the United States . . . were the first paleontologists who tried to create a synthesis between the discoveries of phylogenetics and of genetics. Unlike Simpson, Schindewolf underestimated . . . the impact of population genetics and the role of selection. Because of this some of his theses can no longer be upheld today . . . As far as his typostrophic doctrine goes, I am of the opinion that we must truly give up the concept of typolytic degeneration as a general and regularly-occurring phase. The process of over-specialisation as depicted by Schindewolf has certainly taken place in some cases . . . but not with that regularity which alone would justify raising it to the level of a separate phase. The changes in form associated with typostasis could well be described as 'normal' evolution, as generally observed. It could only be spoken of as an independent phase, then, if the concept of a further separate phase, namely that of typogenesis, were justified.'

This, however, according to the critic, is not the case. Schindewolf's phase of 'explosive upheavals' in the evolution of all types is not acknowledged.

'I would therefore like to suggest that, while not invariably, still generally, the saltation in types is only rendered apparent by gaps in the records.'

If 'typogenesis' and 'typolysis' are not general phenomena, but processes which took place only within a few phylogenetic lines, then these processes are better explained causally. As far as typogenesis goes, this would look like this:

'Each split of the evolutionary tree of the ammonoids was preceded by a general extinction in the previous phylogenetic line. In each case these extinctions suddenly set free previously occupied environmental niches. The branching of the subsequent new types therefore invariably signified a conquering of the vacuum left behind, an explosive invasion of suddenly available biotopes.'

'Typogenesis' here does not appear to be 'autogenesis', steered teleologically from within (something which is steered requires a steerer, or pilot, with a destination in mind), but rather a causal result of random mutations and of the environmental niches set free by the extinction of other species.

The other phase of directed evolution, extinction, is also explained causally by this critic of Schindewolf's hypotheses.

'Changes in environmental conditions are balanced by changes in adaptation as long as this is permitted by the gene reservoir of the population. Where, however, the phylogenetic line had adapted to such an extent that its path had become a one-way street, the phylogenetic dependence of the features automatically increased while the adaptability of the genetic reservoir decreased. Where a change in environmental or living conditions then occurred, suitable mutants were no longer available to bridge the crisis. Only then, when the phylogenetic line had in this manner gambled away the adaptability of its genetic reservoir, only then did its last population stand as a sorry delinquent before its doom: as judge, heredity had already passed sentence, as hangman, natural selection executed that decree.'[42]

6.2.3 *Laws of states*

Laws of states are the commonest type of law in physics. In contrast to causal and developmental laws, they relate not to processes, but to states of the empirical world which result from and which can be changed by processes. In contrast to laws of determinate properties which are governed by time-independent parameters (properties), states are considered to be potentially changeable systems. Laws of states are therefore functional laws because the parameters characterising given states are changeable variables. In form, the laws of states correspond to concepts developed in mathematics since the time of Descartes and Leibniz of the functional relation between two or more changeable variables.

According to the mathematical definition of a function between two variables x and y, y is called a function of x when according to an explicit rule, every value of x can be assigned a certain value of y. The explicit rule according to which values of y are assigned to values of x is the functional rule which unites x and y. The rule need not be expressed by a mathematical formula in order for y to be a function of x. Functional laws can also appear verbally or graphically. The mathematical form, however, is particularly well suited to demonstrate the properties which pertain to all functional laws.

The functional dependency between two variables x and y can be presented in the following general form:

$$y = f(x)$$

where f is the functional sign and stands for the mathematically or verbally expressed rule through which each value of x is assigned a particular value of y. This shows the main difference between causal and functional laws. The causal law relates directly to the process of change and defines the effect's dependency on causal origins. The functional law takes the state to be a lasting one in which the parameters describing the state stand in a particular relationship to one another. A verbally and qualitatively formulated law on the state of calcareous solutions in contact with a carbon-dioxide containing atmosphere, for instance, runs as follows.

'In a solution containing solid calcium which is in contact with an atmosphere containing CO_2, the amount of dissolved calcium (y) increases as more CO_2 (x) is found in the atmosphere'.

This law, which can also be formulated quantitatively, describes the amount of calcium (y) in a solution as a function of the CO_2 content (x) in the atmosphere above the solution. It does not refer to the process, which, following a disturbance of the equilibrium described in the functional law, brought this condition about, that is, it does not refer to any sequence of steps by which an effect ensues from given causes, but only to the relationship between the given parameters x and y when the process has come to rest and no grounds for change exist.

In the formula $y = f(x)$, the variables x and y are distinguished: x is called mathematically the 'independent variable', or 'argument' and y the 'dependent variable' of the function. This asymmetry gives rise to the tendency to speak of cause and effect even in the case of functional laws. When comparing two solutions I and II, in contact with atmospheres containing differing amounts of CO_2, it is usually said that the higher CO_2 content in the air above solution II causes the higher content of dissolved calcium carbonate relative to solution I. The dependent variable y is as the effect, the independent variable as the cause. This tendency to blur the distinction between causal and functional laws is brought about because in considering the equilibrium conditions of I and II, one anticipates the experimental possibility of changing the state of solution I into that of solution II, as can be brought about by increasing the CO_2 content in the air above the solution. In this experiment, the increase in the CO_2 content would be the cause, and in the new state of equilibrium, the increased calcium carbonate content of the solution would be the effect.

While causal laws are characterised by the essential asymmetry between cause and effect (the cause → effect relationship is irreversible), in laws of state, the asymmetry apparent in the expression $y = f(x)$ is a formal one only. In laws of state there is a basic symmetry between the variables x and y, the mathematical expression of this being that for every function $y = f(x)$, there is also a reverse function $x = g(y)$, in which y is the independent, and x the dependent variable. Thus, $x = \sqrt{y}$ is the reverse function of $y = x^2$. The symmetry between the functionally related variables is particularly clear when they are graphed. In the right-angle, two-dimensional coordinate system where one axis represents the variable x and the other the variable y, the functions $y = x^2$ and $x = \sqrt{y}$ appear as one and the same parabola, without any distinction between dependent and independent variable.

Formulae such as $y = f(x)$, in which there is a distinction between the independent and the dependent variable are called the explicit representations of a function. The symmetry of the function corresponds to the implicit formula $f(x,y) = O$, where the two variables have equal status. For one function with two variables there are two explicit formulae, each of which is the reverse of the other, and one implicit form, e.g.

implicit formula: $x^2 - Bx - y = O$
explicit formulae: $y = x^2 - Bx$, and
$$x = B/2 + (y + B^2/4)^{\frac{1}{2}}.$$

What we have shown for simplicity's sake using a function with two variables, is, of course, true for functions with more than two variables. For a function with n variables there is one implicit formula $f(x_1, x_2, \ldots x_n) = O$, which expresses the mutual relationships between all the variables, and n possible explicit formulae $x_n = g_n(x_1, x_2 \ldots x_{n-1})$ each of which expresses one variable as being dependent on all the others.

Functional laws are invariably used in many branches of geoscience in cases where conditions of natural systems can be characterised by variables conditional upon each other in a lawlike manner. Following the model of the mathematical formula, a distinction can be made between implicit and explicit laws, though there is no sharp boundary between the two groups. An implicit law can be transformed into an explicit one, according to the situation at hand, and *vice versa.*

The best known example of an *implicit functional law* is the law of the state of ideal gases, which can be formulated as follows:

$$pV/T - R = 0.$$

This law describes a functional relationship between three variables, mole volume V, pressure p of a gas, and absolute temperature T. The gas constant R remains the same for all ideal gases.

Another implicit law pertains to the elastic state of any rock. The functional relationship between the compression modulus K (governing the reduction in volume, given equal pressure from all sides), the shear modulus G (the measure of deformation under stress–strain), and the Poisson constant s (relationship of relative deformation in length and diameter of a distorted cylinder) is:

$$\frac{3(1 - 2s)K}{2(1 - s)} - G = 0$$

where s, K and G are constants for each rock type. Looking at the complex of all rocks (or all solid bodies), the magnitudes involved change within very

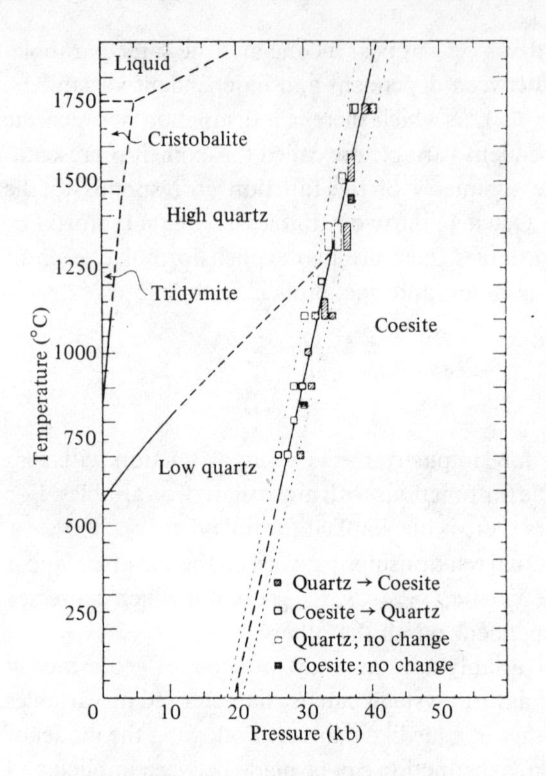

Fig. 6.3. State diagram of the substance SiO_2 in the pressure–temperature field. (Boyd & England, 1960.)

wide boundaries, but their relationships are always equivalent to this function.

Functional laws in geoscience are frequently presented not as mathematical formulae, but as graphs. An example of this is the phase diagram of carbon (Fig. 6.1), which was explained in the context of developmental laws. It can also be understood as a representation of a functional law, determining the boundaries for the stable existence of the crystal types diamond and graphite as a function of pressure and temperature. Another example is the phase diagram in Fig. 6.3 of the chemical substance SiO_2, showing the functional dependency of the stability of the crystals coesite, low quartz, high quartz, tridymite, and cristobalite on temperature and pressure.

In *explicit functional laws*, a single variable is chosen as the one whose value will be determined by the freely chosen values of the other variables. An example of this consists of the laws valid for the normal gravity field of Earth (that is, that gravity field thought to be free of local irregularities caused by the distribution of heavy masses). The dependent variable here is the gravitational acceleration g at a given spot. The independent variables are distance from the middle of the Earth (r), or geographic latitude (β). The following functions can be identified:

I. The gravitational acceleration $g(r)$ outside the Earth's body (M = the Earth's mass); γ = the gravity constant; R = the Earth's radius; r = distance from the centre of the Earth, where $r > R$):

$$g(r) = \gamma \frac{M}{r^2}.$$

II. Gravitational acceleration $g(r)$ within the Earth ($r < R$)

$$g(r) = \gamma \frac{M}{R^3} r.$$

III. Gravitational acceleration at the Earth's surface, dependent on the geographic latitude β:

$$g(\beta) = 9.780490\ (1 - 0.0052884\ \text{sub}^2\ \beta - 0.0000059\ \sin^2 2\beta).$$

Another explicit functional law is the relationship in loosely consolidated sands between the three variable parameters, permeability k, porosity e, and relative inner surface S:

$$k = \frac{e^3}{5\,(1 - e^2)\,S^2}$$

A further example is provided by the following function. It is a theoretically well-founded rule that the porosity of clay sediments will decrease the deeper they are buried beneath the surface. Empirically the functional law of the following form has been found to hold true.

$$E = E_1 - b \cdot \log T$$

where E = relative pore space, or volume of pores/volume of solid matter, T = depth of burial in metres, and where E_1 and b are both constants.

The nomological necessity of laws of state occurring in the theoretical discourse of the geosciences is based on the laws of nature of physics and chemistry. This is obvious in all the examples we have discussed. Further examples would show that the functional laws are either, like the laws on the gravity field of Earth, nothing more than an application of general laws of nature (theory of gravitation) to special configurations (mass and shape of the Earth), or as in the case of the phase diagrams for the substance SiO_2, deal with empirically derived states of affairs whose validity for all cases is established by laws of nature. For just as the theory of crystal structure explains why the light refraction measured on one sample of fluorite is true for all samples of this material, so does it follow from the theory of the behaviour of crystalline substances that the temperatures and pressures of phase transitions are necessary characteristics of phases, so that the data obtained in the individual case have universal validity.

6.2.4 *On the problem of probabilistic lawlike statements*

The examples dealt with in the above sections show that geoscientific explanations are often based on lawlike statements which do not adhere to the criteria for strict universal laws developed at the outset. Such probabilistic laws[43] from which only likely, not certain, statements regarding specific phenomena can be derived, can occur in all classes of laws distinguished above.

Probabilistic laws of determinate property apply, for instance, to the contingent properties of minerals. The shear strength of halite is given as the frequency distribution of values measured on many samples, and represents the probability of observing particular hardness values on any given halite crystal.

Causal lawlike statements can also be probabilistic in type. The reduction in size of bedload in a river caused by transport (p. 150) is shown by an inductively derived curve representing the most probable relationship between transport distance (cause) and diameter (effect). That tin deposits are often found near granites, a regularity seen as a causal relationship between the crystallisation of granite and enrichment of tin, is merely a statement of the probability of finding such ore deposits.

Many developmental laws have a probabilistic character. This is particularly true of phylogenetic laws which predict no more than the most common or probable direction of species evolution. Laws on the development of inorganic systems such as soils or landscapes relate only to changes which are the most frequently observed in many samples and hence the most likely to be expected.

Laws of state are probabilistic when they concern relationships measured on parameters of many samples. The relationship discussed on p. 199 between porosity, internal surface area, and permeability of porous rocks is a derived one, representing the most frequent and therefore most probable relationship.

Probabilistic laws proposed within Earth science are inductive regularities. When applied to a large number of isolated phenomena, they can be trusted on the basis of theoretical plausibility (indirect evidence) and in accordance with the statistical probability on which they are based. In relation to an isolated phenomenon, however, they have no lawlike character. Nevertheless they are often applied to individual cases as well. There are probabilistic, if not quantitatively formulated laws, for instance, on the occurrence of useful natural resources. Oil deposits are often found alongside salt domes, nickel deposits in many gabbro massifs. Although this says nothing about the occurrence of oil next to salt dome A, or of nickel in gabbro massif B, a search will nevertheless be made along the sides of A and in and around B, based on the respective regularities of the probable occurrence of oil or nickel where the searches themselves are justified on the grounds of subjective probability.

ABDUCTION AND DEDUCTION

6.3 Systematic foundations

When introducing the idea that explanation is the main goal of theoretical knowledge we distinguished between three forms of scientific inference: induction, abduction and deduction. The problems of inductive inference have already been treated extensively in Chapter 5. We now wish to examine in detail the procedures of abductive and deductive inference, so indispensable for setting up, refining, and testing hypotheses of all kinds. To begin with we must distinguish between the *logical* form and the *pragmatic application* of these two kinds of logical inference. Let it be remembered that only deduction is a form of inference in the strict sense of formal logic, for in it alone does the conclusion *necessarily* follow from the stated premises. Abductive inference, going from the resulting state of affairs to the controlling state of affairs, must basically remain tenuous since in principle the same results can be produced by any number of premises.

Deduction: if p, then q
p is given
―――――
q must be the case.

Abduction: if p, then q
q is given
―――――
p may possibly be the case.

The inference from q to p is not compulsory according to formal logic. Only the inference from not-q to not-p is compulsory.

if p, then q
$\sim$q is given
―――――
$\sim$p must be the case.

The only case where a positive abductive conclusion is compulsory is where a biconditional law is involved:

If and only if p, then q
q is given
―――――
p must be the case.

Apart from this exception, we agree with Peirce that abductive inference is really a form of guesswork.[1] Abduction can only be justified in connection with deductive and inductive inferences:

> Deduction proves that something *must* be; Induction shows that something *actually* is operative; Abduction merely suggests that something *may* be. Its only justification is that from its suggestion deduction can draw a prediction which can be tested by induction . . .[2]

We will discuss this interaction between the various forms of inference in more detail later.

It is reasonable to distinguish between deduction and abduction on the basis of their differing status in formal logic, but this distinction becomes less marked when their practical application is examined. Deductive inference becomes 'less certain', because neither the unfailing validity of the law 'if p then q', nor the completeness of the premises can be guaranteed. Abductive inference on the other hand becomes 'more certain' when, instead of a multitude of possible controlling states of affairs, only a few need actually be considered, some of which can moreover even be weighted as being either more or less plausible. Once the contrast between deduction and abduction is put into such a pragmatic perspective, even logicians are prepared to accept the abductive process as 'inference', though it is not universally valid.[3]

As types of inference, deduction and abduction are each characterised by specific relations between propositions. As applied modes of scientific inference, a relationship to concrete states of affairs is needed in addition, represented by those assertions. For the purpose of analysis we will use symbols, representing the controlling state of affairs as 'a' and the proposition describing it 'A'. The resulting state of affairs will be called 'e' and the proposition describing it, 'E'. The underlying law in each case (which may also be an inductive generalisation), will be expressed by the letter 'L'. In deduction, A is always the starting point of argument. Given A, we use L to *deductively infer* statement E describing the resulting state of affairs e. In *abduction*, the starting point of the argument is E. Using L we *abductively infer* from E the statement A which describes the controlling state of affairs a.

The move from the logical to a pragmatic analysis of deductive and abductive inference is critically hinged on the consideration of their *temporal relationships*.[4] First we must ask what can be taken as 'given' at the time of argumentation. At the *propositional level* in the case of deduction this is invariably A and L (in abduction, E and L). The deductively inferred statement E (abductively inferred statement A) is therefore 'uttered' later than the premising statements A and L (E and L); yet their real temporal relation at the *object level* can be regarded as 'simultaneous.'[5] The pragmatic analysis therefore hinges on the actual temporal relations of states of affairs a and e to each other and to timepoint t_s of the scientific argument (S), characterised by the deductive or abductive relations between A, E, and L. If we call the timepoint of state of affairs a (premised or inferred) 't_a', and that of state of affairs e 't_e', and indicate the relationship 'earlier than' as '<', 'later than' as '>' and 'simultaneous to' as '=', we may note the following possibilities.

Relation a to S: $t_a < t_S$
$t_a = t_S$
$t_a > t_S$
Relation e to S: $t_e < t_S$
$t_e = t_S$
$t_e > t_S$
Relation a to e: $t_a < t_e$
$t_a = t_e$

We will first examine the timepoint of the object level forming the *starting point* for the inference. In deduction, this is t_a, in abduction, t_e. The temporal relations between a and e in deductive inference can be $t_a < t_e$, or $t_a = t_e$, while in abductive inference it can be $t_e > t_a$ or $t_e = t_a$ (the direction is determined by which state of affairs is the starting point for the inference). This brings us to a basic distinction concerning the temporal relations as they actually obtain and as they appear in our inferences.

We speak of *prediction*[6] when the inferred state of affairs will or should take place in reality later than the initial state of affairs. We speak of *retrodiction*[7] when the inferred state of affairs happened or should have happened in reality before the state of affairs used as the starting point. We speak of *codiction*[8] when the inferred and initial states of affairs are simultaneous.

Remembering that deductive inferences always go from a controlling to a resulting state of affairs, while abductive inferences always draw conclusions from the resulting as to the controlling state of affairs, it will be clear that by their logical structure, predictions must always be deductive, retrodictions abductive. Codictions can involve either form of inference, and we therefore distinguish between *deductive codiction* and *abductive codiction.*

Methaphorically speaking prediction represents a 'look forwards', retrodiction a 'look backwards', and codiction a 'look sideways' with regard to synchronicity and functional interaction within the range of objects under investigation. The point from which this look forwards, backwards, or sideways is cast can for its part lie in the past, the present, or the future. That is, timepoints t_a and t_e, related to each other through prediction, retrodiction, or codiction, are now being regarded from the timepoint of the argument itself, that is, from t_S. This again provides us with three possibilities: we will here use t_a/t_e to indicate t_a in relation to t_e, that is, predictively, $t_a < t_e$, retrodictively $t_e > t_a$, or codictively $t_a = t_e$ or $t_e = t_a$).

1 $t_a/t_e < t_S$
2 t_a, t_e, or $t_a/t_e = t_S$
3 $t_a/t_e > t_S$

In the first case everything takes place in reality in the past, outside of the temporal range of the investigator. In the third case everything takes place in reality in the future, again out of the temporal range of the investigator. Only in the second case are either a, e, or both synchronous with the researcher's time period, and only in this case can scientific inference be based on a factually existing, theoretically observable state of affairs, or aim towards such a state as an immediate test of its validity. Of course there are cases where the state of affairs serving as the origin or aim of argument may or should be contemporaneous but cannot be observed at the time of the inference. Nevertheless the theoretical possibility exists that an improvement in the range and precision of observational techniques should make it possible to relate such inferences to directly observable states of affairs.

It is clear that predictions, retrodictions, and codictions which are directly empirically based (when the initial state of affairs is present and observable) or which can be tested empirically in a direct manner (when the inferred state of affairs is present and observable) generally provide more forceful arguments than inferences which have factual givens neither as their basis nor as their goal. Such inferences must 'dangle' unless supported by further inferences drawn from present and observable states of affairs. We have already mentioned that independently of this a theoretical asymmetry exists between the force of deductive and abductive inferences in argumentation. Even when based on factually given states of affairs, abductive inferences remain uncertain and rely for support on additional deductive consequences which are empirically testable. We will return later to this interplay between deduction and abduction.

Finally, looking at the differences in prediction, retrodiction, and codiction relative to the timepoint of the inferred state of affairs, we can distinguish between inferences relating to the past, (*historical inferences*), to the present (*contemporaneous inferences*), and to the future (*anticipatory inferences*), each characterised by their own temporal relations between a and e. This provides us with three types each of prediction, retrodiction, and codiction. Not all of these are of equal practical significance, however. Retrodictions are almost exclusively historical retrodictions, for which reason we will omit the term 'historical' from here on. Retrodictions always involve inferences concerning past states of affairs, and therefore form the core of geoscientific reconstructions of Earth history.

In predictions, anticipatory predictions clearly dominate. These we will call *prognoses* below, following the general usage according to which a prognosis relates to a future event which can only be observed at a point in time after t_S.[9] In Earth science, however, historical and contemporaneous predictions also play a major role, reversing the direction of retrodictive questioning and making it possible, through deduction of possible consequences, to confirm or refute abductively generated hypotheses.

In codiction, finally, it is the contemporaneous type which dominates, for there is always the hope that new observational techniques will retrospectively support hypothetically assumed states of affairs. In the case of prognosis this becomes increasingly difficult the greater the time gap between the prognosis and the prognosticised event. When we speak below of codiction without employing additional modifiers, we will always mean the type occurring in the present, for which $t_a = t_e = t_S$ holds.

To summarize once again: we have distinguished three respects in which abductive and deductive inferences can be subdivided within the context of scientific argumentation.

1 According to the temporal direction of the inference with reference to the state of affairs on which it is based (predictive, retrodictive, and codictive inferences).
2 According to the temporal relations between the inferred state of affairs and the timepoint of the investigator (historical, contemporaneous and anticipatory inferences).
3 According to the relation to the material actually given, whether as the empirical basis for the starting point of the inference, or as the direct empirical testability of the inferred conclusion.

In conclusion we have attempted to show graphically the various possible abductive and deductive inferences in Table 6.1. The horizontal line represents an oriented time axis running from left to right, while the vertical line divides the past from the future. Their intersection marks the point in time of a given inference, t_S, and therefore the timepoint of the investigator. Predictions (P) are shown above the time line and retrodictions below it, while deductive codictions (dC) are to the left and abductive codictions (aC) to the right of the point marking the simultaneous existence of a and e. The

Table 6.1 *Possible types of abductive and deductive inference*

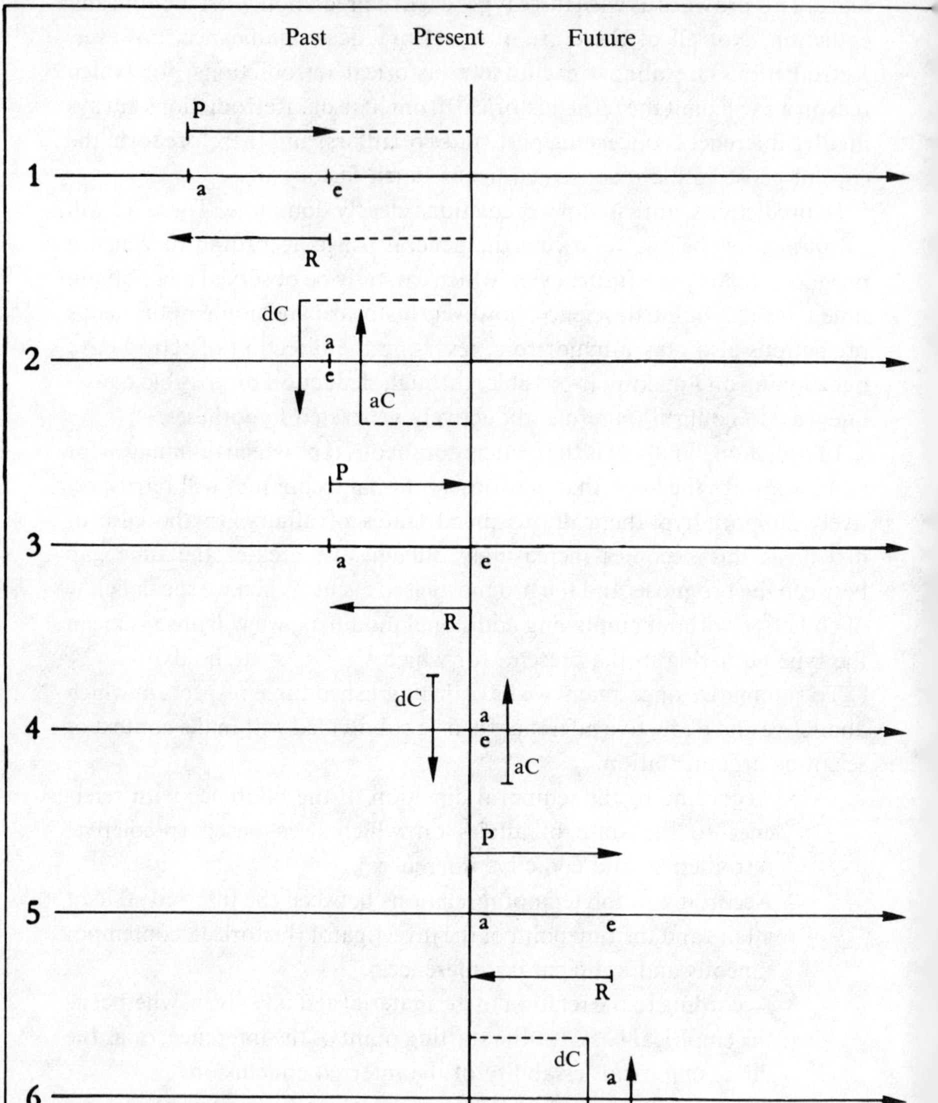

Past Present Future

P

1 a e

R

dC

2 a e

aC

P

3 a e

R

dC

4 a e

aC

P

5 a e

R

dC

6 a e

aC

P

7 a e

R

direction of inference is shown by arrows. Dotted lines show where inferences need support through observations of present states of affairs which may serve as their foundation.

As has already been mentioned, not all the possible deductive and abductive inferences shown have the same practical significance. We will discuss only those which are particularly common or important in the framework of geoscientific research. The complex of possible combinations of various types of retrodiction, prediction, and codiction exemplified in the illustration will be discussed under Section 6.6 below.

6.4 Abductive inference in Earth science

6.4.1 *Introduction*

In order to illustrate the amount of leeway for the validity of abductive inference in Earth science, we choose two examples which differ considerably in the certainty of their conclusions.

(1) Resulting statement E: The crystallised mineral X, whose name is not yet known, is optically isotropic.

Law L: All crystals of the cubic crystal system, and only these crystals, are optically isotropic.

Controlling statement A: Mineral X belongs to the cubic crystal system, that is, the arrangement of atoms in its crystal lattice corresponds to the conditions of one of the cubic symmetry groups.

(2) Resulting statement E: Certain rocks of the crater-shaped structure of Lake Manicouagan in Canada contain a glass whose chemical composition corresponds to a plagioclase (feldspar).

Law L: If a plagioclase is heated above its melting point, and the resulting melt is cooled quickly, a glass of plagioclase composition is always produced.

Controlling statement A: The plagioclase glass of Lake Manicouagan was therefore produced by the melting and rapid cooling of a plagioclase-bearing rock.

Both examples are abductive in structure. A statement (E) supported by an observation is attributed via a crystallographic or physico-chemical law (L) to a cause or certain conditions (A). Yet there is a major difference between the two examples. In case (1) the inference from E to A is inevitable since mineral X, if it is optically isotropic, must belong to the cubic crystal system, for all cubic crystals and only cubic crystals are optically isotropic. The inference from E to A in case (2), however, is not at all necessary, but at best probable. There is a valid law that rapid cooling of silicate melts always produces glasses, but none that all glasses are produced in this manner alone. Glasses may also be produced out of crystalline material without melting, e.g. through high energy radiation or through the action of shock waves of sufficiently high pressure. In fact we know today that the conclusion in example (2) is (probably) false. The crater structure and the feldspar glass of Lake Manicouagan were very probably produced by the impact of a major meteor and by the shock waves which this created.

The two examples show that there are differing degrees of reliability in abductive inferences, depending on the form of the law on which the inference is based. In the first case the law is *biconditional*, stating that: if and only if A is true, then E is true. This form implies that the inference from E back to A is inevitable, and that the conclusion is a certainty. In the second case the law is simply *conditional*, stating that: if A is true, then E is true. The conclusion from E back to A is not unequivocal then, and the result of the abduction is not a certainty. It can only be relied on with a greater or lesser degree of probability.

There are many gradations of probability between our extreme examples illustrated in (1) and (2) above, ranging from practical certainty (in the case of biconditional laws) to relatively large uncertainty (in the case of poorly attested conditional laws). Even biconditional laws cannot guarantee absolute certainty of an abductive inference, as their biconditionality may be open to doubt. It is assumed to this day, for example, that the following law is true in biconditional form:

Coesite is produced out of SiO_2-rich material if and only if it is exposed to pressures of over 30 kbar.

It can be abductively inferred from this law that rocks containing coesite were at one time exposed to pressures of over 30 kbar. But since we know that other minerals can be produced in so-called metastable phases outside of their thermodynamic stability fields (aragonite, for instance, ($CaCO_3$), is only stable at pressures of over 4000 atmospheres but can also be produced at normal atmospheric pressures) it is conceivable that the coesite law is not invariably biconditional, and that under special conditions this mineral might be produced at lower pressures.

We therefore conclude that the results of abductive argument are basically hypothetical and differ significantly from inductively and deductively obtained knowledge. The deductive inference develops with logical necessity that which was already contained in the premises. If the premises are true, then the conclusion must also be true. In abductive as in inductive argument, however, the truth of the premises does not in any way guarantee the truth of the conclusions. Nevertheless this does not rule out holding the results of abductive and inductive inference to be true within the context of scientific discourse (as we held inductive generalisation to be, see p. 153). It means that we attribute a certain degree of probability to abductively or inductively produced statements, whether we do so explicitly, implicitly or altogether tacitly. In this respect induction and abduction are similar; where they differ is in the results they must achieve and at which they aim. Inductive inference proceeds from statements about known phenomena to amplified statements about phenomena which resemble the known. Abduction involves a search for hypothetical conditions which are linked with suitable laws in such a manner that the two premises will necessarily yield a statement exactly describing the states of affairs to be explained. Something new must therefore be found in order to explain the known, without the guiding light of any generally applicable *ars inveniendi* as dreamed by Leibniz. Its rules were supposed to make it possible to infer all possible contexts which might be held accountable to explain any given state of affairs. Abductively obtained explanations are therefore only more or less probable, like inductive inferences, and can only be accepted with the proviso that later revision may be needed since any actual state of affairs may result from various causes or various laws. Moreover the success of an abductive argument depends on an act of creative imagination on the part of the researcher: this is not open to philosophical analysis, and can at best be noted as a psychological phenomenon.

It is in this sense that Peirce states:

> But every single item of scientific theory which stands established today has been due to abduction. But how is it that all this truth has ever been lit up by a process in which there is no compulsiveness nor tendency toward compulsiveness? Is it by chance? Consider the multitude of theories that might have been suggested. A physicist comes across some new phenomenon in his laboratory. How does he know but that the conjunctions of the planets have something to do with it or that it is not perhaps because the dowager Empress of China has at that time a year ago chanced to pronounce some word of mystical power or some invisible jinnee may be present. Think of what trillions of trillions of hypotheses might be made of which only one is true.[10]

How is it then that after only a few false starts the researcher generally does hit upon the 'right' hypothesis, that is, the one which is usually corroborated in the further course of research? We can only register the fact that

> However man may have acquired his faculty of divining the ways of Nature, it has certainly not been by a self-controlled and critical logic. Even now he cannot give any exact reason for his best guesses.[11]

Popper speaks similarly about the activity of the researcher:

> We have characterised the activity of the scientific researcher as setting up and testing hypotheses. The first half of this activity, the setting up of theories seems to us neither to need nor to be capable of logical analysis. The question, how does it happen that someone gets a new idea, whether a musical theme, dramatic conflict, or scientific theory, is of interest to empirical psychology, but not to the logic of discovery.[12]

The researcher who employs abductive inferences not only depends on explanatory insights which may or may not occur to him, but must also (and this is a real problem in the philosophy of science), come to terms in theoretical discussions with the more or less hypothetical nature of the results of abductive inferences. We will see below that non-observable states of affairs which can only be abductively inferred play a major role in the Earth sciences. The Earth scientist must therefore master the skill of dealing with conjectures and hypotheses.

Some critics attribute the use of vague statements, conjectures and mooted possibilities to the supposed immaturity of the Earth sciences at present, and to a corresponding lack of exact methods and laws. Although the increased development of quantitative methods and the more thorough application of physical and chemical knowledge may bring about substantial improvement, no amount of quantification can eliminate the basically hypothetical character of abductively obtained knowledge. The events of the geologic past, the origin of the oceans and continents, the course of Earth's development, the structure of the deep zones of the Earth's body, and most other topics of research in the Earth sciences can only be studied abductively, since they involve either events which are long past, or conditions which are inaccessible. The Earth scientist will always remain dependent upon abduction, the weakest of all modes of inference.

Let us look somewhat more closely at the preconditions of abductive argument. The investigator is faced at the outset simply with state of affairs e. He asks: why is e the case? Two things must be done in order to answer this question according to the abductive model. First, state of affairs e must be placed in a hypothetical context from which explanatory laws can be derived. Then, with the help of such laws L, condition a must be derived, a hypothetical state of affairs which represents state of affairs e.

Given, for instance, a stratum of sand e observed in nature, whose bedding position, structure, grain size distribution, etc., are all known. To explain this phenomenon we must first decide whether the hypothetical context from which we are to derive the formational conditions, a, of the sand, should be 'deposition through flowing water,' or 'deposition by wind.' Each of the two hypotheses is represented by a physical model implying the special laws, L, namely the law of transport of sand grains in water, L_W, or of transport in air, L_A. Having chosen one or the other hypothesis, laws L_W or L_A may be applied, and from them we infer the causally effective conditions a_W or a_A in water or wind, which may have led to the deposition of this particular stratum of sand, e, with its special features.

The use of abduction to explain a real state of affairs thus necessitates not only that the phenomena constituting this state are assembled, but also that they are ordered within a hypothetical system in which they have their own place and significance. Hanson[13] speaks here of 'patterns' which must be developed, Peirce[14] of 'colligations' which must precede abductive explanation.

The creation of such a hypothetical structure tentatively unites phenomena which as isolated data had had no previous connection with each other. We are therefore not talking about simple listings, data tables, or other collections from which inductive generalisations can be derived: rather we are talking about the creation of contexts which imply that the data can be interpreted from a supra-ordinate viewpoint and along the lines of certain expectations. The state of affairs to be explained, represented by a number of singular statements (data), must first be seen differently in terms of a characteristic context or pattern. Only once such a pattern has been formed can the second step be taken, and the laws inherent in the pattern be used to derive those conditions which can be seen as grounds sufficient for the existence of the phenomenon to be explained.

The investigator when dealing with a problem does not generally have both tasks of abductive argument simultaneously in mind, i.e. the tasks of setting up a hypothesis and of reconstructing the explanatory conditions. We can therefore differentiate two intentions.

The first intention may be to use a given state of affairs e either to develop a new hypothesis or model, or to refine an incompletely formulated hypothesis. This is the *abductive proposal of hypotheses*.

Alternatively the investigator may start from already accepted hypotheses or theories which in the given context are not disputed, and may on the basis of laws L belonging to that group of theories, attempt to reconstruct conditions a in order to explain situation e. This is what we have called *retrodiction* or *abductive codiction*.

Both aims of discovery and of explanation can be served at the same time. Generally, however, the emphasis will be on one or the other, and in many cases the distinction is quite clear.

An example of retrodiction is when grain size distributions, mineral contents, structural, textural, or other features of Buntsandstein samples from various localities in Central Europe are used to form a picture of the water depth, current directions, salinity, etc. in the Central European basin in which the Buntsandstein Formation was deposited. A codiction is involved in the reconstruction of the shape and position of an iron ore mass in the basement of the basis of magnetic anomalies at the Earth's surface. In both cases, specific hypothetical models and their laws are used which in these contexts are not questioned.

A hypothesis is abductively invented when measurements on the increased rounding of cobbles along a river course are used to develop a physical model on the mechanisms which round cobbles during transport along a river bed. An existing hypothesis is refined when observations on mineral compositions at different horizons in a soil profile are used to identify more precisely the process of chemical solution of minerals in gradually sinking soil solutions. Abductive conception of hypotheses may either take place in the light of previously accepted theories, or may form the first step towards the formation of new theories, which are to be understood as systems of related hypotheses and laws deduced from them about certain object domains. The development, structure, and function of geoscientific hypotheses and theories will be discussed in a later chapter. We will therefore postpone a look at the abductive invention of hypotheses until then, and turn now to look at retrodictions and codictions.

6.4.2 *Retrodiction*

Two types of retrodiction may be distinguished with regard to their importance for Earth sciences.

1 The resulting states of affairs exist at present (Table 6.1, Case 3), and either observable facts or non-observable hypothesised states of affairs are codictively inferred.

2 The resulting states of affairs belong in the past, and are for their part also retrodictively inferred (Table 6.1, Case 1).

Before we discuss some examples, we would like to refer once more to the general significance of retrodiction for the Earth sciences. A major task of geoscientific research, a task which can be taken over by no other natural science, consists of using the phenomena present on the Earth's surface today to draw conclusions as to conditions, events, or processes of the geological past. This can only occur through retrodictions based on laws derived from accepted hypotheses and theories. Retrodictions can function pragmatically in two ways. They can be made in order to explain isolated discoveries which for some reason seem unusual and in need of explanation. The discovery of a lump of nickel-containing iron on the surface of the Greenland ice sheet, for instance, may be grounds for the retrodiction that a meteoritic fall took place in the past. Retrodictions can also be made in order to use appropriate current facts to reconstruct conditions and events of the Earth's past. Here interest is directed primarily at elucidating greater or smaller segments of Earth's history, rather than explaining isolated phenomena.

We will illustrate the intentions and procedures of retrodiction by outlining the thought processes involved in four special investigations of largely retrodictive character.

(1) Using the laws of radioactive dęcay, the amounts of isotopic rubidium, strontium, potassium, and argon found in selected rocks and minerals from the Central Alps of the western Hohe Tauern and the Ötztal crystalline massif served as a basis on which to reconstruct the events of the geologic past which the given rocks and minerals had undergone. The abductive explanation of how the isotopic proportions present today came to exist was used to retrodict a sequence of rock-forming events whose age could be dated. These events were the solidification of magma at a certain time into granite rock, and the datable phases of warming and cooling of the rock complexes during the Alpine orogeny and metamorphism. The explanation of the samples' isotopic content only interested the author insofar as it provided data relevant to reconstructing a segment of Earth's history in the Alpine region. The number of samples investigated was therefore limited, and the analyses were broken off as soon as the author believed that sufficient data were available for the intended retrodiction.[15]

(2) The facts to be explained consist of measurements of magnetisation in rocks of various geologic ages from India. These data are abductively explained according to the laws of Earth's magnetism by the position of the Indian continent relative to the Earth's magnetic pole at the time the rocks were formed. The positions obtained hypothetically indicate a drifting of the Indian continent upon the surface of the Earth's sphere during the long period from the Permian to the present. Here again the author is not interested in explaining the magnetic properties of the individual samples, but in elucidating a geohistoric process. This interest determines the choice of samples, according to their geologic age and geographic locality.[16]

(3) The investigation is empirically based on measurements of microscopic fission tracks created by spontaneous nuclear fission of uranium in the minerals garnet, epidote, vesuvianite, and apatite, and on measure-

ments of the uranium content of these minerals. The samples were taken from rocks collected in various localities of the Damara Range in Southwest Africa. The intensity of the fission tracks can be explained with the aid of physical laws, i.e. by retrodiction to the last time that the given mineral reached a certain threshold temperature. From the retrodictions for the individual minerals and from the position of the samples, the author reconstructs a history of cooling for the rocks of the Damara orogeny. Again, the author is not concerned with explaining why this or that sample had a certain density of fission tracks, but with reconstructing a geohistoric process.[17]

(4) The investigation is based on observations set down in numerous collections of maps on the distribution of various rock formations, particularly of the characteristic breccias in the region of the Nördlinger Ries. The many isolated facts are explained by means of the impact hypothesis, whose validity is not questioned in the context of this work. The authors state:

> The creation of the Ries Crater through the impact of a cosmic body may be taken as assured today on the basis of mineralogical, petrographical, and geophysical finds. . . . Nevertheless, the known rules on crater formation and mass ejection upon the collision of a projectile into a solid body have not yet been sufficiently applied to a precise interpretation of the regional geological finds in the area of the crater and its ejecta.

This is precisely what the authors now wish to do.

> In the following we will attempt, based on the rules of impact mechanics, to interpret the distribution and properties of the ejecta of the Ries Crater.

This involves a retrodiction, which in contrast to examples (1), (2) and (3), is intended to explain an isolated instance, the Nördlinger Ries, by reconstructing its formational conditions in the geologic past. This is carried out on the basis of a certain hypothesis, the impact hypothesis and its related laws. The result of this retrodiction is a narration of the formational conditions.

> The main ballistic ejection of the crater (diameter c. 23 km) created a continuous blanket of 'Bunte Breccia', 90% of which outside of the crater consists of weakly-stressed fragments of pre-Ries sedimentary rock, 750 m thick. This blanket of ejecta is discordantly overlain by Suevite Breccia. . . . Today's asymmetric distribution of the ejecta with respect to the crater rim was created by Pliocene and Pleistocene erosion. It was very probably originally distributed symmetrically within a circle of 40 km radius around the Crater centre.

The 'laws' involved in reconstructing the ejecta around the crater, are supported, wherever possible, by quantitative data. It is shown that these 'laws' may be derived qualitatively from the physical processes involved in crater formation, and can be simulated by ballistic experiments. These 'laws', which we place in quotation marks, are in our nomenclature inductive generalisations, derived from observational data. The authors show that these empirical regularities can be explained qualitatively by the impact hypothesis, which is supported both physically and experimentally.[18]

The four examples clarify the two functions of retrodictions. Studies (1), (2) and (3) are not primarily concerned with explaining isolated facts, but with reconstructing certain episodes of the geologic past as segments of Earth history as a whole, whose clarification is one of the goals of Earth science. It cannot be the intention of geoscientific research to uncover the cause of every single phenomenon which is as yet unexplained. The number of phenomena is infinite, so that such random research would achieve neither a tentative nor a final conclusion. Instead, an empirical basis for abductive argument is created by selecting these possible facts which, on the basis of a presupposed hypothesis, can be expected to pave the way towards an abductive explanation and thus towards the controlling conditions or processes which should be known in order to reconstruct the past. This is why only representative samples of minerals and rocks are chosen in the first three studies, not arbitrary masses of material. The researcher's horizon of expectations is limited right from the start by the hypothesis underlying the retrodiction, and by his intention of explaining an episode in Earth's past. This is an important insight for the purposes of critical evaluation, for we can now determine whether the author has planned his investigation using hypothetical expectations as a guide, or whether he has simply accumulated (sometimes meaningless) reams of facts in the usually vain hope that from this alone something of retrodictive value will emerge. Studies of this kind are at best attempts to obtain inductive generalisations.

Example (4) is a case of retrodiction used to explain isolated states of affairs which were considered striking and anomalous even before the impact hypothesis had been suggested. Various explanations had previously been proposed based on other hypotheses, interpreting, for example, the breccia masses in and around the Ries Basin as products of volcanic events or as the deposits of a Ries Glacier.

The four studies may be taken as retrodictions of our first kind insofar as conclusions as to processes and conditions of the past are based on facts observable today. A closer look, however, shows that, as in most cases of geohistoric retrodictions, such conclusions also imply retrodictions of the second kind. When present isotopic contents of minerals are used to make inferences about the processes of magmatic cooling and metamorphism which took place long ago, it must be assumed that the minerals which formed at that time have not meanwhile been protected and conserved from all further external influences. The processes which have occurred in the meantime must be taken into consideration. As a first step, therefore, one makes a retrodiction of the first kind, drawing inferences from the present facts as to the conditions which existed immediately after the termination of the rock-forming processes long ago. Only from this hypothetical state of affairs can a retrodiction of the second type be made as to the processes which are of interest to the researcher, namely those of the various phases of magmatic solidification, warming, and cooling. Examples (2) and (4) are analogous. The magnetisation of the Indian rocks today is not necessarily the same as that of their formation. To make retrodictions about their original position in the Earth's magnetic field, it must first be determined if and how their magnetic properties have changed in the meantime. From the present distribution of breccia (interpreted as ejecta) around the centre of the Ries Basin, a retrodiction of the first kind must be made, taking into account the erosion which has presumably taken place and inferring the distribution of the outflung material immediately after the Ries event. Only from this hypothetical state of affairs can the desired retrodiction of the second kind be made as to the ejection processes during the impact. When looked at closely, retrodictions are nearly always both retrodictions of the first (direct) and second (indirect) kind, often with several of the latter linked together. These form so-called *chains of explanation*,[19] using which one may work step by step from the present back into the past. Since interest and attention are usually directed toward the 'first causes' of a given geohistoric process, the problems which can arise in constructing such chains of explanation are often overlooked. One example is the reconstruction of formational conditions of rocks which have undergone diagenetic or metamorphic alteration. Sandstones which at first appear to represent the practically unaltered structure of a shallow wave-cut beach, may in actual fact have undergone significant changes in structure, texture, and mineral content through diagenesis (alteration at low temperatures and pressures). Before even thinking of inferring the past conditions on that marine shore, the first steps of retrodiction must work from the present mineral composition, texture, and structure of the sandstone back by means of certain laws of diagenesis to the composition of the sand as it was when deposited by the waves.

If we attempt to identify the abductive components in the above examples, we find that in such cases the resulting state of affairs to be explained is clear. In each example it is provided by descriptions of singular facts: data on the content of certain kinds of atoms in minerals in (1); magnetic properties of rock samples in (2); fission tracks in minerals in (3); and distribution of breccias in and around the Nördlinger Ries in (4). The explanation should consist of laws and statements describing the hypothetical conditions which have been retrodictively inferred. The laws involved in example (1) are of various kinds, firstly the physical laws governing the decay of rubidium and potassium, and secondly the experimentally established inductive generalisations determining the temperatures at which the gas argon is released during potassium decay. In example (2) the former position of the Indian continent relative to the Earth's magnetic pole is inferred using the physical law that ferromagnetic minerals segregated out during magmatic solidification will orient themselves such that their magnetic moments will reflect the direction of the magnetic field pertaining at the time and place of the rock formation. In example (3) the retrodiction is based on the physical law concerning the radioactive decay of uranium, and on inductive generalisations on how heating to various temperatures repairs the scars created on minerals by radiation during decay. The retrodiction in example (4) is based on the physical laws of the impact process and on the ballistic transport of accelerated particles, as well as on inductive generalisations which were obtained by experiments with projectiles at high speeds.

The underlying laws in the argument are not always explicitly revealed in retrodictions. This may be for two reasons. Some laws are too trivial to be mentioned, others are implicit in the descriptions of the empirical basis.

Trivial laws are tacitly assumed in ascertaining the temporal relations of the geologic past. The 'law of superposition' is one such triviality which is rarely stated outright. It asserts that in a sequence of bedded or sedimentary rocks, if the original position has remained undisturbed, each bed was formed later than the one beneath, and earlier than the one above. Equally trivial is the observation that a rock identified by certain features as an igneous intrusive is always younger than the rocks surrounding it, or that the substance of the pebbles in a conglomerate was created by processes which took place before the rounding of the pebbles. For tectonic events the trivial rule holds that a given deformation of rocks was created by a phase of movement younger than any deformation it distorts.

A case where a law is implicit in the description of the empirical basis is provided by the following example.

(5) A variety of geomorphologic phenomena are described as moraines, glacial valleys, cirque-type valley heads, arretes, roches moutonnées, erratics, glacial polishing, and glacial striations, that is, they are defined from the outset as glacially caused, while various deposits (fluvio-glacial coarse sediments, morainal marls) are also described as being of glacial

origin. From these phenomena and deposits the authors conclude that a Pleistocene glaciation took place in a certain part of Venezuela. The laws underlying this retrodiction are implicit in the use of genetically defined terms of geomorphology and petrography to describe the observations.[20]

In this context we should refer back to the discussion of non-empirical assumptions in empirical statements (p. 99), where it was shown that many geoscientific terms used in describing observed phenomena are not purely descriptive, but themselves already imply explanations and laws.

6.4.3 *Abductive codiction*

The abductive codictions we are primarily interested in are those which, within the framework of an accepted theory and its implied laws, can help attribute current phenomena to coexisting states of affairs which have either not yet been observed or are non-observable and perhaps cannot realistically be expected to become observable within any reasonable time period. These abductions correspond to case 4 in Table 6.1.[21]

The following cases can pragmatically be distinguished.

1 The resulting state of affairs which is the starting point of the argument is
   (a) an observable fact, or
   (b) hypothetically assumed.
2 The abductively inferred, controlling state of affairs is
   (a) empirically directly testable, or
   (b) not directly testable by any available methods.

States of affairs inferred through abductive codiction are tested empirically by deductive codictions. In 2(a) it is deduced that specific observations will either support or refute the claim that the state of affairs factually obtains. In 2(b) the inferred but not observable state of affairs is used to deduce statements about further states of affairs which are open to empirical testing.

We will illustrate these cases of abductive codiction through some examples.

1(a) The resulting state of affairs is an anomaly in the Earth's magnetic field, established by geophysical measurements in the area of the Nördlinger Ries. The intensity of the magnetic field is sharply reduced relative to that of the surroundings, a phenomenon which is to be explained in reference to the present distribution of basement rock. According to an older theory, the Ries was produced by volcanic processes during the Upper Miocene. Since volcanic rocks of the same geologic age in other areas have been found to have negative magnetic anomalies (due to the reversed magnetic field of the Earth in the Upper Miocene), it was inferred within the framework of the volcanic theory that basaltic masses must be present in the basement of the Ries Basin. Within the context of the newer impact theory the magnetic anomalies have been used to infer the presence of so-called 'suevite', a breccia created by the impact, in the basement. Where such breccia occurs at the surface, it has been found to be negatively magnetised, that is, its magnetisation is reversed with respect to the Earth's magnetic field today.[22]

1(b) According to the theory of plate tectonics as generally accepted today, it is assumed that the North American and European continents are drifting apart, yet the speed of this movement is so slow that it cannot at present be empirically established. It is inferred from this hypothetical state of affairs that corresponding convection streams exist within the upper mantle which are driving the European and North American plates apart.

2(a) Abductive codictions of this kind form the basis for using various geophysical methods to search for ore deposits. The presence of a local positive magnetic anomaly may be used to infer the existence of an iron ore deposit, while a negative gravity anomaly suggests a buried salt dome. In both cases, the presence or absence of the abductively inferred state of affairs can be tested by drilling, which for its part is carried out on the basis of deductive codiction.

2(b) Abductive codictions of this kind include, e.g. the explanation of so-called 'hot-spots' (abnormally high thermal flow from depth to the surface at certain points on the deep sea floors) in terms of local upwelling of hot mantle material, or 'mantle plumes' from great depth.

Further examples include, indeed, all inferences from geophysical observations as to the inobservable structure of the Earth's body, based, for example, on observations of heat flow in the Earth's crust, and the paths and propogational speed of elastic waves during natural earthquakes or during experimental seismic explosions.

In some cases interest in codictions coincides with an interest in converting hitherto inaccessible states of affairs into directly observable ones by developing new technologies. An international programme, for instance, is discussing the possibility of developing technology for deeper coring to drill into hitherto unexplored depths. This would permit abductive means to be used to develop new concepts on the structure and petrography of the Upper Mantle.

Anticipatory codictions, shown as Case 6 in Table 6.1, like anticipatory retrodictions, play little role in Earth science. This is not the case for historical codictions, shown as Case 2 in Table 6.1, which often appear mistakenly as contemporaneous codictions. For strictly speaking, the abduction in such cases refers to the simultaneous past existence of the resulting and controlling states of affairs. Nevertheless it is tacitly assumed that conditions in the investigated object domain have not changed, so that the resulting and controlling states of affairs are evaluated as manifested at present.

The nearly unaltered lateritic soils (consisting mostly of kaolin and of iron and aluminium oxides) of Miocene age on the Vogelsberg, for instance, are used to infer that tropically humid climatic conditions must have existed there when the soil was being formed.

As in the case of retrodiction, we can distinguish two functions of abductive codiction. The intention may be to explain a striking singularity, such as the magnetic anomaly in the Nördlinger Ries, the high thermal flow at a 'hot spot', or a certain earthquake attributed to local tensions and fracturing processes in the Earth's crust. Alternatively the intention may be to enable direct observations of inaccessible or as yet inaccessible general relationships. Thus, observations on the path and propogational speed of elastic waves released by a single earthquake give rise both to inferences about the epicentre of this event, and to the mechanical properties and structures of the deeper zones of the Earth's sphere.

Summing up we find that abductive codictions are no less hypothetical than retrodictions. There are always an indeterminate number of ways in which observed phenomena or inferred states of affairs may be linked to current controlling conditions and processes. In geophysics one speaks of *models* which explain observed phenomena according to accepted theoretical relations or preconceived hypotheses. But there is no logical procedure for discovering the 'true' model, and the abductive construction of such a model is therefore never final. New measurements, refinements in theory, or entirely new theories may cause older models to be changed or entirely recast.

There is a major pragmatic difference between retrodictions and abductive codictions. Retrodictions concern that which has irretrievably vanished. Codiction concerns the present existence of certain conditions or processes. If these are observable, direct tests may establish whether or not they actually exist. If these are not observable, it is always possible that improved observational techniques will enable this to be done at some time in the future. So long as this is not the case there is always the possibility of testing indirectly by deducing observable facts from the inferred state of affairs. The connection between abduction and deduction, an important one for research, will be discussed in more detail in Section 6.6 below.

6.5 Deductive inference in Earth science

6.5.1 *Introduction*

We have already explained that deduction is the only mode of inference which can be called 'conclusive' in the strict sense of formal logic. The deductively derived conclusion alone is invariably true, provided that the stated premises (laws and assertions as to the controlling state of affairs) are true and actually 'contain' everything expressed in the conclusion. Leaving aside for the moment the fact that in practical application to empirical states of affairs, these conditions of truth and completeness of the premises can never be regarded as being absolutely satisfied, we may contrast the advantage of logical certainty with the disadvantage that the deductive inference is basically nothing more than a tautology. Unlike induction and abduction, deduction may develop something not yet known or thought out, but it can never discover anything wholly new. The deductively inferred statement says neither more nor less than what is already analytically implied in the premises, as can be shown in the following example, where A and L are used to deduce E.

Controlling statement A:	The mineral substance calcite belongs to the ditrigonal scalenohedral crystal system.
Law G:	All crystals belonging to the ditrigonal scalenohedral class have one optical axis.
Resulting Statement E:	A light ray will be broken into two rays with differing refractive indices or velocities in all directions other than that of the crystallographic c-axis of calcite.

The deduced statement E says nothing more than what is contained in statements A and L, although this may not be immediately obvious to every beginner in crystallography.

Deductions play a major role in Earth sciences, as they do in all science, and this role increases in importance with the growing usage of universal laws of physics and chemistry to explain geoscientific phenomena. These laws are applied to a large part by deriving from them what must exist or happen in special conditions or contexts. The use of general physical or chemical laws to explain geoscientific phenomena often presupposes the deduction of special laws applying to special geological conditions, substances, or bodies. Stokes' law on falling spheres in viscous media is used to derive laws on the sedimentation of mineral particles in water and in air. The Hagen–Poiseuille law on flowing liquids in capillaries gives rise to laws on the flow of gases, oil, and water through the complex capillary systems of porous rocks.

We will investigate below the most important applications of deductive inferences following the scheme worked out above; we will distinguish between prediction and deductive codiction and subdivide each of these according to whether the deductively inferred state of affairs existed in the past, present, or the future. Depending on whether or not the state of affairs on which the inference is based is factually 'given', i.e. observable, we distinguish further between *factually based* and *hypothetically based* predictions or codictions. *Counterfactually founded* deductions also play a role in Earth science when the initial conditions are explicitly formulated contrary to the actual circumstances. These have the linguistic form of *conditionals* such as: 'if a were the case (but we know or at least suspect that a is not the case), then on the basis of L, situation e would have to follow'. Such inferences are particularly important for rejecting competing hypotheses, as they attempt to make plausible what 'cannot or could not have been the case' (see Section 6.7).

6.5.2 *Deductive codiction*

Codictions generally imply the simultaneity of states of affairs a and e on the basis of certain laws (corresponding to Case 4 in Table 6.1). In the case of abductive codiction the argument goes from resulting state of affairs e to controlling state of affairs a, while in deductive codiction the argument goes from controlling state of affairs a to resulting one e. Abductive and deductive codictions can only be distinguished, therefore when the law which links the coexisting states of affairs is 'asymmetrical', so that the one state of affairs, a, controls the other, e, and not vice versa. Coexisting states of affairs, however, are states, and the laws linking states have been described as laws of states on p. 195. We stated there that in their general or implicit form they are characterised by symmetry of linkage, meaning that there is a functional relationship between two states, which unlike in causal laws, allows no distinction between resulting and controlling states of affairs. From this viewpoint there is apparently no definitive boundary between abductive and deductive codiction, since in functionally mutually dependent states it is possible to choose either state as the 'starting point' or argument.

In the context of geoscientific explanations, however, functional laws can be applied pragmatically in their explicit form, that is, they are interpreted 'quasi-causally', so that one of the two states is posited as the independent stimulus of the dependent state. Here the rule applies that that state is considered to be the dependent or resulting state which can be functionally associated with various different controlling conditions. A negative gravity anomaly at the Earth's surface, for instance, can be functionally related to various specific light rocks in the basement, therefore in the context 'salt dome – negative gravity anomaly' the salt dome is regarded as the controlling state of affairs. In practice, therefore, the distinction between abductive and deductive codiction is usually clear. One argues abductively from a state to a coexisting 'stimulus', deductively from one state to another 'responding' to it.

Like abductive codictions, deductive codictions are most interesting when they relate to presently coexisting states, as it is only these which, at least in principle, can be based on observation. We will therefore limit ourselves below to this most important type. The argumentative combination of abductive and deductive codiction is particularly worth stressing. Abductive codiction infers the existence of a state of affairs not accessible to observation; as the controlling state this becomes the goal of the argument. That same state can also be made the basis of a deductive codiction inferring the existence of a functionally controlled state which can be confirmed by observation. Deductive codiction thus reverses the direction of the abductive codiction, just as historical and contemporaneous predictions reverse the direction of retrodictions. By deducing observable consequences, deductive codiction may help solidify the corresponding abductive codiction. This, of course, can only be the case if it is not deducing the very state of affairs upon which the abductive argument was based, for this would be circular reasoning. The deduction must involve different states of affairs which, however, are homogeneous and thereforc confirmatory. An example of this is the use of geophysical evidence in prospecting. Geophysical data are used to infer the existence of a salt dome flanked by steeply dipping sandstone beds: an abductive codiction. Observations gathered throughout the world have led to the inductive generalisation that sandstone strata along the flanks of salt domes in sedimentary basins frequently contain accumulations of petroleum. The oil geologist will therefore risk the deductive codiction that as a consequence of this regularity, a core drilled alongside the salt dome will probably strike an oil deposit. This codiction is not based on a law, but on an inductive regularity, and the existence of the inferred state of affairs can therefore only be asserted with a certain degree of probability, despite the fact that the conclusion itself is inescapable in terms of formal logic, since it follows the pattern of deductive inference.

6.5.3 *Prediction based on retrodictively inferred past states of affairs*

An especially characteristic use of deduction in Earth science is the deduction of consequences from retrodictively (i.e. abductively) reconstructed states in the Earth's past. The inferred state of affairs itself may in turn be a hypothetically assumed past state of affairs. This brings us, in some cases via long chains of reasoning, to predictions which aim ultimately at some observable present state of affairs. Deductively inferred past states of affairs are often treated as though they were presently observable, namely when the state of affairs in question can reasonably be assumed to have 'survived' unchanged into the present. Such historical prediction, however, cannot be made with the same degree of certainty as those predictions which can be confirmed or refuted through direct observation. An example of this is the discovery of the mineral coesite in the breccias of the Nördlinger Ries.

This resulted from the retrodiction that the Nördlinger Ries and its breccias were created by the collision of a major meteor. This impact hypothesis entails that very high temperatures and pressures would be locally created during such an event. From the laws of alterations caused by rising pressures in mineral substances consisting of SiO_2, it follows that in those breccias which underwent particularly high pressures during the impact, the quartz will have been altered to the high-pressure mineral coesite. From these premises, namely the conditions during an impact and the law of pressure transformation of quartz, it was deduced that coesite occurs in certain breccias of the Nördlinger Ries. This prediction was confirmed by a careful investigation of the rocks concerned. That the presence of the mineral remained unknown till predicted and actually found is because coesite is present only in tiny amounts and in microscopically small crystals, so that chemical enrichment and x-ray techniques had to be used to find it. The predicted state of affairs was so hidden that it could only be discovered after deliberate search.

Strictly speaking this is a case of a historical prediction, in which the deduced state of affairs is past and has to be considered only as a hypothetically postulated state. Nevertheless, the resulting state is assumed to have lasted to the present, so that the historical prediction can be confirmed as though it were a contemporaneous one. The prediction could still be false, even though it appears to be 'confirmed' if, for instance, it could be shown that the quartz alteration could have taken place later and under different conditions. On the other hand, the prediction could be historically quite correct even if the empirical test turned out negative, if it were conceded that the deduced past state of affairs had not survived to the present as originally postulated. This example shows the complexity of the assumptions in what at first seems a simple scientific inference.

Another example of an expressly historical prediction is the following. Retrodictions of conditions during the earliest periods of Earth's history imply, among other things, that the terrestrial atmosphere around 3×10^9 years ago contained little or no oxygen. The laws of water and atmospheric action on primary minerals would indicate that weathering products from this period of Earth's history must differ from later weathering products. The historical prediction therefore states that sediments of this time must contain small quantities of undecomposed sulphides and compounds of metals such as iron, manganese and uranium in low states of valence. In a further step we then infer the contemporaneous prediction that ancient pre-Cambrian rocks surviving undisturbed since their formation (escaping metamorphism or orogenies) will still contain these products of weathering under anoxic conditions.

Empirical testimony of the weathering products at least indirectly confirms the original historical prediction. But even direct evidence available for a contemporaneous prognosis is qualified by the constraint that the transition from past state of affairs a to present state of affairs e must have proceeded without any intervening 'disturbing influences'. In practice this qualification limits the certainty of deductive conclusions, since though the formal logical necessity of the deduction can be assumed, the material necessity that a given event will obtain on the grounds of a prediction cannot.

6.5.4 *Prognoses*

Those predictions targeted on a state of affairs which will not be observed until the future, we have named prognoses. We divide these into those which are *factually based* on a presently observable state of affairs, and those which are *hypothetically based*, whose initial state of affairs is only postulated as possible and whose realisation therefore depends on conditions which will not be met until a future time or which are at least for the time being inaccessible to observation. Another pragmatically significant question for hypothetically founded prognoses is whether or not one can manipulate the conditions on which the prognosis depends. The argumentative status of a prognosis finally depends on the time-span between the stating of the prognosis and its possible empirical confirmation.

Prognoses are best tested and have their greatest predictive strength in experiments. Here they are either based on facts, or their initial conditions (insofar as these are hypothesised) can be least be manipulated precisely, and the prognosis can be confirmed or refuted within reasonable periods of time. Researchers are therefore constantly attempting to reconstruct within an experimental framework the states of affairs brought together in a prognosis. In Earth science this is usually done by designing appropriate models. As we have already discussed the problems involved in experimentation and model building, further discussion of these is unnecessary here. We will therefore largely confine ourselves below to examples involving observation of geoscientific states of affairs outside experimental contexts, even though isolated aspects may concurrently be treated experimentally. Our first example is typical of a factually based prognosis.

One example is founded on the observation that increased concentrations of lead occur in many anthropogenically influenced river waters. Based on the laws of the behaviour of dissolved lead in clay suspensions under the conditions pertaining where river- and sea-waters are mixing, the authors come to the following prognosis:

> If river waters are slightly acidic upon entering the coastal zone, significant pH changes occur as the fresh river waters mix with the saline, normally more alkaline, oceanic waters. Soluble lead, brought to the coast by slightly acidic river waters, will precipitate after mixing with coastal waters. Lead, adsorbed to suspended clays . . . accumulates in the fine grained bottom sediments, and frequently reaches concentrations greatly exceeding lead concentrations in the surrounding water mass.[23]

Hypothetically founded prognoses are of particular practical significance when they contain at least some initial conditions which have not yet been realised. They usually start off with some conditions known today, but envision that these conditions may change in time along with certain parameters of the basic laws involved. Since a variety of such changes are usually possible, such a hypothetically founded prognosis does not generally claim that a particular event will take place in the future, but offers a cluster of partial prognoses, projecting a range of possible 'scenarios' for what may be expected in the future.

Hypothetically founded prognoses which are important for practical purposes are either those whose conditions cannot be influenced by man, or those concerning processes in which people have an influence in that they themselves are a geologic factor. The first kind includes predictions of unavoidable natural events. These may be sudden natural catastrophes such as earthquakes, spring floods, or volcanic outbursts, or very slow processes such as climatic changes, displacement of coastal lines, or the movements of dunes or glaciers. Geoscientific research has in recent times increasingly been trying to achieve some progress here. The difficulty of such prognoses is that the initial conditions are very complex and only incompletely understood, and exact laws are lacking. The statistical frequency of major earthquakes in certain regions is known, for instance, as are the type and frequency of past eruptions of certain volcanoes. This, however, indicates at best how many earthquakes or eruptions can be expected in the course of the next 100 years. For any prediction of such practical significance as a specific event to be expected in the immediate future, however, there is insufficient detailed knowledge of observable phenomena which could 'warn' of such an imminent event and provide the basis for a sure prognosis. That is why prognoses as to the dates of individual natural catastrophes are as yet extremely unreliable. This is not true of prognoses on the course and effects of a given catastrophe, be this an earthquake of particular intensity in a certain region, or the eruption of a volcano of known magmatic type and eruptive mechanism. For these it is possible to make prognoses which can be used to design protective measures.

The prospects for predicting slow processes such as changes in sea level or movements of glaciers and dunes are more favourable, since inductive extrapolations of changes observed in the immediate past can often be used for these. Here again, however, the general observation holds that geoscientific prognoses become more difficult and harder to test the greater the range of initial conditions and the longer the time period between the initial and predicted states of affairs. The precision and testability of such global, long-term prognoses is improved if they are split up 'horizontally' into prognosis clusters, (separating the initial state of affairs into its related partial aspects) and 'vertically' into prognosis-chains (dividing up the entire chronological course of the change into its inter-related phases).

Attempts to prognosticate the climate of the future provide an example of global prognoses which involve many partial aspects and long periods of time, and which also involve historical predictions based on retrodictively inferred states of affairs of the Earth's past. These prognoses often centre around the question 'when is the next ice age coming?' Some researchers are certain that the current 'postglacial' is actually an interglacial. This is expressed in titles such as the following: 'The end of the present interglacial'[24] or 'When will the present interglacial end'.[25] The journal, *Quaternary Research*, dedicated its entire November 1972 issue to the theme 'The present interglacial: how and when will it end?' Prognoses of an imminent ice age are based on retrodictively obtained knowledge on the duration and interglacial segmentation of earlier ice ages and on climatic developments during the Pleistocene interglacials and the postglacial. These data are used to infer inductive regularities on the periodicity of glacials and interglacials, which are then supposed to warrant inferred prognoses on future climatic developments. These prognoses, whatever their results, are equally as uncertain as the inductive generalisations on which they are founded.[26]

Hypothetically founded prognoses which relate to processes and events that can be influenced by human actions and behaviour have of late increased greatly in significance. This is because humans are becoming increasingly effective geologic agents, so that present and future natural events are no longer determined simply by combinations of 'natural' conditions. Geologic conditions and processes of the Earth's surface, waters and atmosphere and the physical composition and quantity of material resources necessary for human civilisation are increasingly dependent on how mankind behaves with respect to his environment. In order to control these influences in a desirable manner, a more conscious style of 'rule' over the physical world is needed, one which is not possible without geoscientific prognoses.

> Since our actions can control and influence the future, but not the past, prognoses aimed at control over nature are of considerably more interest than 'backward looking' applications of theories.[27]

Prognoses of this kind relate in the first place to all unintentional side-effects of technological activities, such as the releasing of pollutants through technological processes of various kinds into the earth, air or waters, where these materials were previously absent; local or global changes in the thermal balance of the Earth's surface through the introduction of warm sewage into rivers or the increase of CO_2 levels of the atmosphere; changes in conditions governing transport and deposition of sand and clay due to the construction of retaining ponds, harbours, and regulatory structures along the coasts; earthquake-inducing changes in local tensions in the Earth's crust, caused by damming up large water masses; increased soil erosion caused by the destruction of protective forests; increased deflation of the upper soil horizons through the use of monocultures in agriculture; influences on ground-water movement and ground-water reservoir capacities by drainage and canalisation of rivers and streams. In all these cases, hypothetically based prognoses can be used to ascertain the measures by which one can presumably avoid undesirable consequences or bring about desirable ones.

Equally as important, in the second place, are hypothetically founded prognoses concerning the effects of our increasing demands on lithospheric reserves of ground-water, raw materials, and energy sources. Predictions on the growing demands and dwindling reserves of these resources, some of which will probably be used up in the course of decades rather than of

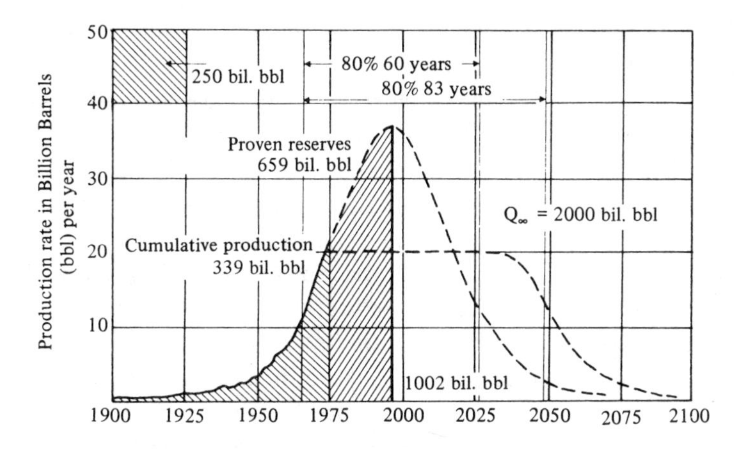

Fig. 6.4. Hypothetical prognosis (from 1975) of the future course of world oil production. Annual production is shown in barrels/year, assuming that around 2000 × 10⁹ barrels (254 × 10⁹ tons) of petroleum are commercially available in the Earth's crust. The solid line shows the past increase in oil production. The dashed lines show two possible prognoses: one assumes a continued increase in annual production, the second assumes that production rates are held constant at 1975 levels. (Hubbert, 1977.)

centuries, are based to a large extent on geoscientifically founded inferences. We will give two examples of this.

Fig. 6.4 shows the hypothetically-founded prognosis on the expected future of global petroleum reserves.[28] The prognosis rests on quantitative estimates of the total quantity of existing liquid carbohydrates and of the technologically accessible quantity. The underlying assumption of a total of 2000×10^9 barrels (254×10^9 tons) includes the contents of the deposits known today plus an estimate of the quantities yet to be discovered. This estimate is hypothetical, and is based on geoscientific knowledge and theories of the formation and occurrence of oil deposits. The underlying laws on the depletion of deposits are not geoscientific, and indeed not scientific at all, but are a description of the politically, economically, and technologically determined behaviour of the people exploiting the deposits. The hypothetically founded prognosis in Fig. 6.4 limits itself to two of many possible scenarios. The exponentially rising curve describes the increase in annual production up to the year 1975. If this trend continues, annual production will reach a maximum shortly before the year 2000 and then fall off sharply with the exhaustion of the reserves. The curve running horizontally from the year 1975 shows the projected course of annual production assuming that planning measures succeed in stabilising yearly production at

1975 levels. In this case the prognosis states that the reserves will last around 25 years longer than in the first case.

A second example concerns the development of ground-water reserves in a desert area.[29] Investigation of the geologic structure of the Murzuk Basin in the Central Sahara has shown the following: The ground-water reserves stored in the porous basement rock derive, according to radiometric dating, largely from a period of increased precipitation or pluvial, 20 000 to 40 000 years ago. The bedding structure is such that any renewal of ground-water from the rainy mountains is practically out of the question. These are the established conditions which must form the basis for any prognosis on the future of the ground-water reserves there. According to the trivial law that any depletion of a closed reservoir must ultimately result in its exhaustion at a date depending on the speed of depletion, i.e. from a 'law' established by the people who use the water, the authors arrive at the prognosis:

> Any withdrawal of water from the Murzuk Basin, as from any other Saharan Basin, will deplete the reservoir. This is particularly true of the artesian aquifers which are not being replenished at all today, and which appear to have been replenished little or not at all in the post-pluvial humid periods. Within the Murzuk Basin, the Wadi Shati seems to be the most endangered.

The foundation of geoscientific prognoses, particularly those of practical benefit, is still relatively uncertain, for research has only just dared to enter this new field. Geology *sensu strictu* has so far largely been concerned with retrodictions, geophysics with codictions. The practical demand for the increased use of factually and hypothetically founded prognoses requires a complete turn around in the geologist's accustomed point of view. This exposes him to a danger he does not encounter in making retrodictions: each prognosis will at some time be proven right or wrong, and he who made it must reckon with the possibility that the course of events or the very measures he proposed may 'refute' it. The backward-looking geologist is safe from such dangers, for the ambiguity of abductive inference and the impossibility of 'direct' proof make it impossible to disprove 'once and for all' certain retrodictions or abductive codictions, possibly setting long years of research at nought.

6.6 Scientific inference as an interplay of abduction, deduction, and induction

The interplay between abuduction, deduction, and induction has been stressed above as a structural feature of scientific explanation.[30] Inductive inferences play a special role outside of explanatory argument:

the inductive organisation of empirical data was discussed in Chapter 5. Abduction and deduction are tied far more closely to the concept of explanation, and in their application display almost mutual symmetry. The connection between controlling state of affairs a and resulting state of affairs e can be represented using either deductive or abductive inference. The starting point of the one inference is the end-point of the other, and vice versa. Viewed in this manner, retrodiction and prediction, as well as abductive conjecture. This testing of hypotheses is often also referred to as then, an entailment-relationship obtains here, since an initially abductively inferred state of affairs can serve as the basis for inferring deductive consequences, whose empirical manifestation may help confirm the abductive conjecture. This testing of hypotheses is often also referred to as 'induction'. We will treat the problems connected with this later on when discussing the testing of hypotheses and theories, and will concentrate here on the relationship between abduction and deduction.

In general the application of these modes of inference to geoscientific states of affairs serves to expand our knowledge 'backwards' through retrodiction, 'forwards' through prediction, and 'sideways' through abductive and deductive codiction beyond the already accepted knowledge on which the argument is based. As a genuinely explanatory inference, abduction answers the question '*why* is something the case?' and deduction, '*what follows from this*?' In reality, abductive conclusions in the Earth sciences serve principally to increase our (though in principle always hypothetical) knowledge about the past states of affairs hidden from direct observation, while deductive inferences in the form of prognoses increase our potential knowledge (subject to future confirmation) about that which has not yet happened.

While abduction has far greater explanatory power than deduction, it is as we have explained, far weaker in terms of the certainty of its inferences. Retrodictions and abductive codictions must remain hypothetical if only in terms of formal logic, since in principle one initial state of affairs can be related to a variety of different controlling states of affairs.

As we have already mentioned, however, the formal superiority of deductive inference becomes relative in terms of the pragmatic context of research. We can never be sure that the basic premises (conditions and laws) are complete and that they correspond to the facts in all respects. The law used can never be definitively verified as universal. Where universal laws are absent, deductions are often carried out on the basis of regularities to which can be attributed only a certain degree of probability. Such at best 'probable' inferences underlie much ore prospecting using deductive codictions, such as drilling for tin in the region of granites or searching for nickel deposits in connection with extremely silica-poor rocks. The conditions chosen as 'relevant' may turn out to be irrelevant, and vice versa. Complete control of all the initial conditions is at best possible in closed systems such as laboratory experiments. Natural systems are always open; complete knowledge of all conditions which may influence the outcome can never be achieved.

It is therefore vain to hope that when the prognosticated event has actually occurred and can be observed, the correctness of the premises of the deductive inference can at least be determined retrospectively. The occurrence of the prognosticated state of affairs is in no way definite 'proof' of the correctness of the prognosis, or the correctness of its assumptions, for false premises can also be used to infer statements which can empirically be confirmed.[31]

That empirical confirmation is no 'proof' of the correctness of an inference becomes much more obvious when justifying retrodictions or abductive codictions by deriving testable consequences. The arguments of water diviners are an example of this and repeatedly entrap officials and geologically uninformed laymen. From certain movements of his instrument, the water diviner 'infers' the existence of 'water veins' in the ground, which according to him can only be found by drilling in very precise locations. This is an abductive codiction in form. The phenomenon of the dipping of the rod is interpreted by 'laws' known to the water diviner to suggest the existence of 'water veins' which govern the dipping of the rod. Should drilling actually reveal a water-bearing horizon the diviner chalks up this success as 'proof' of his method. In fact, however, he has argued on the basis of false premises, for every geologist knows that the 'water vein' hypothesis was invented for the diviner's convenience. Water rarely occurs in the ground in narrow 'veins' or pockets: water-bearing rocks are nearly always porous rock strata of considerable lateral breadth.

This example is not meant to belittle the cognitive value of empirical confirmation for abductive and deductive inferences. Such confirmation is actually crucial for at least tentative support for such inferences, though it can never prove them. We should mention here that the absence of expected empirically testable consequences also represents a gain in knowledge. Since, logically speaking, the conclusion of a deductive inference must be true if the premises are true, the reverse therefore also follows that if the conclusion of a deductive inference is not true, then one or both of the premises must be false.

In conclusion we return to the combined use of deductive and abductive inferences. This best becomes apparent in Earth science's central task, *the reconstruction of Earth's history*. Here retrodictively inferred states of affairs are used to deduce consequences which also lie in the past. These are either recognised as events which have already been retrodictively established by independent means or they are used as a basis for further prognostic steps leading up to currently observable states of affairs. In this manner a network of mutually supportive abductive and deductive argument is built up encompassing the geological history of the Earth (or in a broader context, of the planets), at first in segments, and finally *in toto*. This is done within a historical context in which retrodictive chains of explanation lead from present facts back to past states of affairs, and deductively derived chains of predictions lead from the past back up to the present.

HYPOTHESES AND THEORIES

6.7 Structure, construction, and function of hypotheses and theories

6.7.1 *Definitions*

According to the hierarchical model on which our study is based, the highpoint of geoscientific research is the conception of hypotheses and theories. What the individual disciplines discover at an empirical level through phenomenological and experimental observation concerning the composition and behaviour of the Earth, its components, and those of other planets should not merely be enumerated. It does not suffice to describe the 'face of the Earth' or to recount the Earth's natural history. Geoscience as a natural science wishes to achieve more than this. In accordance with the idealised deductive-nomological model of scientific explanation, we wish to construct connections in time and space ordered by natural law. These we call hypotheses and theories; they encompass a wealth of empirical facts and are the most concentrated and informative of all forms of geoscientific knowledge concerning the natural world as it surrounds us today, as it was in the past, and as it will be in the future. Hypotheses and theories, as approximative models which can never definitively attain the goal of 'complete' knowledge of reality, are what guide the researcher in his choice of problems and in evaluating the relevance of facts; in this way they, by a feedback process, determine the form and type of experience upon which they themselves are based.[1]

A preliminary example will illustrate how a hypothesis is raised to the status of a more comprehensive theory.

We designate as hypotheses the various retrodictions which have for some time been made concerning the origin of the morphologic, tectonic, petrographic, and mineralogical phenomena in the area of the Nördlinger Ries. Various Ries hypotheses were suggested, including a volcanic origin, creation by a steam explosion, glacial and tectonic origin, till in recent times it has become generally accepted that it was formed by a meteoritic impact.[2]

The weakness of the old hypotheses was that they could scarcely be integrated into any more comprehensive theory. Each stood more or less alone, explaining the Ries as an isolated phenomenon. The impact hypothesis on the other hand, could be integrated into the context of a general impact theory. It confirmed the claim that there are impact structures on Earth, and the phenomena observed in the Ries made it possible to refine and improve the general model of an impact process underlying the theory. In this way the Ries phenomenon lost its singular character, and the hypothesis of its origin through impact gained greater support than the other hypotheses by being incorporated into the impact theory in general.

The intimate connection between hypotheses and theories, and the fact that hypotheses, though isolated at first, often contain the germ from which later theories evolve, justify examining the problems of hypotheses and theories together in the following sections. By way of introduction, the next section will present two geoscientific examples of the structure of theories, showing how they are constructed from interlocking hypotheses.

The terms 'hypothesis' and 'theory' are not precisely defined in ordinary linguistic usage. It is generally agreed that a theory is higher ranking than a hypothesis in the range, significance, and reliability of its contents. To define this ranking more precisely for our purposes we will first attempt to determine what we mean and have so far implicitly understood by hypothetical propositions and hypotheses.

In the presentation of the results of geoscientific research, three types of propositions and sets of propositions can be distinguished. First, there are logical or mathematical propositions, whose validity is not open to doubt, at least not in the context of Earth science. Second, there are assertoric propositions of the empirical basis. These can be reduced to so-called protocol sentences which describe singular events and which are certified at least to the extent that in principle anyone can reproduce the basic phenomenological and experimental observations. Third, there are hypothetical propositions which transcend experience and whose validity, while not certain, is assumed.

Our discussions on the problems of the empirical basis have shown that there is no fixed boundary between assertoric and hypothetical statements. Observational methods and descriptive terminologies both depend on nonempirical assumptions based on hypotheses and theories, so that the boundary is conventional only. Historically it depends on the state of research. Statements once recognised as indubitable descriptions of phenomena may in the light of new insights prove to be hypothetical interpretations. The boundary is not necessarily fixed at any given time, however: the same state of affairs may be treated in one investigation as an empirical fact and used as a basis for further argument, while in another study it may be questioned as hypothetical and subjected to new methods of testing.

Such cases aside, there is usually no doubt in actual research as to which statements can be accepted as descriptions of fact, and which are hypothetical. The claim that the terrestrial atmosphere contained no oxygen three billion years ago will be called a hypothesis, while the assertion that the peak of the Wasserkuppe in the Rhön Mountains consists of basalt (that is, of a volcanic lava) is an assertoric statement of fact, even though the naming of the rock as 'basalt' is based on a hypothetical explanation of its origin.

We will not, for purposes of pragmatic delineation, describe all statements which are in any way open to doubt as hypothetical. Rather we will categorise as hypothetical assertions and hypotheses only those propositions and sets of propositions which go beyond experience (which is based on phenomenological and experimental observation and consists of descriptions of the empirical basis) and which in the actual course of research may be questioned as more or less problematical and in need of further testing for their refinement, corroboration, or refutation.

We therefore characterise as *hypothetical* first off, all inductive generalisations and regularities, and all statements based on retrodictions, predictions, or codictions. Laws are also basically hypothetical, even though many of them, being general laws of nature and having a firm basis in recognised and proven theoretical systems, are accorded a degree of subjective probability bordering on certainty. Complex constructions relating to larger contexts and uniting several hypothetical components, such as regularities, laws, and codictively, retrodictively, or predictively inferred states of affairs are also classed as hypothetical here.

Hypotheses[3] relate to empirical states of affairs which, though often fairly wide-ranging are nevertheless well delimited. They stand alone at first, even when supported by indirect evidence, that is, when in agreement with other accepted hypotheses and theoretical discoveries. Hypotheses may gain a different character and above all, a greater degree of certainty when they are linked up with other hypotheses in the systematic contexts which we call theories.

Theories,[4] as will be made clear in the examples of the next section, are hierarchically ordered systems of hypotheses interlocked by a network of deductive relationships. True, it is usually possible to express such a thing as the quintessence of a theory using a few basic hypotheses from which the more specialised partial hypotheses may be developed. Nevertheless, that which is expressed by a geoscientific theory cannot be stated simply in terms of these basic hypotheses: it is nothing less than the totality of all that is conveyed by all its partial hypotheses.

6.7.2 *Examples of geoscientific theories*

Theory of the formation of igneous rocks through fractionated crystallisation

The *empirical basis* from which this theory was developed and upon which it still rests consists of a large quantity of data on the mineralogical and chemical composition of igneous rocks of various geologic ages and from varying regions of the Earth. Later these data from phenomenological observation or inventorying were supplemented by experimental data from laboratory investigations on the behaviour of silica melts under given conditions.

Data on the composition of naturally-occurring igneous rocks was first ordered by *inductive generalisation.* It was found that the rocks of a given region which were formed either at depth during a given geologic epoch or at the surface through cooling of a magma, were 'related' by similarities in chemical and mineralogical composition, and differed both from other series of igneous rocks of other areas and from rocks formed in the same area at a different time. The concept of 'petrographic province' was formed through inductive generalisation, each province defined by certain characteristic features (e.g. greater frequency of sodium than potassium, or absence or abundance of aqueous minerals). The rocks of a province often formed a continuous series of changing composition, e.g. a series showing a gradual increase in SiO_2 content. Many smaller provinces were defined in this way, such as the 'rocks of the Oslo graben', or the 'Thule province', as were larger supergroups, such as the 'Pacific' and the 'Atlantic' group.

Outlines of explanatory hypotheses generally entered into these generalisations, whether expressly or not. They formed the first step towards a theoretical interpretation of the relationships expressed by inductive generalisation. There was a widespread hypothetical assumption that the spatial proximity, chronologically correlated formation, and overall resemblance of rocks of one province or group could be explained by origin from a common parent magma. This hypothesis was expressed in terms such as 'consanguineous' rocks or 'rock tribes' (e.g. 'Pacific tribe', or 'Atlantic tribe'). Other hypotheses assumed that major groups such as the 'Atlantic' and 'Pacific' rock series each had their own parent magmas. Finally attempts were even made to trace all igneous rocks back to a single original magma. For such hypotheses it was necessary to assume that certain differentiation processes had produced rocks of differing compositions (such as series of rocks with continuously increasing SiO_2 content) out of one parent magma. The concept of differentiation of parent magmas presented researchers with the task of developing a model of processes which, working in accordance with physical and chemical laws, could create the observed heterogeneous rock series out of a hypothetical, homogeneous original substance. Of various models created to solve this problem during the first decades of this century, one led to the theory of differentiation through fractional crystallisation.

The theory rests on two *basic hypotheses*:

1 All rock series of the different petrographic provinces developed out of a single parent magma of basaltic composition.
2 Differentiation took place according to a certain physico-chemical mechanism called 'fractional crystallisation'.

Both basic hypotheses started off as nothing more than plausible assumptions. The first hypothesis both agreed with general monistic concepts and was empirically and theoretically plausible. It was empirically plausible since rocks of basaltic composition are found in most rock series of the petrographic provinces, and theoretically plausible since it had been clear from the outset on physico-chemical grounds that the other main types of igneous rocks could only be derived by mechanisms of fractional crystallisation from an initial magma having highest temperatures and lowest SiO_2 content. Of all the widespread igneous rocks, the basalts are the only ones which fulfill these conditions.

The second basic hypothesis was plausible on theoretical grounds, since the model it sketches is based on known and accepted physical and chemical laws which hold universally for crystallisation from melts. This model of fractional crystallisation applied to the behaviour of silicate melts to which the magmas belong can be characterised as follows. A melt consisting of a simple mineral substance, e.g. NaCl, solidifies when cooled at a well-defined temperature (the melting point) into a homogeneous mass of crystals of one kind, in our example, into a mass of sodium chloride crystals. The cooling of a complex silicate melt on the other hand, will involve several steps of crystallisation within a given temperature interval. When an upper temperature threshold is crossed (the so-called liquidus temperature), crystals of mineral species belonging to the early crystallisation series appear (e.g. SiO_2-free minerals such as spinel, or SiO_2-poor minerals such as olivine). With continued drop in temperature, further mineral phases successively replace each other: these belong to the main reaction series and include minerals such as plagioclase, pyroxene, hornblende and biotite. The last to crystallise are the most SiO_2-rich minerals of the remnant reaction series, such as quartz, alkali feldspars, and muscovite. During this process the volume of the melt decreases, until finally at the lowest temperature boundary (the solidus temperature) only solid phases remain. The composition of the melt changes constantly during this process of fractionation while the composition of the solid phases which have already crystallised out (e.g. the olivines, feldspars, and pyroxenes) also changes due to chemical reaction with the melt. Should the precipitated crystals and the melt remain together until the final remnants of melt have disappeared, the composition of the resulting rock will exactly correspond to that of the parent magma. Manifold variations, that is, rocks and melts of differing compositions are formed when the remnant melt and the precipitated crystals are separated at some stage during the cooling and solidification process. Of all the common rock melts, basaltic magma has the highest liquidus temperature, and it is from basaltic magma therefore, that the greatest number of differing rock and magmatic types can be developed provided that a separation between crystals and melt occurs at some temperature between the liquidus and the solidus temperatures.

Further development of the theory occurred through two interrelated ways. The first involved establishing the physico-chemical and mineralogical–crystallographic laws governing the crystallisation of melts of the appropriate composition. This was done experimentally in the laboratory, starting with simple chemical systems having few components, and gradually approaching the compositions of the magmas and rocks of concern by adding further components. The laws discovered in this manner established how changes in pressure and temperature would affect the chemical composition and quantities of the liquid and gaseous phases, and the kind, quantity, and composition of the solid mineral phases. Since these laws describe the sequence of changes occurring, e.g. during cooling, they are developmental laws (see p. 175).

Once the laws of the model of fractional crystallisation are known there are various methods by which the compositions of differentiated rocks may be deduced: as accumulations of crystals which were precipitated early by mechanical separation from the melt with which they were in equilibrium; as crystallisations out of a melt which at some stage of cooling was separated from previously precipitated crystals; and as various degrees of incomplete separation of melts and precipitated crystals. Since the melt and the crystals generally have differing specific gravities, the most important factor for separating them appears to be gravity: heavier crystals tend to collect on the floor of the melt chamber, while lighter crystals rise. Tectonic forces can also bring about separation movements of the melt and the crystal sludge.

Decades of systematic laboratory work have built up and continue to elaborate on a complex model of the behaviour of silicate melts, aimed at deducing how different rocks and rock series are produced out of a parent magma. This model is based on many general and specialised proposed laws of physical, chemical, and mineralogic-crystallographic type.

The second manner in which the theory underwent further development was through increasing investigation of naturally occurring igneous rocks and rock series in the light of the theoretical model, leading to a rapid and significant expansion of the empirical basis. The theory focussed attention on those properties most relevant to these investigations. The most theoretically relevant properties included: chemical relationships (of both main and trace elements) between rocks viewed as 'consanguineous'; microscopic texture, from which can be read the formational sequence of the minerals; and the geologic depositional situation and circumstances surrounding the production and siting of the magmas, etc. The accumulation of empirical material was and is carried out, from the point of view of theory development, with two chief goals in mind. On the one hand, confirmation for predictions deduced from the theoretical model is sought, and on the other, new empirical discoveries which will modify and refine the theoretical model are also looked for. An improved model can provide prognoses which are more specialised, exact, and more frequently quantitatively and empirically testable, as well as retrodictions which are more detailed and probable. A classic presentation of the theory by its most significant author and promoter, N. L. Bowen[5] includes the following description of the application of the theory in research. This forms the introduction to a chapter which develops the series of most common igneous rock types from the crystallisation of a basaltic magma.

> We shall now endeavour to discuss the fractional crystallisation of basaltic magma. By way of anticipation it may be stated now that we shall first develop the thesis that normal subalkaline series of rocks (the most frequently occurring group of igneous rocks on the continents) could be developed from basaltic magma by fractional crystallisation. In doing so, much use will be made of the data of investigated systems, but since the investigated systems are always much simpler than magmas, it is not possible to use directly the actual quantitative values of concentrations, temperatures, etc. of the experimental results. The principal service of these must be rather to point the road. We find in a certain investigated system a definite course of crystallisation and definite possibilities of differentiation through fractional crystallisation. We turn then to an actual magma as near as may be to some liquid of the investigated system and ask ourselves whether the crystallisation of such a

magma presents any parallelism with that of the investigated liquid. To answer this question we may consider what indications there may be in physico-chemical theory as to the expected extent of departure from the simple system. We then turn to the evidence of the course of crystallisation of the magma, as determined from rocks, and see to what extent there is parallelism with the simple liquid and to what extent the departure is of the expected kind. If the correspondence with our expectations appears to be sufficiently good we may then proceed to deduce the results of *fractional* crystallisation of the natural magma, using the evidence from the simple liquid with such modifications as may be appropriate in the light of physico-chemical theory. The deduced course of fractionations is then to be checked against actual rock series and if the rocks are of the anticipated kind, there is considerable likelihood that they have been formed by fractional crystallisation. There is a little in this of the nature of circuitous reasoning but it is the kind of circuitous reasoning upon which all scientific generalisations are based. It is the common procedure of science, given indications that a certain general relation is true, to *assume* that it is true, to push deductions to their ultimate consequences in all directions and to make the degree of correspondence of observation with deduction the measure of the probable truth of the original assumption. This does not mean of course that one should not entertain alternative hypotheses. Nevertheless the alternatives must be checked in precisely the same manner, and it is but a poor recommendation for a hypothesis that it can be checked against observations to such a limited extent that it is difficult to prove wrong.

Of all the hypotheses of differentiation of magmas none except the hypothesis of fractional crystallisation can be checked against observation in any detail.

The theory of the formation of magmas through fractional crystallisation is no longer the only theory of petrogenesis today, being supplemented and replaced by the theory of the formation of magma through partial melting (anatexis) of solid rock through pressure release and/or through the raising of temperature. It is also based on laws obtained experimentally in the investigation of silicate systems under controlled conditions of pressure and temperature.

Theory of plate tectonics

As a second example of a geoscientific theory we will attempt to describe the basic features of the theory of plate tectonics.[6] This is the most comprehensive theory ever produced in the field of Earth sciences, and it is largely due to this theory that one can speak in pragmatic terms of a unified geoscience today. Given acceptance of this model of Earth's development, most geoscientific problems of any import can no longer be tackled within the framework of a single discipline, as was possible and customary just a few decades ago. The theory of plate tectonics is a very complex construct in terms of the structural interconnections between its empirical and hypothetical elements and the range of its empirical basis, and we can describe only its basic features here. Even that cannot be done in the allotted space in the same way as in our first example of a theory of more limited scope. We therefore provide a schematic sketch of the theory's structure in Table 6.2, distinguishing between an empirical level, a level of partial hypotheses, and a level of basic hypotheses.

The *empirical level* includes more than just 'facts', encompassing all states of affairs which in this context are considered non-problematical and which anyone can in principle establish by empirical means. It also includes many states of affairs discovered by interpreting observations or by inductive generalisations, statements, that is, which strictly speaking contain various hypothetical elements. By no means all the empirical components which played or still play a role in discussions on the theory of plate tectonics have been mentioned; we have simply chosen the most important empirical elements for understanding the theory.

Among the theory's *partial hypotheses*, varying levels of generality or different degrees of deductive capacity can be distinguished. Hypotheses shown higher in the sketch (H_5 and H_6) have a higher deductive capacity than those shown beneath them (H_1, H_2, and H_3), as the latter can be deduced from H_5 and H_6. Highest of all stand the *basic hypotheses* H_b of the theory, from which all the partial hypotheses and all the facts of the empirical basis should be derivable. The lines connecting them with the partial hypotheses H_1 to H_6 and also with the components E_1 to E_9 of the empirical basis lay the foundation for the theory's external consistency (p. 255), meaning that the basic hypotheses can be derived from, or at least are not contradicted by, independent hypotheses concerning the theory of the solid Earth.

Our sketch can now be elucidated by summarising the contents of the empirical and hypothetical components and the relationships between the various levels.

Components of the empirical level

E_1 *Geological findings.* Geological formations of similar age and structure, and tectonic structures (orogenies) of the same age are found on continents separated today by deep and wide oceans.

Table 6.2 *The theory of plate tectonics: structural scheme*

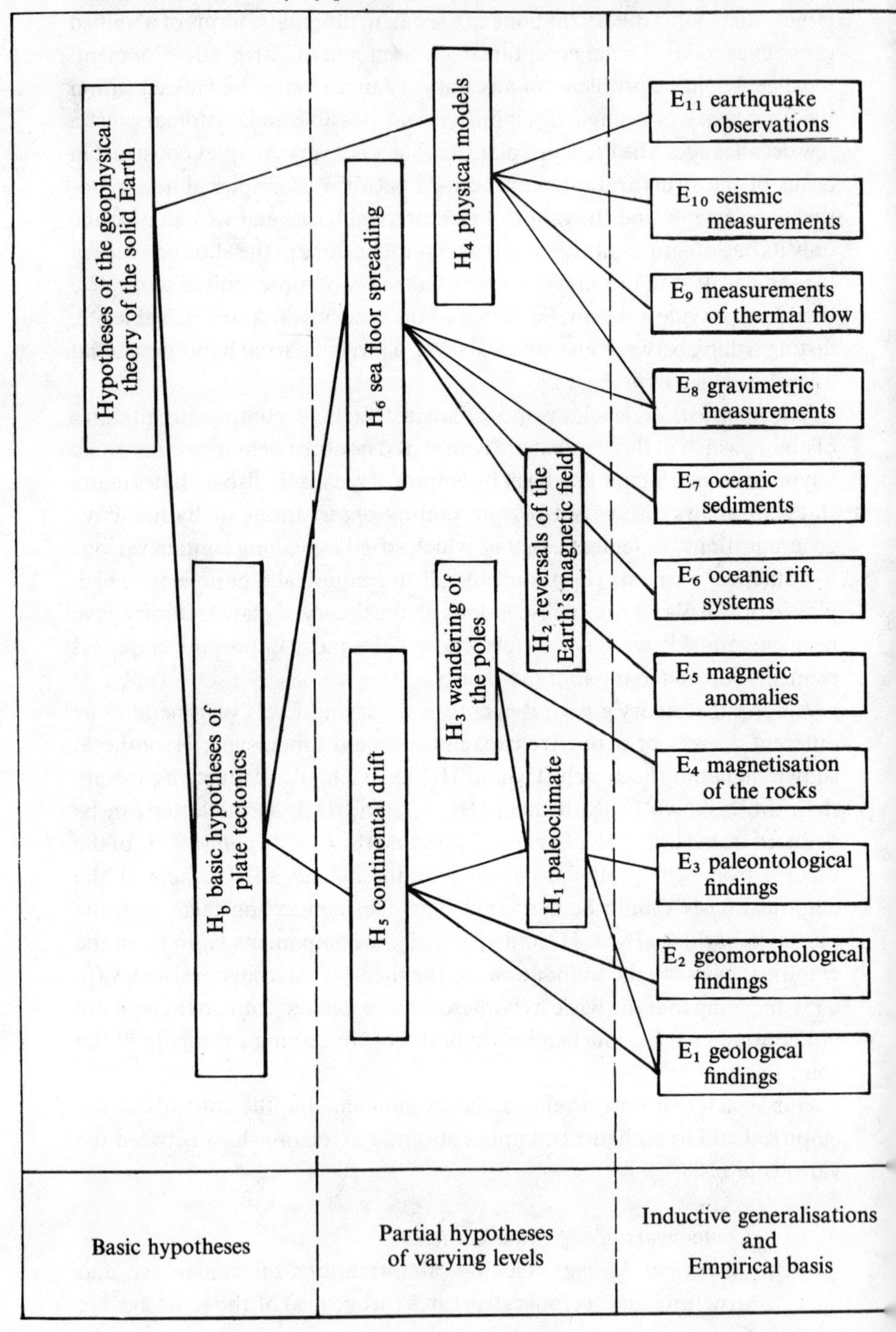

Examples include formations indicating an Upper Carboniferous glaciation in Australia, India, Africa, South America, and Antarctica; Precambrian gneiss complexes of the same age in West Africa as in Northeast Brazil; and petrologically, structurally, and chronologically analogous Early Paleozoic Caledonian and Late Paleozoic Variscian mountain ranges in Western Europe on the one hand, and in Newfoundland and Nova Scotia on the other.

E_2 *Morphological findings.* The shelf margins of the continental masses of South America and Africa, North America, Greenland, and Eurasia, all separated by the South and North Atlantic, and of India, Madagascar, Australia, New Zealand, and Antarctica, all separated by the Indian Ocean, fit together so well that they could be joined practically without a gap into the single land mass 'Pangea'.

E_3 *Paleontological findings.* Until late in the Mesozoic, faunas evolved identically in Brazil, Australia, India, and Madagascar. After this time, separate faunas developed on the continents now separated by oceans.

E_4 *Magnetisation of the rock.* Some rocks containing magnetic minerals (especially volcanic rocks) are magnetised, that is, they show natural residual magnetism. Assuming that the rocks were magnetised by the Earth's magnetic field at the time they were formed, the magnetic rock can be used today as a fossil compass indicating the direction of the magnetic field at the time and place the rock was formed.

E_5 *Magnetic anomalies.* In thick series of volcanic rocks, and later in cores from the ocean floors, many sequences of reversals in the polarity of the residual magnetism were discovered. The rocks are either magnetised positively (with the magnetic field today) or negatively (opposite to today's magnetic field).

E_6 *Oceanic rift systems.* A system of broad mountain ridges consisting of igneous basaltic rock runs through the oceans. A deep graben-like rift runs along the centre of these ridges, interrupted by so-called transform faults which horizontally displace the crust at right angles to the rifts. The oceanic rift systems are distinguished by volcanic activity producing volcanic islands such as Iceland on the mid-Atlantic ridge.

E_7 *Oceanic sediments.* The sediments lying on the ocean floor are geologically young, even deep cores yielding no material older than the Lower Cretaceous (c. 135 million years old). In the Atlantic Ocean the age of the sediments immediately above the underlying basalt increases systematically with distance from the mid-Atlantic ridge, and the same is true in the other oceans as well.

E_8 *Gravimetric measurements.* The continents consist of lighter rock (usually of granitic composition) than the ocean floors which under their veneer of sediment consist of heavier basaltic rocks; yet the gravitational pull on the continents is on the average the same as in the oceans. Negative gravity anomalies coincide clearly with the deep sea trenches, such as those on the seaward side of the island arcs of Southeast Asia.

E_9 *Measurements of thermal flow.* The heat flow from the Earth's centre towards the exterior is on the average identical per unit time and area in the continents and the oceans. The oceanic ridges, however, form local areas of higher thermal flow.

E_{10} *Seismic measurements* indicate that the Earth's crust is divided up into layers in which elastic waves travel at different velocities. Beneath a low density 'crust' in which waves travel at low velocities lies the denser 'mantle' where waves have higher velocities. The crust and the mantle are separated by a discontinuity interface, the Moho or Mohorovočic Discontinuity. This lies at a depth of around 35 km beneath the continents, under the oceans at around 12 km. Under the oceans the velocities of seismic waves indicate three layers of crust: layer I is around 1 km thick and consists of sediments, layer II is around 1.7 km thick and consists of consolidated sediments or modified material from layer III, and layer III is basaltic rock. Within the mantle a layer of reduced seismic velocities at about 100–200 km down is called the 'asthenosphere'.

E_{11} *Earthquake observations.* The oceanic ridges are characterised by seismic activity whose epicentres lie exclusively at shallow depths. Another zone of high earthquake activity is the system of deep sea trenches and island arcs surrounding the Pacific, which has many volcanoes as well. The epicentres of earthquakes in this zone are at depths of down to 700 km.

Partial hypotheses

H_1 *Paleoclimate.* From the rocks being formed at a given time and place on the Earth's surface, and from the coetaneous animal and plant types living there, retrodictions can be made as to the climatic conditions existing there at that time. Coal formations, for instance, suggest a moist climate, while gypsum and halites point to arid, hot conditions; certain coarsely clastic sediments called tillites are interpreted as glacial formations, and therefore reflect a cold climate; large reptiles indicate warmth, as do reef-forming corals; annual rings in tree trunks suggest seasonality, their absence indicates tropical conditions. It therefore follows that during the Carboniferous, those areas of Europe and North America where coal was being formed were tropically moist, while areas where rocks were being formed at the same time in South America, South Africa, and India were undergoing glaciation under arctic conditions.

H_2 *Reversals of the Earth's magnetic field.* A comparison of the direction of magnetism of rocks from differing rock sequences in different places shows that magnetic reversals took place simultaneously on a global scale. Assuming that the Earth's magnetic field has always been a dipole field (having a magnetic north and south pole), the hypothesis is then suggested that the Earth's field has changed direction several times, and that each such reversal represents a specific date in Earth's history. The reversals have been followed back into Cretaceous times, while radiometric dating has permitted precise calibration of the geomagnetic time scale back into the Tertiary. The average time-span between reversals is 450 000 years, and the last reversal took place about 50 000 years ago.

H_3 *Wandering of the poles.* The position of the magnetic poles at different times in the geologic past, as calculated from the magnetisation of the rocks of a given continent, has led to the construction of a hypothesised continuous curve for the changing position of the North and South poles with respect to the present position of that continent. The hypothesised curves of 'polar wandering' for the various continents (e.g. for Europe and North America) generally do not coincide.

H^4 *Physical models*. The results of geophysical measurements of gravity and thermal flow, and of the propagation speed of both artificially caused elastic waves and natural earthquake waves, viewed in the light of petrological, geochemical, and geological insights, have led to the construction of physical models to interpret how the empirical phenomena came about. These models pertain to the distribution of temperature, pressure, density, and other physico-chemical features, and, although we cannot discuss these here, to the petrographic and mineralogical composition of the deeper zones of the Earth's body. The interpretation of the results of seismic experiments described under E_{10} as components of the empirical level actually belongs to this group of hypotheses.

H_5 *Continental drift.* In contrast to the conception underlying all geoscientific hypotheses up to the middle of this century that the continents have not changed their position relative to one another during the geologic past, the hypothesis first conceived in 1912 by Alfred Wegener claims that the continents have moved, and that up to the end of the Paleozoic, the continents now separated by oceans formed a single land mass, 'Pangea'. Starting in the Mesozoic, that is, during the last 200 million years of Earth's history, this paleocontinent began to break up into individual blocks and drift apart. South America and Africa separated during the Jurassic, forming the South Atlantic between them. Later the North Atlantic opened up, North America splitting off from Eurasia, and Greenland from both of these major blocks. India moved northeastwards, and was pushed up against the Asiatic block, creating the mountain range of the Himalayas. Similarly the Alps were produced by the impact of Africa against Europe. Australia and America separated from Africa during the Cretaceous, Australia later divided from Antarctica, and New Zealand from Australia.

Table 6.2 shows the empirical evidence and hypothetical constructs which this hypothesis is founded on and/or explains. The geological and morphological similarities between continents now separated by oceans become intelligible; the at first uniform, later divergent evolution of life forms becomes something to be expected through the hypothesis. By considering the changing position of the continents through time with respect to one another and to the rotational poles of the Earth, the past climatic zones of the continents, as reconstructed by geologic and paleontologic evidence (paleoclimatic hypotheses) fall into global climatic zones for each timepoint of the geologic past. The apparent 'wandering of the poles' relative to the continents nearly disappears when the movement of the continents with respect to one another and to the rotational axis of the Earth is taken into account. The hypothesis, indeed, has a 'Copernican' character in this regard: it is not the poles which have 'wandered', but the continents which have moved relative to the fairly constant position of the poles (by 'position of the poles' we refer to the position of the magnetic pole relative to the rotational axis of the Earth).

H_6 *Sea floor spreading.* According to this hypothesis the rifts running through all the oceans are systems of seams along which basaltic melt wells up and accumulates on both sides of the rift in the submarine mountain chains of the oceanic ridges. These oceanic ridges with their central, graben-like rifts are caused by tangential tensions, created by the hot streaming of convection cells rising up in the Earth's mantle below. The ocean floor is carried away from the axis of the oceanic ridge as on a conveyor belt toward the deep sea trenches, where the cooled stream of the convection cell sinks again into the depths. In this way the ocean floor grows by means of the basaltic material emerging at the seams, while older ocean floor is being fed back into the mantle in the zones of the deep sea trenches.

In some places, as along both sides of the Atlantic, the continents are being thrust apart by the enlarging ocean floor. In other places, as around the Pacific Ocean, the continents are thrust over the sinking ocean floor, so that, as a result of the tangential pressures, ranges of mountains such as the Andes are formed along the continental margins parallel to the coast.

The most important empirical facts and hypotheses on which this hypothesis is based or which can be deduced from it are shown in Table 6.2. The hypothesis of sea floor spreading explains the ocean rift systems, and makes it clear why today's ocean floors are covered only by young sediments, being themselves therefore geologically young formations; the high thermal flow under the oceanic ridges reflects the rising stream of hot mantle material there; the low gravity values in the deep sea trenches result from the light material being carried down into depth by the sinking convection streams there. The most brilliant confirmation of the hypothesis was the discovery of a striped pattern of magnetic anomalies along both sides of the oceanic ridges. Narrow stripes of basalts of alternating normal and reversed magnetisation are symmetrically arrayed parallel to the rifts. These stripes correspond to the reversals in the Earth's magnetic field through the geologic past, and represent the growth or spreading of the ocean floor. Using the geomagnetic time-scale and the width of similarly magnetised stripes as a basis, the chronology of the ocean floor spreading can be determined. The rate of spreading in the Atlantic is an average of 2 cm per year, while higher values have been found in the Pacific.

Basic hypotheses

We can characterise the basic hypotheses of plate tectonics as a system of propositions bringing together the two partial hypotheses of continental drift (H_5) and sea floor spreading (H_6) and setting these into meaningful relationship with the geophysical theory of the solid Earth.

According to the geophysical theory of the solid Earth, the Earth's crust should be considered rigid down to a depth of about 100 km. Beneath this rigid lithosphere lies the asthenosphere, characterised by lower velocity of elastic shear waves, in which the rocks are in a nearly molten state. The lithosphere is divided up into at least six larger and about twice as many smaller plates, which can move about laterally upon the somewhat less viscous asthenosphere. The larger plates consist of the American, Eurasian, African, Indian, Pacific, and Antarctic plates, all of which with the exception of the Pacific plate include both continental and oceanic areas. Due to the movement of the plates against one another the plate boundaries form narrow zones of particularly intense seismic and volcanic activity. Three types of plate boundaries may be distinguished.

1 Along constructive plate boundaries new crust is produced by basaltic magma welling up out of the mantle. These are the oceanic rifts with their ridges.
2 Along destructive plate boundaries crustal material is dragged down into the depths along mountain chains or in the area of deep sea trenches and island arcs.
3 Along conservative plate boundaries the crust is neither used nor destroyed; these are the transform faults, where the plates slide laterally past each other.

Since it is assumed that the Earth's radius has remained constant through time, the newly produced oceanic crust along the constructive plate boundaries must be compensated by an equal amount of crustal rock disappearing along the destructive plate boundaries.

We cannot describe here the extent to which the theory of plate tectonics touches all areas and disciplines of geoscience. It is equally important for petrology as for the theory of the formation of ore deposits, historical geology, tectonics, volcanology, paleontology, and of course geophysics. What is particularly and increasingly noticeable in this theory is a function characteristic of all great theories: the individual disciplines, which since the end of the nineteenth century had become increasingly specialised and distant from each other, are so united at a higher level by this theory, that new insights of any kind are only possible now through interdisciplinary cooperation. This, of course, also has significant effects on the practical application of Earth science.

The theory is still being developed and will reach a higher degree of completion when a model has been worked out explaining the forces in the interior of the Earth which put the plates in motion. In its present form, the theory asserts that thermal convection in the mantle (particularly in the region of the asthenosphere) is the only motor driving the movements on the

Earth's surface, and that this motor is powered by the decay of radio-active atoms. What still remains to be analysed in detail is the obviously complex relationship between convection in the mantle and the phenomena at the Earth's surface.

6.7.3 *The real world and the explanatory model*

The statements of the empirical basis describe isolated observed facts of the present natural world. Inductive generalisations, regularities, laws, and abductively and deductively inferred propositions also have a descriptive content, for they not only explain empirical facts, but also describe the past, hidden, and future states of affairs. The reconstruction of extinct animal species and their life styles, the extrapolation of climatic features of past periods from explanations of current landforms, the petrologically-based pressure and temperature values under which garnet crystals were formed in a garnet mica schist outcrop today, the nickel/iron core of the Earth inferred from magnetic measurements, all such states of affairs are hypothetical and less certain than empirical facts of which anyone can convince themselves through observation; yet they, too, belong to the universe of geoscientific research. The reality of geoscientific theories and hypotheses is a 'richer' reality, including not only the facts of the empirical basis, obtained or obtainable by observation, but also hypothetical states of affairs inferred through induction, abduction, and deduction.

Few of the hypotheses and none of the theories of Earth science are mirror images of this expanded reality. They are not meant to describe (or not merely to describe), but to explain the empirical and hypothetical components of reality, and are thus only indirectly related to reality. They have a direct relationship to conceptual entities which we call *models*.[7] We will show this by means of some examples.

The theory of differential crystallisation deals with conceptual systems of simple melts behaving under varying conditions of pressure and temperature. The actual processes which at various times produced differentiated series of igneous rocks in the lithosphere form an unlimited body of mediated referents thought to have proceeded in accordance with the model of Bowen's Theory. The theory of plate tectonics deals with the movements and interactions of rigid, spherically curved plates of rock carried on a viscous stream. The actual movements of the Eurasian, American, African, and other plates since the Mesozoic forms a body of mediated referents which is to be interpreted according to the ideal model of plate tectonics. At present this model is being tested to see if it can also be applied to the geotectonic processes of earlier periods or to the structures on the surface of Mars. The theory of evolution operates with the abstract mechanisms of mutation and natural selection. Evolutionary steps such as the transition from dinosaurs to birds are subsumed as actual, geohistoric episodes under the model of evolutionary theory. The impact theory works out the physico-mathematical model for the spread of a shock wave caused by a projectile's collision with a target and propagated semispherically around the target material. All impacts which have actually taken place of planetary bodies into the surfaces of planets are interpreted according to this model, which is immediately related to (is the immediate referent of) the impact theory.

A model set up by an explanatory hypothesis or theory is to a certain degree independent of the real systems which exist in nature. Phenomena exist and processes occur in the real world quite independently of whether or not there are models to explain them. Correspondingly, models may 'exist' in the mind of the researcher as ideals, even when (for the time being) no phenomena are known for certain which fit the model. It would be possible, for instance, to create the idealised model of an impact process before any actual impact or its results have been observed, but such constructs *in vacuo* are rare. As a rule, an explanatory hypothesis or theory is related to a real object requiring explanation. The creation of theories would be a senseless game of cat's cradle if it were not assumed that somewhere in the world, phenomena or processes did, do, or will exist which fit the model.

Three features should be stressed in characterising the model as the direct referent of theories or explanatory hypotheses.

1. The most important feature is the difference between a 'picture' of reality expressed by a description of observed or hypothetical states of affairs, and an explanatory model of real systems provided by a theory or hypothesis. Obviously a description cannot provide a picture in the strict sense since any linguistic or semiotic representation of real phenomena must be selective, but such a representation should be as true to nature as possible: the investigator concerned with description is governed first and foremost by the object. The investigator concerned with explanation, on the other hand, creates a model with the intention of explaining a large quantity of individual real phenomena according to a uniform hypothetical scheme. For this reason the model, in contrast to the picture, must ignore all features which have nothing to do with the essentials of the explanation; the impact model does not include all the morphological, petrographical and other features to be observed in the Nördlinger Ries or in any other particular impact structure; the model of plate tectonics does not incorporate all details of the structures of the Andes along the boundary of the Pacific plate and the westward drifting South American plate.

Compared with the descriptive 'picture', the model is simpler and easier in the sense that it can be described by fewer parameters; it is more variable; in short, it is idealised such that it fits many actual cases, each of which may have many contingent peculiarities which are irrelevant in light of the theory. Idealisation through simplification is particularly obvious where it is possible to express the model mathematically. We will illustrate this with a simple example.

In sands transported and deposited by water or wind, large light quartz grains are found associated with smaller grains of specifically heavier minerals. The problem is part of a larger one in the theory of fluid transport of sedimentary particles, and consists of explaining these size differences quantitatively. The explanation is based on a very simplified physical model, which ignores all special features of sand transport by water in rivers or along sea coasts, or of aeolian transport in deserts, beaches, etc. Sand grains upon a solid substrate are set in motion as soon as the shear stresses at work upon the grain's centre of gravity overcome the friction proportional to the weight of the grain. According to physical laws, the following holds true for the state of equilibrium which must be exceeded for the grain to be set in motion.

$$D = \text{const.}\ [(s_M):(s - s_M)] \cdot f,$$

where D is the diameter of a mineral grain with the specific weight s and s_M is the specific weight of the transporting medium exerting shear stress f upon the substrate. From this formula it follows that in a sand being transported by rolling, the size difference between specifically light quartz grains and grains of a heavier mineral depends upon the specific weights of the two minerals and upon the specific weight of the medium (water or air). Quantitatively the following relationship between the diameter D_a and D_b of mineral grains with the specific weights s_a and s_b holds:

$$\text{in air:} \quad D_a:D_b = s_b:s_a,$$
$$\text{in water:} \quad D_a:D_b = (s_b - 1):(s_a - 1).$$

Measurements of the size distribution of quartz and heavy mineral grains in dune and river sands have shown that the statistical mean of the size relationships derived from the theory do in fact obtain.[8]

2. A theory and an explanatory model must be consistent both internally (within themselves) and externally (they must not contradict any recognised theories of natural science).

3. Finally, observable facts must exist by which the theory, hypothesis, and explanatory model can empirically be tested.

In the following section we will discuss the conditions of internal and external consistency of theories, the refinement of theories through phenomenological and mechanistic models, the basic features of conceiving and developing theories, and the most important functions which theories can fulfill.

6.7.4 *Internal consistency of theories*

The problem of the internal consistency of theories has a semantic and a syntactic aspect.[9] In the *semantic* sense a theory is consistent if the same scientific language is used throughout its various branches and partial hypotheses. Categorical differences must be eliminated. The evolutionary theory based on causal laws, for instance, has no place for teleological concepts and these must be interpreted or translated as indicated on p. 184. Semantic problems crop up when higher level theories unite disciplines which have hitherto evolved independently, each developing its own terms and terminologies. Thus the theory of plate tectonics has brought together geophysics and geology in the narrower sense. This has necessitated mediating between the language in which geophysicists describe rock units as physical entities, and the language in which geologists customarily describe rocks according to their genetic, morphological, or mineralogical–petrographic criteria. Semantic problems also arise through the increasing use of mathematics, physics, and chemistry to construct the explanatory model. A mathematical model may be constructed, for instance, to describe the propagation of compressive shock waves and the following decompressional waves during the impact of an object into an abstract material described by physical parameters. To apply this model to an actual impact structure, however, is difficult, since the physical concepts of the model must be translated to apply to the actual features of the given rock substrate. Finally, semantic problems still linger from the fact that until quite recently the geoscientific disciplines evolved relatively independently in different countries, at least as far as the terminologies of petrography, stratigraphy, and other disciplines are concerned. Geoscientific research is now an international undertaking and the terminologies of different nations have to a large extent become uniform. We have referred elsewhere (p. 46) to those rudiments of earlier national terminologies which still survive.

Semantic consistency is prerequisite for testing *syntactic* consistency. This is present when the partial hypotheses are not a random agglomeration, but are related within a hierarchical structure. As we have seen, the most general basic hypotheses stand at the top, and cannot be justified within the framework of the given theory. From these derive the partial hypotheses of decreasing generality, the hypotheses of the higher levels explaining those of the lower levels. The higher their level in the deductive order, the more empirical facts they will embrace. At the lowest level, finally, are the empirical generalisations and the single empirical propositions.

This characterisation is subject to certain qualifications. Were the syntactic consistency of geoscientific theories actually as strict as this, there would be no difference between them and the axiomatised theories of formal logic, mathematics, and mathematical physics. In fact, however, there are significant differences, firstly with respect to logic and mathematics, and secondly with respect to the physical sciences.[10] Fully axiomatised theories are based on axioms (unproven and unprovable assumptions) and contain those and only those statements as so-called theorems which can be deduced from axioms. The possible contents of such a theory, that is, all known and yet to be discovered theorems are contained in the axioms and the rules of deduction. Mathematics and logic, as formal sciences which do not answer to reality, build their own constructs. Theories in mathematics and in logic are based on arbitrarily established axioms: a mathematical system can be constructed just as well on the axiom that parallel lines cross, as on the axiom that they do not cross; formal logic can be built up just as well on the axiom that there are only two truth values (true and false), as on the axiom that there are more than two truth values.

The natural sciences by contrast discover real structures, and present these in the models of natural scientific theories. True, the basic hypotheses of natural scientific theories are as little subject to direct proof as the axioms of mathematics and logic, but, even allowing considerable leeway for convention,[11] they cannot be created arbitrarily. The basic hypotheses of natural scientific theories must be constructed in such a way that it is possible to derive from them the partial hypotheses and generalisations supported by observations and ultimately the broad spectrum of singular facts.

Within the natural sciences, the theories of physics can be axiomatised to a particularly high degree. Special hypotheses and singular statements can be derived and the entire deductive system can be presented in mathematical form from the basic hypotheses alone without reference to limiting background conditions at lower levels.[12] The theories of geoscience on the other hand cannot be fully axiomatised: although the statements of geoscientific theories are founded on basic hypotheses, they cannot be derived *in toto* from them. Partial hypotheses are indeed consequent upon basic hypotheses, but contain certain additional empirical data which do not simply follow from the basic hypotheses. We will illustrate this with two examples.

When a basic hypothesis of plate tectonics asserts that the terrestrial lithosphere is divided into plates which move about upon the convection streams of the mantle within the Earth, we may derive from this assumption various statements of the partial hypotheses, but not their entire content. The existence of constructive, destructive, and conservative plate boundaries could be deduced, but not the number of independent plates or where in fact their various types of margins are to be found. Here empirical data enter the scene, data which cannot yet be derived from the basic hypothesis, and which will never completely disappear during future development and refinement of the theory. This empirical, non-deducible content of the partial hypotheses determines the open nature of the theory. Research may refine current knowledge or produce new discoveries on the number, margination, type, and direction of movement of the lithospheric plates without toppling the overall structure of the theory. As we descend from partial hypotheses to lower hierarchical levels the process is repeated: the next lowest partial hypothesis contains not only the logical consequences of the higher level but additional empirical data as well.

The theory of gravitational differential crystallisation is founded, in part, on the basic hypothesis that magmas will differentiate out according to the mechanism of fractional crystallisation. This sets up a model according to which each magma, according to its chemical composition, will develop as laid down by the laws of the crystallisation sequence. These laws themselves, however, are not contained within the model. The theory is open in this respect, the actual laws involved being external to it, derived either from existing knowledge of physical chemistry or obtained and constantly improved by focussed experimentation.

These examples show that the structure of a geoscientific theory rests on two differing foundations: *basic hypotheses*, and *subsidiary premises*. The latter contain both empirical facts which enter at all hierarchical levels and so-called *lemmata*, laws and hypotheses derived from other contexts. *Lemmata* in geoscientific theories are drawn largely from the discoveries of physics, chemistry, and biology.[13] Due to these subsidiary premises which are not contained in the basic hypotheses but which enter at the various levels of the theoretical system, geoscientific theories at no stage of their development form a closed deductive structure. Some theories, particularly those based on physically derived models may contain an axiomatised or axiomatisable core. This is true of the impact theory. The general features of energy transmission and distribution during the collision of a projectile, the formation and propagation of shock waves, material flow, etc., can all be presented mathematically and axiomatised. To apply the theory to actual events, however, this axiomatised core must be supplemented by empirical data and discoveries relating to the characteristics and behaviour of the rocks of the planetary surface, the impacting body, etc. Even geoscientific theories with an axiomatised core, therefore, contain subsidiary premises which cannot be eliminated, and they therefore have an open character which does not occur in this manner in fully axiomatisable theories.

6.7.5 *External consistency of theories*

A theory must not only be a uniform and non-contradictory structure in itself (semantically and syntactically consistent), it must above and beyond this also be externally consistent. Every theory should bear a meaningful relationship to other known theories pertaining to the same realm of reality. This means that statements of theory T_1 should not imply contraditions to statements of another theory T_2, T_2 being a theory generally accepted at the time. It also means that some sort of mediation between T_1 and T_2 should be possible, such that assertions from T_1 can at least be understood in terms of T_2; at best statements of T_1 can be derived from T_2 and vice versa.

The question of the external consistency of a geoscientific theory has both broader and narrower contexts. Specific external consistency relates to theories and hypotheses recognised at the time within Earth science, while general external consistency relates to theories of other natural sciences. External consistency of geoscientific theories postulates a homogeneous and non-contradictory geoscience, even more, a uniform natural science. This postulate is never completely fulfilled by theories; it is and must repeatedly be broken by newly-conceived theories. We will deal in later chapters with the role of external consistency in testing theories and with the advancement of geoscientific inquiry. Here we wish simply to note that both specific and general external consistency are justifiably required of any theory. External inconsistency in a theory is a serious flaw which researchers must mend either by changing the theory or by making changes among the theories surrounding it.

The need for *general external consistency* with the laws of nature of physics and chemistry is particularly and indispensably important for all the Earth sciences. Every geoscientific hypothesis and theory must harmonise with the theories and laws of physics and chemistry. Any discrepancies which occur are usually due to geoscientific theory, although the reverse can also occur. Thus the hypothesis suggested by the geologist Carey and the physicist Jordan,[14] that geological discoveries supported a model of Earth evolution according to which the Earth's sphere has expanded constantly during the Earth's past, contradicts the recognised laws of physics. To get around this inconsistency the authors suggested the radically new hypothesis that the gravitational constant is not in truth a constant, but has steadily decreased through the geologic past.

There is general agreement, however, that all laws, hypotheses, and theories of Earth science must be consistent with the theories and laws of physics and chemistry, and the development of the geoscientific disciplines since the last century has largely been marked by efforts to bring about this external consistency. This has gradually brought about a fruitful rapprochement between geology, which had previously been something of an outsider in scientific discourse, and the exact sciences. The very interesting story in this context of the conflict between the physicist Kelvin and the geologists who believed in constant, very slow cooling of the Earth's body, will be discussed in a later chapter (p. 283).

Theories, hypotheses, and laws can be called 'geological' only because they relate to 'geologic' objects, not because they operate in a vacuum wherein matter can behave without regard to chemical and physical laws. Geologic objects are components of the material world, and are therefore subject to chemical and physical laws. In contrast to objects dealt with in the laboratory, however, they are more complicated and 'impure' in construction and composition, they frequently take up larger volumes of space, and are often subject to processes which unfold in time periods which are experimentally not controllable: usually these are very long time periods, but they can also be extremely short. When for instance, 'geological' laws, hypotheses, or theories are proposed on mountain building processes, such as the model of nappe structures developed in the Swiss Alps and applied to many other mountain ranges, such a theory may be called 'geological' because it deals with the very slow movements of enormous units of the Earth's crust, made up of differing rock types. Based on observational statements obtained by research in various different mountain ranges, regularities have been found for the formation of huge recumbent folds moved hundreds of kilometres by lateral pressures and compressed into masses several kilometres thick. The model for such movements are 'geological' or 'tectonic' and not 'physical' so long as the complex structures of the rock masses moved, the slow movements, etc. *still* make it impossible to describe the model with sufficient precision using physically defined parameters. Nevertheless, the postulate of external consistency with physical theories still applies to such 'tectonic' or 'geological' theories: the theory may not contain any statement contradicting physical laws. The aim of any further development of the theory must be to develop a model which will correspond not only in qualitative, but also in quantitative detail as closely as possible with the laws of plastic deformation of solid matter.

Specific external consistency is present when a theory harmonises with all other currently recognised geoscientific theories, hypotheses, and hypothetical states of affairs. Since geoscience is constantly in a state of progress complete harmony can never exist between all theoretical and hypothetical conceptions developed in all the various fields and disciplines. The attempt to reach such a state and to eliminate contradictions between theories, hypotheses, and hypothetical states of affairs is nevertheless an important regulative factor contributing to the advance of geoscientific inquiry. A new theory proves itself when earlier hypothetical explanations which apparently contradicted it can be corrected in its light. It gains in explanatory power when it can be modified or expanded in such a way that contradictions disappear.

In the geoscientific literature of the last years, a considerable part has been played by discussions of the inconsistencies which exist or seem to exist between the newer theory of plate tectonics and older or independently developed theories, hypotheses, and geohistoric retrodictions. It has been pointed out that the model developed since the 1920s of periodic worldwide upheavals of long-sunken crustal portions (the theory of orogenic cycles or inundations), a model which has often proven useful in explaining orogenies, appears to contradict the model of plate tectonics.[15] Another author believes that the occurrence of marine halites, coal, desert, and glacial formations developed on the present continents since the Devonian can be used to derive a distribution of paleoclimatic zones upon the Earth contradicting the lateral movements of the continental plates proposed by the newer theory.[16]

6.7.6 *Empirical and theoretical hypotheses as building blocks of theories*

A striking difference between many of the Earth sciences and the physical and chemical sciences is that the hypotheses incorporated into geoscientific theories operate more frequently with empirical concepts relating directly to observable entities, while hypotheses in physics and chemistry deal to a greater extent with theoretical concepts such as energy, acceleration, entropy, or chemical valence, all of which relate to entities which cannot directly be observed.

To take account of this peculiarity of Earth science, we distinguish between empirical and theoretical types of hypotheses. We refer to these as types, rather than as mutually exclusive sets of hypotheses because it is obvious that there can be no sharp boundary between the two, as it is impossible to draw a clear division between empirical and theoretical concepts.[17] Hypotheses used as building blocks of geoscientific theories can therefore only be assigned to one or the other of these groups according to whether empirical or theoretical concepts predominate in them.

Empirical hypotheses primarily include inductive generalisations so long as these do no more than express relationships between observable quantities and qualities. Other laws can also be classed as empirical hypotheses, such as the laws of determinate properties which relate to observable qualities, and causal and developmental laws which set observed events and changes in relationship to one another. Empirical hypotheses also include codictive, retrodictive, and predictive hypotheses which describe past, hidden and future phenomena and processes in the same terms used to describe currently-existing objects.

Theoretical hypotheses operate largely with theoretical concepts describing conceptual entities which cannot be observed. Such concepts are usually physical or chemical. Therefore, all laws, generalisations, codictive, retrodictive, and predictive hypotheses may be considered as theoretical hypotheses in so far as they involve physical and chemical concepts which may be derived from observed phenomena, but are not themselves directly observable.

The distinction between the two types of hypothesis is particularly useful for understanding the development of hypotheses and theories. Hypotheses are created in order to go beyond the simple description of observations by finding explanations, recognising relationships, discovering the hidden, and looking into the future or the past. Hypotheses fulfill these functions better the more effectively their theoretical concepts can elucidate observable phenomena by going beyond single empirical instances. For this reason it is characteristic that as scientific inquiry progresses, empirical hypotheses become increasingly shot through with theoretical concepts. The closer a hypothesis approaches to the theoretical type, the greater its precision, the stronger its explanatory power, and the wider its domain of application. Only through theoretical concepts can relationships be established between at first unrelated hypotheses, linking them together into comprehensive theories.

We will illustrate this with the development of the hypothesis explaining metamorphism in sedimentary rocks, metamorphism being the alterations undergone by rocks deposited at the Earth's surface after they have been buried at depths where they are subjected to higher pressures and temperatures. To begin with there were empirical hypotheses: in mountains such as the Alps where metamorphic rocks crop out, it was possible to distinguish different 'metamorphic zones': the epizone, mesozone, and catazone, each characterised by certain combinations of minerals. These zones were interpreted as a comparative series of rising temperatures and pressures in which alteration reactions had taken place. It was not possible at that time, however, to provide quantitative values for the corresponding pressure and temperature fields. Further development of the theory, which still continues today, was brought about by the introduction of the basic theoretical hypothesis that mineral assemblages of metamorphic rocks in most cases, i.e. at a first approximation, represent equilibrium states in the thermodynamic sense. On this basis petrologists succeeded in explaining the observed phenomena with increasing precision in the light of the concepts, laws, and models of thermodynamics. This meant that theoretical concepts were increasingly incorporated into a hypothesis which had initially been predominantly empirical. One important consequence resulting from the basic hypothesis was the application of a universal law, the so-called phase rule, pertaining to all systems in a state of thermodynamic equilibrium. This law permits certain conclusions as to the number of minerals which may form out of a material of given chemical composition under varying conditions of pressure and temperature. For every rock of a given chemical composition, the two-dimensional pressure/temperature continuum is divided into fields in which the number of coexisting minerals equals the number of independent chemical components of which the material consists. The fields are bounded by lines representing those pressures and temperatures at which the number of coexisting phases equals one more than the number of chemical components; any three such lines join at the point representing the unvarying pressure/temperature conditions at which the number of coexisting mineral types is two greater than the number of chemical components. This provides a universal formal framework for all metamorphic rocks. It can be rendered more concrete by experimentally determining the pressure and temperature values of the field boundaries for certain chemical compounds, and the types of minerals which exist within the fields. These experimental data can be used to derive the pressure/temperature values at which naturally-occurring mineral assemblages of metamorphic rocks were formed.

Another, even more theoretically-dominated development in petrological hypotheses of metamorphism was brought about by efforts to calculate the pressure/temperature conditions of certain mineral reactions according to thermodynamic principles, based on the caloric properties of the substances involved. We have already discussed these methods above (p. 115).

The development of petrological hypotheses sketched here, from their initial empirical stages to more theoretical forms, has had the consequences described above: the hypotheses became more precise, containing not only qualitative and comparative predicates, but quantitative predicates as well. Their domain of application has increased through the establishment of laws, leading to a growth in their explanatory powers as well. Based on theoretical hypotheses, a theory of rock metamorphism has been created on physical and chemical foundations.

Similar developments can be described in many other areas of geoscientific research, as in the development of the oft-cited theories of plate tectonics or of impact processes. These examples show the significance of the frequency of empirical hypotheses in geoscience and of their development into theoretical hypotheses. There is a tendency to incorporate more and more theoretical concepts into hypotheses which at first were wholly empirical, causing them increasingly to resemble the theoretical type of hypothesis. This reduces their original singularity and permits several hypotheses to be linked into theoretical systems. The significance of empirical hypotheses in the Earth sciences, as we have stressed above, must at the same time not be overlooked. Whenever empirical hypotheses are set up in retrodictive, codictive, or predictive reconstructions of non-observable events, an indispensable and never-ending task of geoscience has been performed. Presently available phenomena, objects, and processes which are directly accessible to observation are augmented by the world of past, hidden, and future phenomena, objects, and processes. All of these can be described in empirical terms, and only together can they form the basis for the creation of theories. For this reason the frequency of empirical hypotheses is not a sign of any 'immaturity' of the Earth sciences compared to physics, but is a necessary consequence of the fact that the object of geoscientific research is not simply that which is presently observable.

6.7.7 *Phenomenological and mechanistic models as building blocks of theories*

Every theory is intended to explain empirical circumstances by reducing the multiplicity of phenomena to a few schemes or models. Two types of models can be distinguished according to their claim to help establish new states of affairs or relationships 'underlying' or 'producing' directly observable phenomena: these two types are the *phenomenological* and the *mechanistic* models.[18] The classic example for the representation of the same set of objects using both models is provided by the thermodynamic and kinetic theories of gases.

The thermodynamic theory is phenomenologically based: it systematises and explains the behaviour of gases using concepts derived from observable phenomena such as pressure, temperature, volume, density, internal energy, specific heat, enthalpy, and entropy. The model created by the thermodynamic theory is an abstract construct, a system of functions of quantified phenomenological concepts or so-called state variables representing each gas and the conditions to which it is subjected.

The kinetic theory of gases is mechanistically founded. It reaches 'deeper', so to speak, and attributes the observable phenomena to inner mechanisms on a micro-scale, namely to the speeds and masses of atomic or molecular particles and to the exchange of kinetic energy between them through their collision and through the effects of these collisions on the walls surrounding the gas. The phenomenological property of temperature, for instance, is explained as the kinetic energy of the particles, and the phenomenological pressure upon the walls is attributed to the collisions of the particles. The mechanistic model corresponding to this theory is a graphic construct representing a mass of moving particles subjected to statistical laws of mechanics.

Phenomenological and mechanistic models occur in the Earth sciences as well, nor do the latter by any means simply represent the mechanisms of physical mechanics in its narrower sense. Phenomenologically based models are provided by, for instance, hypotheses explaining the stability or alteration of mineral substances after changes in external conditions in terms of thermodynamic concepts such as free enthalpy or conversional heat without recourse to the atomic mechanisms which actually create the alterations in the crystal lattices. Another phenomenological hypothesis is the theoretical explanation of the transportation of suspended particles in a turbulent flow of water. This relates the quantity of particles carried in the cross section of the current per unit time to other phenomenological quantities such as the mean velocity of flow or the turbulence or speed at which the particles sink in quiet water, without regard to the movement of the individual particles and the eddying structure of the water in which they are carried.

Various laws may function as phenomenological components of theories, such as geochemical laws which, as developmental laws, express the behaviour of chemical elements during the formation and disintegration of rocks; regularities of geotectonic development of parts of the Earth's crust; and some paleontological laws on the phylogenetic evolution of life forms.

A mechanistic model is involved in the theory of crystallography, which attributes the optical, elastic, thermal, electrical, and other physical properties of crystals to the arrangement, size, and electronic shells of the atoms in the crystal lattice, as well as to their inter-atomic interactions. Other models which are mechanistic explain the diagenetic hardening and compaction of sediments through pressure solution at the grain contacts, or attribute the alterations during rock metamorphisms to models of ionic diffusion along the grain boundaries or through pore solutions. Darwin's theory of natural selection has a mechanistic basis, since the phenomena of species evolution are explained through the mechanism of selection of particularly well-adapted individuals.

Phenomenologically and mechanistically founded theories each have their specific advantages and disadvantages. The advantage of phenomenologically-founded theories is that they are less 'hypothetical' than mechanistic ones, which are clearly more problematic. The latter, however, have the advantage that they are simpler and may be used to explain other phenomena, becoming the basis for constructing more comprehensive theories. Where certain phenomena or processes must be explained, there is often a choice of whether to try this by phenomenological or by mechanistic means, so that it is important to weigh up their advantages and disadvantages before making the decision.

The historical development of knowledge in a given domain usually runs from an early phenomenological stage, towards a later mechanistic stage, much as empirical hypotheses are replaced by theoretical ones in the progress of inquiry. The theory of rock metamorphism seems to be developing in this direction at present. The phenomenologically founded hypotheses and theories based on the principle of thermodynamic equilibrium are increasingly being interpreted kinetically, e.g. through mechanistic models relating to atomic processes in which equilibrium is approached. In other cases the sequence may be reversed, and a mechanistic model may serve as a sketchy outline from which to conceive a phenomenological theory. V. M. Goldschmidt found his phenomenological laws on how chemical elements became distributed in rocks and minerals during the various processes of their formation and disintegration by working on the basis of a mechanistic model in which the uptake of atoms and ions in the crystal lattices of the minerals depended on the radius and electrical charge of the atoms and ions, which he conceived of as spheres.

6.7.8 *The conception and development of theories*

A new geoscientific theory usually begins with a concrete problem. Its primary aim is generally to provide a better and more complete explanation for empirical findings, a greater or lesser number of singular phenomena which do not seem to fit well into the existing explanatory schemes. The improved explanation develops by conceiving a vague idea or 'founder hypothesis',[19] in a creative act which defies epistemological analysis. This creation does not take place in a vacuum, however, but in a pre-theoretic situation controlled by objective and subjective factors.[20]

The *objectively valid factors* include such things as the kind and quantity of pre-theoretically known empirical facts, the state of techniques and instruments available for observations in the laboratory and in nature, the discoveries and accepted theories in Earth science and in other natural sciences, and industrial demands, (e.g. a correct theory of petroleum formation is needed to develop rational methods of finding and exploiting oil deposits).

At any one given time the objectively valid factors of the pre-theoretic situation are more or less the same for all researchers. This is why theories sometimes seem to be 'in the air', and are conceived independently by different researchers 'when the time is ripe'. An example from the more recent geological history is the hypothesis proposed independently in 1962 by the American researchers Hess and Dietz of sea-floor spreading, from which the theory of plate tectonics was born (see pp. 246, 312).

The role of purely *subjectively valid factors* in the pre-theoretic field should not be underestimated.[21] They include particular knowledge or lack thereof in special geoscientific disciplines and in other natural sciences, experience in locations have a certain geological–petrographic make-up, special knowledge of certain geological processes, technical experience, obligations towards certain authorities or schools of thought, and ties to certain regulative principles (p. 314) or even to certain dogmas. Some subjectively valid factors defy rational analysis. Others become more objectifiable in the course of inquiry, become amenable to empirical testing, and thus become objectively valid factors. Because of the special role of abductive inference in forming geoscientific theories, subjective elements of many kinds play a larger part in Earth science than in other natural sciences. It is due to subjective conditions that researchers with differing personal backgrounds often explain the same problem situation by setting up different theories.

The outline of a theory is created when, out of the matrix of objective factors and subjective conditions, a possible hypothesis to explain as yet unexplained phenomena begins to emerge, providing a 'pattern'[22] according to which the observational data may be sifted and organised. This pattern is as a rule pre-determined by accepted theories, for these are important components of the pre-theoretic situation. To this extent it may be said that 'theories are generated by theories.'[23] At any rate, even in the pre-theoretical stage of a theory its foundations are not *bruta facta* or arbitrary data represented by singular assertions; instead the pattern allows certain data relevant to the intended explanations to be selected. They are viewed not as what they at first seemed, but as elements of a characteristic context. The model to be conceived on the basis of these selected and pre-arranged facts should achieve the following:

First, the model should do more than explain the singular configuration which triggered the formulation of the theory. After all, the model should relate not only to a single object, or even to a few objects, but to an indefinite number of analogous objects. The specific object which gave rise to the theory thus loses its unique character and becomes an example, an element of the class of all objects that can be explained by the model.

Second, the known facts take on new meaning in the context of the theory. They are no longer simply facts, but are meaningful, and those important to the consolidation and refinement of the theory can be selected from them. This brings order and form into a previously unbounded and amorphous mass of random observational data.

Third, the theory and model explain relationships between observed facts whose conjunction or concurrence previously seemed inexplicable or purely coincidental.

Fourth, the theory throws light on previously ignored or as yet undiscovered states of affairs. This increases the quantity and type of empirical findings beyond what was already pre-theoretically known. In this way a new theory creates its own base on which it stands. This is why it is impossible to arrive at a theory purely by inductively generalising from pre-theoretically known data.

Fifth, a theory will often integrate pre-theoretically existing hypotheses into a higher level. This may sometimes occur in the initial stages of conceiving a theory, but will at any rate certainly occur later in its development. By this means hypotheses lend each other mutual support; their incorporation into a theoretical system makes them more precise, and they achieve a wider range of application.

While theories may often seem to burst into existence suddenly, revolutionising the methods and goals of research, no theory ever actually sprang to life full-blown like Athena from the head of Zeus. The first prototype of a theory coalesces gradually out of a pre-theoretical matrix and subsequently undergoes many changes. The very striking stage of a theory's first conception is followed by a long period of gradual refinement through laborious research. This process of refinement and completion of a theory can never end. For one thing, as a theory is applied, new data continually arise which require additions or improvements to partial hypotheses and which add scope to the model's details. For a second thing, as more singular cases differing in detail become incorporated into the model, it gains universality. The basic hypotheses become more firmly grounded in universal laws of nature, and the deductive structure, ranging from general and special hypotheses (including defined subsidiary additions) to empirical facts, becomes structurally more compact. This requires, thirdly, that the theory becomes increasingly mathematical in expression, and that the parameters characterising the model are quantified. Fourthly, changes in the theories of neighbouring disciplines and sciences, the proliferation of physical and chemical knowledge, and improvements in phenomenological and experimental methods of observation must be dealt with, so that the theory's empirical basis and the evaluation of its external consistency are constantly subject to change. All this means that no theory accepted in Earth science can ever be regarded as 'finished' or closed. Even the apparently most secure theories should not be examined in terms of their static structure of the moment, but must be regarded as dynamic constructs which must continue to prove themselves as they develop. We will come back later to this aspect of continued corroboration of developing theories (p. 276).

6.7.9 *The function of theories*

In our description of geoscientific inquiry, theories form the highest rank of a hierarchy, apart from principles which as such are not the object of actual research but form a background or 'world view'. Seen from this perspective, the construction of theories using material from the empirical basis, inductive generalisations, and hypotheses, forms the ultimate aim of research, and the theories themselves would seem to be of no further use. Such a picture is inadequate to a complete understanding of the role of theories in Earth science. Though they are the ultimate result of research and synthetic efforts, theories are not merely the results of research, but also the instruments of the growth of knowledge, for we cannot observe, draw conclusions beyond the observed, find hypotheses, or develop further theories without already possessing some theories. The instrumental character of theories can be described by listing the functions which theories fulfill, both within science and outside the realm of pure research.

It is useful to organise the functions of theories according to three aspects. Statically, theories represent the given state of geoscientific knowledge. Dynamically they are indispensable for the progress of inquiry. Practically, they serve to apply geoscientific knowledge to non-scientific purposes.

1. *Statically*, theories serve at any given moment as the actual bearers of geoscientific knowledge. They contain in concentrated and conceptual form the information derived from many observations. Recognised theories form a thesaurus of the knowledge available to tackle problems, whether these be questions internal, or applications external to science. As long as observational results remain simply verbal lists of descriptions stored in data tables or computer banks, they are dead capital which can only bear interest when it has been theoretically reworked, i.e. incorporated into the ordered structure of a theory.

But theories do more than just sum up masses of empirical data, systematising them and making them comprehensible. What is more fundamental to the growth of knowledge is the *reduction in theories*:[24] the number of separate hypotheses and independent theories is reduced when these are integrated into a system of more general and comprehensive theories as partial hypotheses or partial theories. The most convincing example of such an integrative force is provided in recent geoscience by the theory of plate tectonics, whose success and rapid recognition in the scientific community is due not least to the reduction in theories which it has already brought about and which it promises to continue doing in the future.

2. In their *dynamic* aspect, theories function as bearers and motors for the proliferation of knowledge, and serve the growth of scientific knowledge in a critical sense as well.

By 'the proliferation of knowledge' we mean the exhausting of possibilities provided by the laws belonging to a theory for abductively and deductively expanding directly experiencible reality. We have already discussed elsewhere (p. 249) how this uncovering of the past, hidden, and future states of affairs significantly enriches the universe of research in relation to the specific goals of Earth science. This can only take place with the aid of theories.

Theories serve the progress of inquiry in a critical function since new theories and hypotheses can only be tested with the help of theories and can only be created against the background of already accepted and proven theories. We will come back to this important role of theories in more detail in the following chapters on the testing of hypotheses and theories and on the growth of geoscientific knowledge.

3. The significance of geoscientific theories for *practical application* has grown rapidly since the beginning of the nineteenth century. It will continue to grow, for all useful knowledge of the Earth's material composition is contained in theories, and the intensive and extensive increase in technological civilisation concurrent with the rising world population mean that the demands upon Earth's material capacities will become a problem upon whose solution the future of mankind is becoming ever more dependent. Geoscientific theories form an indispensable basis on which to critically evaluate the potential energy and material available on Earth to the coming generations, and for an evaluation of the conditions in endangered ecosystems which after all are governed not only by the behaviour and food chains of living things, but also by inorganic components and geological and geochemical processes. Prognoses which are based on geoscientific theories and which are factually and hypothetically justified will be needed in order to frame our conduct with nature so as to fulfill our obligations to future generations.

6.8 Testing hypotheses and theories

6.8.1 *Introduction*

The development of geoscientific knowledge is manifested in all disciplines through a progression of sequentially replaced theories. After being accepted for a time, theories are replaced by improved or completely new ones. Theories are in principle always exposed to critical discourse, carried out in the scientific community through *pro* and *contra* arguments contributed by the individual researchers.[25] This discourse, which may lead to tentative acceptance or rejection of a theory is based on *testing* the theory with regard to its expediency for the intended explanation. A theory is *accepted* if it fulfills certain requirements and has withstood certain tests in its application. A theory is *rejected* when it does not fulfill requirements and expectations, and breaks down in the course of application.

The actual process of critically testing a geoscientific theory is complex and proceeds on various levels; it is not easily analysed in individual cases. We will attempt to break it down into its basic components, taking into account the special problems of the Earth sciences and using the achievements of recent philosophy of science.

The requirements which a theory, as a deductive system of hypotheses, must fulfill, are implied by the function of hypotheses and theories and the structure of theories and their mutual interactions as discussed in previous chapters. As we mentioned there, a theory relates directly to a model and indirectly to reality (see p. 249). This suggests two issues concerning the testing of a theory: *non-empirical tests* of a theory or model examine its internal and external consistency, while *empirical tests* relate to the theory's indirect referents, that is, to the theoretical model's expediency in explaining reality.

In the course of research, criticism of a theory usually starts with empirical tests, which we will therefore examine separately first. Empirical testing of a theory uses phenomenological and experimental observation to check the truth of statements derived from the theory. Since observation can only determine the existence or non-existence of singular states of affairs, it is impossible empirically to test a theory as a whole.[26] Only individual hypotheses belonging to the theory can be checked by observation. We will therefore begin by looking at the empirical testing of individual hypotheses, after which we will investigate what the results of this testing mean for the critical evaluation of the higher-level theory.

6.8.2 *Empirical tests of hypotheses*

To discuss the empirical testing of hypotheses, we must start by distinguishing between *logical* and *pragmatic* conditions of empirical testability. Logical conditions concern the logico-syntactic structure of the statement to be tested, pragmatic conditions are the practical processes which can be applied to the statement to determine its validity. The consequences of these two forms of testing are not necessarily the same. A universal statement, for instance, which on logical grounds can never definitively be proven true[27] can, in pragmatic terms, certainly be accepted as 'true' due to experimentally corroborative evidence. It is therefore reasonable always to distinguish between the logical and the pragmatic validity of a hypothetically introduced statement which is being tested. If a statement can be proved true on logico-syntactic grounds, we refer to it as *verifiable*; if it can be proved false on logico-syntactic grounds, we refer to it as *falsifiable*. If a statement is acceptable as true on pragmatic grounds, we call it *confirmable*, if rejectable as false on pragmatic grounds we call it *refutable*. While logical criteria are strict and unambiguous, pragmatic criteria are more flexible, and depend upon evaluations of the given situation and of the general state of research.

The pragmatic conditions for empirically testing hypothetical statements may be divided into two classes: *directly testable* statements relate to the existence or non-existence of observable facts; *indirectly testable* statements relate to states of affairs which cannot be observed as they belong to the past or to inaccessible areas. Here the objects of empirical testing are, strictly speaking, not the statements themselves, but rather certain statements associated with them representing the observable consequences of the hypothetically postulated state of affairs. The boundary between direct and indirect empirical testability is not always clear-cut. For one thing observational techniques range from simple *demonstratio ad oculos* to very complicated and indirect procedures, while on the other hand an evaluation of the theory's testability depends on how the domain for which a theory is to hold true is delimited. Laws can be regarded as 'directly' testable if it is only their validity here and now which is concerned. Generally, however, their domain is extended beyond the here and now to past and future states of affairs, and in this sense they are only 'indirectly' testable.

Singular hypotheses represent a clear case of direct testability: these relate to states of affairs which can be observed either now or at a specific timepoint t_0 in the future, that is, they relate to all singular statements which follow deductively from contemporaneous or anticipatory predictions or codictions. A particularly important group of these consist of 'short term' prognoses, relating to states of affairs which are directly impending or which will shortly be effected by certain manipulations, e.g. the results of experiments. 'Long-term' prognoses, e.g. on climatic evolution or on geomorphological changes cannot be tested directly, that is, within a relatively short time period. As mentioned above, observability can only be defined relative to the state of available observational techniques.[28] Differences in direct testability at a given time-point t_0 are particularly evident in codictions. A singular hypothesis on the existence of an underground oil deposit can be tested directly by deep drilling. A hypothesis on the existence of a sharp boundary between silica-rich crustal material and silica-poor mantle material at a certain great depth, however, cannot at present be tested directly for technological reasons. This situation may, however, change at some time in the future.

Directly testable singular statements logically presuppose existential statements of the following form:

There is (at least) one object of type X.

By virtue of their logical form, existential statements can only be verified, never falsified. The discovery of a single object of type X verifies the statement, but it can never empirically be proved that there is *no* object of type X in the world.

In the pragmatic context of research we prefer not to speak of verification and falsification even in regard to tests of singular statements, but more circumspectly of empirical confirmation and refutation. All singular, directly testable assertions are confirmable, although under certain circumstances a state of affairs at first accepted as a confirming instance may, after more specific testing be interpreted differently.

Some researchers, for instance, formerly maintained that particular structures in certain meteorites (carbon-containing chondrites) represented the remains of organisms, that is that they confirmed the hypothesis that meteorites containing organic compounds (carbohydrates) contained structural remains of extra-terrestrial life forms. The structures in question were later recognised to be inorganic, an interpretation generally accepted today.

As far as their refutability goes two types of singular hypothetical propositions may be distinguished: refutable, spatially definite statements, and non-refutable, spatially indefinite ones.

Spatially definite singular propositions maintain that certain currently-observable phenomena exist at a given place. Hypotheses of this kind are particularly common in the Earth sciences, in contrast to physics. Like predictions on the result of experiments, they can in principle be either confirmed or refuted. The extent to which they may be considered empirically tested is often limited, however, since few researchers may in fact be in a position to carry out the test *in situ*, whereas an experimental test may be carried out at any time in any suitably equipped laboratory, independent of site.

When, for instance, an author[29] refutes a given hypothesis on regional metamorphism of rocks in an area of East India by stating that 'the dominant aluminosilicates in the rocks of the Sonapet Valley are andalusite and kyanite and not sillimanite', this assertion can only be followed up by somebody who is in a position to visit this remote valley.

Spatially indefinite singular assertions assert the current existence of certain observable phenomena within an area that is either delimited but very large, or (in practical or real terms) unbounded. Such hypotheses can be confirmed by observation but never refuted. Here the case is the inverse to that of laws: proof of one single instance of the asserted phenomenon confirms the hypothesis, while even the most thorough search could not refute it due to the impossibility of searching the entire area exhaustively.

Should the hypothesis relate to a large delimited area which can be broken down into a finite number of completely analysable elementary regions, the spatially indefinite assertion can be broken down into a collection of spatially definite assertions which can collectively be confirmed or refuted.

The theory of the formation of impact craters and the hypothesis that the Ries Basin was formed by an impact which created shockwaves of more than 100 kbar in the quartz-bearing rocks imply the indefinite hypothesis that the high pressure mineral coesite should occur somewhere in the widespread breccias of the Nördlinger Ries. The successful discovery of coesite in various localities of the Ries confirmed this spatially indefinite hypothesis, but had it not been found, as is the case in various other impact craters, this would not have refuted the hypothesis.

Consideration of probability and of the theory of the physical and chemical conditions necessary for the origin and evolution of life imply the indefinite hypothesis that somewhere in the universe there are other planets inhabited by life forms. This would be confirmed as soon as one single such heavenly body were discovered, but it can never be refuted, as it is impossible to investigate the entire universe.

The direct empirical testability of *universal hypotheses* is limited insofar as the totality of states of affairs represented by the hypothesis cannot be checked even when the validity of the hypothesis is restricted to those states of affairs observable here and now. Universal hypotheses have the logical structure of *universal statements*, being statements in the form:

All states of affairs of type (X) have property (a).

Empirical testing cannot verify such a universal statement, since the number of *all* states of affairs of type (X) cannot empirically be exhausted. Such an assertion can be falsified however, for evidence of a single state of affairs of type (X) lacking property (a) would necessitate rejecting the universal statement as false. In practice, however, empirical testing of universal hypotheses is not carried out so rigourously. Let us look at the case of a hypothetically asserted law whose factual validity in the here and now may be regarded as directly provable.

All objects of fluorite have a light refraction index n_D of 1.434.

Due to its logical form as a universal statement this hypothetical law cannot be verified. In practice, however, the impossibility of completely verifying it is logically trivial, and is less important than the fact that the law can be confirmed by measuring the light refraction index of a quantity of objects consisting of fluorite. Empirical testing of this kind of law can therefore lead to confirmation, though it must always remain tentative since the possibility of obtaining contradictory measurements cannot be ruled out.

The concept of falsification also cannot simply be applied in all its logical strictness to the pragmatic context of research. Should someone report the discovery of a fluorite crystal having a light refractive index that does not equal 1.434, this empirical situation does not necessarily imply that the law is false. The measurement requires strict procedures and a certain expenditure on apparatus, and may have been subject to error; the fluorite crystal may have contained trace impurities influencing its light refraction; mechanical influences might have deformed the crystal and affected its optic properties. Additional testing must therefore ensure that disruptive influences can be discounted before accepting an instance contradicting the law. Since there is always a certain amount of room for decision here it is more appropriate in a pragmatic context to speak of a refutation of a hypothetical law rather than of its falsification.

Hypotheses which are empirically indirectly testable play a supreme role in geoscientific research, for as we have frequently stressed it is one of the peculiarities of Earth science that its main goal consists of going beyond the current reality accessible to observation to explore the geologic past and areas which are currently or permanently inaccessible.

Examples of indirectly testable singular hypotheses include all *retrodictions* by which past singular states of affairs are abductively inferred. Indirect empirical testing here involves regarding the abductively inferred assertion as an initial hypothesis from which observable consequences are deduced which can be tested directly.

As we have seen, all singular statements can be confirmed. Logically however, the confirmation of an inferred singular hypothesis by no means implies that the initial hypothesis being tested is true, for true conclusions can be drawn from false premises. The confirmation of an inferred singular hypothesis simply means that the initial hypothesis might be true, not that it is necessarily true. Since indirect testing through deduced singular hypotheses is the only way in which to test retrodictions empirically, however, and since retrodictions are fundamental to the main task of Earth science (the explanation of the Earth's present configuration in historical terms), this kind of indirect empirical testing is very important despite its inherent uncertainty. Even if the confirmation of inferred singular hypotheses cannot certify the validity of retrodictions, practical research is still guided by the basic rule that each new confirmation increases the probability that the retrodiction is true.

It may further be asked if a retrodiction cannot be confirmed with certainty, may it at least be refuted with certainty? Again, retrodictions can only be refuted through deductively derived, singular hypotheses. Only spatially definite singular hypotheses can be empirically refuted, and these must therefore form the basis for refuting a given retrodiction. If it follows deductively from a retrodictively obtained hypothesis that a given state of affairs must exist at a certain place for the hypothesis to be true, then the hypothesis can be refuted if observations show that the given state of affairs does not exist at the expected site. Pragmatically, however, it must be remembered that the correctness of inferring a singular statement from a retrodiction cannot be guaranteed. In the initial stages in developing a hypothesis particularly, the retrodictively postulated states of affairs are still not known in sufficient detail to prevent errors from occurring when deriving testable singular hypotheses. We will illustrate this through two examples.

For an empirical test of the retrodiction that the Ries Basin was formed by impact, the spatially definite assertion may be derived that the glass which was formed by melting of the rock and which is found in many sites in and around the Ries Basin must all be of the same age as the impact event. This was confirmed by measuring the radiometric age of many glasses: all the investigated samples yielded an age of 14.8 million years.

This confirms the impact hypothesis on the formation of the Ries, insofar as the evidence of the age of the glass does not contradict it. It does not, however, rule out that the agreement in ages might be explained in another fashion, even if no alternative hypothesis has yet been suggested.

For an indirect proof of the retrodiction that the Barringer Crater in Arizona was formed by the impact of a large iron meteorite, the definite hypothesis was deduced that a large iron mass, the remains of the meteorite, must lie immediately below the crater floor. Magnetic measurements and cores found no such iron mass. Many geologists took this as a refutation of the retrodiction and concluded that the Barringer Crater was not formed by the impact of an iron meteorite.

Later it was found that the inferred existence of an iron mass in the crater was incorrect. Refined analysis of the physical laws governing impact, i.e. improvements in the theoretical model of the impact theory showed that the temperatures created during an impact are so high that the meteoritic mass evaporates completely at the site of impact. The inference that an iron mass existed in the crater was therefore false, and the negative results of the test, according to current understanding, fully accord with the hypothesis of the impact origin of the Barringer Crater.

The same applies to the indirect testing of *codictions of non-observable states of affairs*. Here the codiction, as an initial hypothesis, can only be tested with the aid of directly testable singular assertions deduced from it. Again, certain confirmation is logically impossible, but refutation is quite possible through spatially definite singular hypotheses.

An example of abductive codiction is the hypothesis based on the existence of a magnetic anomaly, that a large mass of iron ore exists in the inaccessible basement rock. The spatially definite assertion can be derived from this hypothesis that the basement rock in the given region must have increased electrical conductivity, which can be directly tested by setting up electrical voltages in a suitable spatial arrangement. Should such measurements produce a negative result, the initial hypothesis is refuted: where there is no increased conductivity, no iron mass is present and the magnetic field must be explained in another manner. A positive outcome to the test would tentatively confirm the codiction but would not be 'proof' in the strict sense, since the increased conductivity and the magnetic field might have some other explanation.

Finally we also meet in the Earth sciences with *universal hypotheses which can only be tested empirically by indirect means*. Here we are ignoring the problem discussed elsewhere, of the counterfactual validity of 'true' laws, according to which only laws of the inductive generalisation type would be regarded as directly testable hypotheses. We are thinking rather of laws whose postulated validity relates solely to states of affairs of the geologic past inferred abductively through retrodiction. Here again, and for the same reasons as for directly testable laws, only indirectly testable laws can be refuted; in both cases, however, the inherent uncertainties of abductive inference make both forms of refutation argumentatively weak.

The following is an example of an indirectly testable, lawlike statement since it relates to the origin of the given state of affairs in the geologic past. 'All rocks with glassy structures are volcanic'. Due to a large number of confirming findings, each a testable deductive consequence of the initial hypothesis, this postulated law was regarded as valid. It is now refuted, since glassy rocks have been found which in accordance with a generally accepted abductive inference were not formed by solidification from a volcanic melt, but by meteoritic impact (partly through melting and partly in a solid state).

To summarise, it has been found that for logical and pragmatic reasons, the possibilities for empirically determining by test whether hypotheses are true or false are limited. Many hypotheses can be empirically directly tested. Directly testable hypothetical laws can be refuted or at least provisionally confirmed. Directly testable assertions as to the existence of spatially definite singular states of affairs can be confirmed as well as refuted. Where a state of affairs is claimed to exist in a large or unbounded area (spatially indefinite hypothesis), this can be confirmed but not refuted. The hypotheses so important to geology which are not subject to direct testing can be tested by directly testable hypotheses that can be deduced from them. In this manner it is possible at least to refute abductive codictions of unobservable states of affairs, indirectly testable hypothetical laws, and retrodictions, and even to a certain degree (given the proviso that alternative explanations may be possible) to confirm them.

6.8.3 *Testing theories*

Truth as a regulative idea

We have found that most hypothetical assertions, especially laws and retrodictions, can at best be provisionally confirmed, but can be refuted decisively. This has certain consequences for the empirical testability of theories. Since all geoscientific theories contain laws and abductive hypotheses as indispensable components, what is true for the testing of such hypotheses must also be true for testing geoscientific theories. Empirical means can only decide with some certainty that a geoscientific theory is not true in the version which is being tested. This, however, does not determine whether its 'core' is shaken, or whether minor modification of its 'periphery' will save it. In any case, empirical tests cannot be used to form sure conclusions as to the truth of a theory.

We agree here with Popper's criticism of the so-called 'justification model' of scientific kowledge[30] which dominated the early verificationist period of modern philosophy of science. According to this model, scientific research sets up theories which are then collectively supposed to be empirically verifiable. Popper's falsificationist argument maintains that such a model is questionable even on logical grounds. Theories necessarily contain lawlike universal statements with counterfactual import; these can at best be falsified, never verified.

Linguistic usage in the Earth sciences, however, still largely follows the 'positive' model of justification of scientific knowledge questioned by Popper. Geoscientific publications speak as often of 'proofs' *for* a theory, as of 'proofs' *against* a theory. Such wording is incorrect, and at the very least misleading, for strictly speaking there are no 'proofs' for the truth of a theory. Nevertheless, the correctness of the falsificationist criticism of the justification model cannot eliminate the fact that the researchers' motive is to find the 'true' theory. Phenomenological and experimental observations are therefore used in order to find how the Alps were 'actually' formed, or under what physical, chemical, and climatological conditions the pre-Cambrian uranium and gold-bearing conglomerates of South Africa were 'really' deposited. For the Earth scientist, therefore, 'truth' remains the goal of research, even if there is nothing which can give his theoretical results the seal of certainty: in Popper's words: 'We seek the truth, but we do not possess it.'[31]

Recognising falsificationism as an epistemological and scientific principle of criticism does not imply abandoning the ideal of 'truth' as the regulative idea of research. The awareness that a theory's validity cannot definitively be 'proven' does not prevent the researcher from attempting to approach the 'true' theory in an empirically controlled manner.[32] During such an approach it will be necessary to eliminate hypotheses and theories which can be shown on good grounds to be 'false'. It is equally important, however, to find the most comprehensive empirical confirmation possible for those hypotheses and theories regarded as 'true' in order to indicate why they should be accepted. Popper's falsificationism suggests that such 'positive' confirmation-orientated research activity is of little value, and only 'negative' refutation-orientated research should be considered relevant to science. Popper conceives of the proper handling of theories in the form of science envisaged and recommended by him as follows:

> research should not attempt to defend theories; . . . it attempts with all available logical, mathematical and technological–experimental means to refute them.[33]

The oft-deplored alienation of the natural sciences from the more modern philosophy of science may well be due in good part to the fact that probably no science, certainly no Earth science, follows or will follow this rule. Every researcher endeavours to find confirmative evidence for his theory, whether he is looking at Alpine structures in terms of the nappe theory, at lunar craters and basins in light of the impact theory, or at remains of the oldest mammals in view of the theory of evolution. As theoretical inquiry progresses, research success is accomplished not so much by the refutation of a theory, as by the foundation, confirmation, and successful application of a new theory. Were there a Nobel Prize in Earth sciences, it would have been given to persons such as V. M. Goldschmidt for his theoretical foundations of geochemistry, or A. Wegener for his basic conception of the theory of plate tectonics, and not to N.N., who, at least according to contemporary evaluation, conclusively refuted the long-accepted theory of the cooling and shrinking of the Earth.

The history of a geoscientific theory, beginning with its first draft and culminating in its acceptance or rejection by the scientific community is largely equivalent to continuous testing of its basic hypotheses via their deduced consequences leading to confirmation or refutation of the separate hypotheses. The researchers involved do not actually regard this process as a critical test procedure, but rather as a series of attempts to apply the theoretical model on an increasingly broad basis to explain and master real contexts and phenomena. In the course of this 'testing-through-application' the theory becomes corroborated if its component hypotheses are empirically confirmed or if attempts to refute them are met by corrections in observational procedures or by modifications of the theory. On the other hand the absence of empirical confirmation and the successful refutation of particularly relevant partial hypotheses lead to the disconfirmation of a theory. The terms 'corroboration' and 'disconfirmation' are intended to underscore the procedural character and pragmatic orientation of theory testing. Continued corroboration of a theory leads to its *acceptance* by the scientific community, while continued disconfirmation can lead to its *rejection*.

Corroboration of theories

That a theory is corroborated becomes manifest when it fulfills those dynamic functions which a theory must serve as an instrument of the growth of knowledge. These functions were investigated in an earlier chapter (p. 266). A theory is corroborated to the degree to which the theoretical model may successfully be used to explain reality, its success being empirically checked by seeing that the hypotheses derived from applying the theory to reality can empirically be confirmed. This involves singular hypotheses (which alone are strictly speaking confirmable), retrodictions, and laws. During corroboration the *number of real referents to which the model can be applied increases*. Between 1972 and 1979, for instance, the number of terrestrial structures that could be explained by the impact theory rose from 43 to 91. Such proliferation of the referents of a theory implies increasing simplification and generalisation of the basic concepts. Further, with each new application it *gains depth* as new empirical details appear which must be explained. The first sketchy model thus becomes refined along the lines discussed in a previous chapter (p. 264).

In the course of applying a theory more will occur than simply the corroboration of certain hypotheses, for certain expectations will not be fulfilled and some hypotheses initially deduced from the theory will be empirically refuted. The theory will be corroborated to the extent that such contradictions may successfully be eliminated by appropriate *modifications of the theoretical model*, so that the theory increasingly, 'approximates' reality. According to Lakatos,[34] the hypotheses constituting a theory may be differentiated into those that form its 'hard core' and those belonging to its 'protective belt'. A theory is corroborated as long as empirical contradictions are eliminated by corrections in the protective belt, leaving the hard core intact. This corresponds to the exhaustion strategy of conservative conventionalism (p. 299).

Corroboration of a theory also depends on the extent to which it can successfully be fitted into the framework of accepted theories relevant to that aspect of reality. Theories and hypotheses, after all, are not tested and corroborated in a theoretical vacuum. *External consistency* (p. 255) is no less important for a theory than empirical confirmation of its deduced consequences. Critical consideration of the facts is possible only if one can draw on theoretical knowledge which is unquestioned in the context of the examination. During the initial stages of outlining hypotheses to be developed into theories, this foundation of previous knowledge is not fixed (p. 262); later it becomes more clearly defined as its external consistency is tested. The accepted theories which a new geoscientific theory must reckon with or must not contradict primarily include physical and chemical laws and theories, as well as all accepted special laws and theories of Earth science relating to the pertinent aspect of reality.

In the course of 'testing-by-application' the researcher is guided by the concept or idea of a 'true' theory. This 'true' theory resembles the philosopher's stone of the alchemists of old: its possessor would find that it would correspond to reality in all its empirically testable consequences, only useful information would be produced for practical application, and it would agree with all other true theories. In the course of corroboration a given theory acquires more and more of the features which the 'true' theory would have; we may say that in the course of continued corroboration a theory gains in *plausibility*. The fund of theories accepted at any given time by geoscientific research does not consist of demonstrably 'true' theories, but of theories whose proximity to truth is plausible.

To illustrate the process by which a theory becomes corroborated we will examine some incidents in the history of the impact theory.

The development of this theory can be divided into the *elaboration* of the model and its *application* to explaining real structures on the surface of the Earth and other planets. Both developments go hand-in-hand, and it is precisely the joint nature of the non-empirical elaboration of the model and empirical 'testing-through application' which constitutes the *corroboration* of a theory.

Using a model, the theory describes the sequence of events initiated when a mass travelling at planetary speeds of many km/s hits the surface of a planet. In its present form the theory contains the following basic and partial hypotheses relating to terrestrial impacts in particular.

1 In space between the large planets there are meteoric bodies of varying sizes which will with a given probability collide with the Earth, Moon, or other planets.
2 Bodies of this type hitting the Earth travel at speeds of between 11 and 70 km/s.
3 When meteoric bodies collide with the Earth, the decelerating effect of the atmosphere permits only bodies of more than 1000 tons to reach the Earth's surface, travelling at a speed of many kilometres per second.
4 Given the initial conditions (kinds of subsurface rocks and the velocity, mass and composition of the meteor) and physical laws deriving partially from more general laws and partially from experiments with models, it is possible to deduce (a) the laws governing the events occurring during and after the impact event and (b) the observable phenomena surviving as necessary consequences of the impact.
(a) The kinetic energy of the meteor is transformed into mechanical and thermal energy by a compression wave (shock wave) travelling at supersonic speed in the meteor and the surrounding ground. The meteor and the surrounding rock are vaporised and concentric zones of melting, pressure transformation and fracturing of the rock form around the impact point. Thermally and mechanically altered and comminuted rocks are ejected, forming a crater surrounded and partially filled by impact breccias; the shape of this crater may be modified later by compensatory tectonic movements.
(b) Where not destroyed by later erosion, the following traces of an impact can afterwards be observed: crater structures filled and surrounded by impact breccias whose components reveal the effects of shock wave metamorphism (melting, high pressure minerals, and traces of plastic deformation at high pressures).

The testing-through-application of the impact theory began with a hypothesis derived from initial hypotheses, which, though at first applied only to Earth had the character of an indefinite existential statement: there are structures created by impacts on the Earth's surface.

The suggestion that there are impact structures on Earth was first made in 1891 by the American geologist Gilbert[35] to explain the origin of the

Barringer Crater in Arizona. Craters at that time were generally interpreted as products of volcanic activity. Located in sandstone and limestone strata and having a diameter of 2.5 km and a depth of 170 m, this crater differs in two respects from normal volcanic craters. Firstly there are no volcanic rocks or loose debris in or around the crater, and secondly there is a scattering around the crater of large and small fragments of metallic iron whose nickel content indubitably shows them to be of extra-terrestrial origin. Gilbert himself wrote the following on his reasons for assuming that the crater was created by impact.

> Among the various theories that have been proposed for the origin of the planet there is one which ascribes it to the falling together under mutual attraction of many smaller celestial bodies. . . . Speculating on such lines I had asked myself what would result if another small star should now be added to the earth, and one of the consequences which had occurred to me was the formation of a crater, the suggestion springing from the many familiar instances of craters formed by collision. . . . So when Dr Foote described a limestone crater (i.e. the Barringer Crater) in association with iron masses from outer space, it at once occurred to me that the theme of my speculation might here find its realization. The suggested explanation assumes that the shower of falling iron masses included one larger than the rest, and that this greater mass, by the violence of its collision, produced the crater. Here again you will observe that a single theory explains the crater, the iron and their association.

This as yet sketchy outline promised to fulfill four of our conditions (see p. 263) required of a theory. First, an unusual phenomenon was explained not through special assumptions fitting only the individual case, but within a larger, astronomic–cosmologic context. According to the theory, the crater in Arizona is an example of an impact crater, of which there should be more examples upon Earth. Second, the theory throws new light on the empirical facts. In attempts to confirm the theory, the debris masses, crater dimensions, and basement composition would all be investigated for clues relevant to the impact process. Third, the impact theory explains the occurrence of meteoritic iron around the crater. The hypothesis supported by other researchers that the crater originated by steam explosion left this unexplained as pure coincidence. Fourth, the new theory opens a door onto as yet unknown empirical phenomena to be discovered by new observational methods, phenomena which could be inferred from the basic hypothesis of impact.

Gilbert himself attempted to develop his theory further in this last regard, believing that he could decide between his theory and the alternative hypothesis of a steam explosion by an *experimentum crucis*.

> If the crater was produced by the collision and penetration of a stellar body that body now lay beneath the bowl, but not so if the crater resulted from explosion. Any observation which would determine the presence or absence of a buried star might therefore serve as a crucial test. . . . Again it occurred to me that the stellar body would presumptively be composed, like the smaller masses round about, of iron, and that its presence or absence might, therefore, be determined by means of the magnetic needle.

Magnetic measurements carried out in the crater at Gilbert's request failed to indicate the presence of an iron mass beneath the surface. Gilbert therefore considered that the hypothesis of a crater created by impact had been refuted, and that the steam explosion theory of the crater origin was better founded. This meant, naturally, that the highly improbable coincidence of the explosive event and a causally unrelated fall of an iron meteor had to be accepted.

In any event, Gilbert did not consider the question of the origin of the Barringer Crater to be permanently closed.

> The method of hypotheses, and that method is the method of Science, founds its explanations of Nature wholly on observed facts, and its results are ever subject to the limitations imposed by imperfect observation. However grand, however widely accepted, however useful its conclusion, none is so sure that it cannot be called in question by a newly discovered fact. In the domain of the world's knowledge there is no infallibility.

The theory of the meteoritic origin of the Barringer Crater appeared to have been refuted, but the impact theory itself was not, even though no other acknowledged examples of major impact craters were known at the time. It was not until the 1950s that further progress was made and the model became sufficiently conceptually refined (particularly with regard to the physical laws governing the impact event) to show that Gilbert's conclusion had been over hasty. More precise calculations showed that in crater-forming impacts the temperatures immediately around the point of collision are so high that the meteor must vaporise. There could therefore be no meteorite buried under the crater floor, and the iron fragments around the crater in Arizona probably sheared off during the meteor's descent through the atmosphere. Moreover, in view of the high pressures during collision it would be expected that some quartz grains in the sandstone strata would be transformed into the mineral coesite, a high pressure silica modification created shortly before in the laboratory of an American researcher. In 1960, the American geologist Shoemaker[36] successfully discovered coesite in the brecciated sandstones of the Barringer Crater. This supplied a new empirical argument for the impact origin of the Barringer Crater, making the crater after all into one of the acknowledged objects which could be explained according to the model of the impact theory. The mineral coesite, it should be emphasised, occurs in such tiny amounts and minute crystals that its presence can only be demonstrated with certainty through X-ray analysis after special chemical enrichment. Coesite could be found only through directed search, only, that is, in the light of the theory. The theory created its own empirical basis.

Further progress in the impact theory was made by using it to explain the origins of other structures. The Nördlinger Ries was explained as an impact crater after Shoemaker and Chao[37] found coesite in inclusions in its breccias. As soon as the Ries structure and its breccias had been explained as products of an impact event, then all impact-related phenomena observed in the Ries acquired a supra-local significance for the impact model in general and hence for all examples of it, i.e. for all known and as yet undiscovered impact formations on the Earth, Moon, and other planets. What had happened before with the first discovery of coesite now occurred again repeatedly. The new theory opened a window onto phenomena which had previously gone ignored or which could only be discovered when specially looked for by refined methods. A few examples on a microscopic scale include lamellar structures in minerals such as quartz and feldspar indicating plastic deformation due to the high pressures of shock waves; the alteration of quartz and feldspar into glassy substances (diaplectic glasses) by shock waves without melting; the peculiar melting of some minerals and entire rock units by shock waves; the formation of glassy bodies of characteristic structure through the rapid cooling of such melts, etc. No less important are macroscopic observations. The composition and depositional siting of the various breccias and the size and distribution of ejected blocks which previous to the theory had seemed chaotic acquired regularity and organisation. They agreed with the concepts derived from the model on the mechanisms of crater formation, ejection, and deposition of debris, while at the same time they served to refine the concepts both in general and in their application to the particular geologic situation of the Ries.

The same occurred again in applying the theory to further structures, particularly to the many craters discovered and investigated on the Canadian Shield during the 1950s and '60s. These were discovered by searching aerial photographs for basins and lakes of circular shape (a necessary, though not sufficient characteristic of impact craters which have not been subsequently altered); of these structures, those to which the theory could be applied were then sought out.

Applying the theory to an individual case involves the retrodiction that the structure visible today, its rocks, and so forth, were created by an impact event in the geologic past, that is, the theory's application leads to a conjecture which can only be tested indirectly by means of deduced hypotheses which themselves are directly testable. To confirm each retrodiction it is important to test and confirm various mutually independent hypotheses. The accumulation of confirming evidence increases the plausibility of the retrodiction. The fewer mutually independent instances of confirmation are found, the less certain is the retrodiction, and the easier it might be to find an alternative explanation.

Each confirming retrodiction strengthened the general impact theory. But the increase in the number of terrestrial structures of all epochs (and now of lunar structures as well) to which the theory could be applied, and the increase in the number of confirming details observed in each case not only corroborated the model: over and above this the continued 'testing-through-application' led to a *further development and improvement* of the theory, which in turn had a corroborative effect. The structures examined in light of the impact theory led to the discovery of *new facts*. Efforts to explain these new findings built up and refined the concepts of the causative impact event: they improved the retrodictions.

These observations of nature went hand in hand with experimental and theoretical investigations, which collectively served to interpret the model of an impact event according to physical (and particularly quantitative) laws. Experiments, are of course, unable to simulate either the enormous masses or the high velocities and energies of planetary impacts. Nevertheless, a combination of experimental and theoretical work on the energy transformation, the sequence of phases of an impact event, and the shock waves producing certain mineral deformations, transformations, and melts, served to develop laws on the trajectories of the ejected material, whose conclusions agreed with the findings in nature.

Disconfirmation of theories

Theories may be disconfirmed either by showing them to be externally inconsistent, or by empirically refuting some of their partial hypotheses.

A theory is *externally inconsistent* when it conflicts with another belonging to the fund of theories accepted as valid at the time by the scientific community. Where such a conflict occurs, the first attempt will be to eliminate it without giving up the disconfirmed theory. This is done by adjusting either the theory being tested or the accepted theories. Where there are reasons for adhering to the accepted theories and where the tested theory cannot be changed so as to eliminate the conflict, that theory must be regarded as refuted and should be rejected. Such rejection is in fact not always carried out by the scientific community, or at least not unanimously. Pragmatic needs are often stronger than epistemological ones, so that theories which have been and continue to be corroborated within a limited area are adhered to despite ineradicable inconsistencies, since it is believed that for the time being they are indispensable as instruments of research. Since Earth science is not a static structure but changes and progresses continually through research, it constantly uses theories which are inconsistent with others belonging to the ensemble of generally accepted theories. Eventually such conflicts vanish when the inconsistent theory is rejected in favour of a new, externally consistent theory or when changes removing the inconsistency occur in the fund of accepted theories. We will illustrate this through some examples.

The theory of the Earth's cooling was promulgated in the nineteenth and far into the twentieth century: according to this theory, gradual cooling has made the Earth contract, causing mountain folding and other tectonic phenomena. In 1890 this was rejected as externally inconsistent by the physicist Kelvin.[38] Kelvin demonstrated that according to the laws of the physical theory of heat conductivity, a mass the size and composition of Earth would cool from an originally homogenous temperature of 7000 °C to its present condition (established by measuring temperature gradients in bore holes) in not more than 100 million years. This is far too short a time for biological evolution or for any geologic processes. The cooling theory therefore conflicted with many components of corroborated theories of physics and of the Earth sciences, and the condition of external consistency was not fulfilled.

A general principle dogmatically believed in by most geologists in the second half of the nineteenth century and later was also accused by Kelvin of external inconsistency on similar grounds. This was the principle of uniformitarianism on which Lyell, following Hutton, had founded the explanation of all geologic change. According to uniformitarianism all processes have always proceeded in the same way through the geologic past with the same intensity or energy input. Since it was further assumed that the energy necessary for geologic processes in the Earth's crust came solely from internal sources (e.g. from a reservoir of heat remaining from the time of the Earth's origin), Kelvin accused all uniformitarian theories of implying perpetual motion, which is ruled out by the second law of thermodynamics. Uniformitarian theories are therefore externally inconsistent and must be rejected. In contrast to Lyell, Kelvin claimed that Earth's past and future history must be explained using a model of directional process, running from an original state of high energy content and intensive endogenous activity (tectonics and vulcanism), and winding down to conditions of lower energy content and correspondingly restricted geologic activity. This was considered to be completely independent of any conceptions as to the nature of the internal energy sources effective in the Earth's interior.

At Kelvin's time (1862) there was no evidence to counter this demonstration of external inconsistency and the logical consequence should have been to reject the theory of the cooling Earth and the principle of uniformitarianism. In fact, both theories were for pragmatic reasons maintained till long afterwards. Kelvin's criticisms were largely hushed up, particularly as they came not from within the camp, but from another science, namely physics. The geologist Chamberlin (1899) made one of the very few attempts at defence, casting doubts on Kelvin's underlying assumption that the Earth had originally been molten.

Around the turn of the century the situation changed radically with the discovery of radioactivity. The decay of radioactive atoms in the Earth's interior provided an additional constant source of heat compatible with a far longer terrestrial history than Kelvin had reckoned possible with his theory of cooling. This eliminated one inconsistency.

This has still not happened for strict uniformitarianism. The supply of heat-yielding radioactive atoms must surely, if slowly, dwindle due to their decay, meaning that the amount of heat released per unit time and available for endogenous geologic processes must steadily diminish. For this reason, strict uniformitarianism conflicts even with the revised theory of Earth's thermal history as determined by radioactive decay.

Empirically a theory can be disconfirmed by the refutation of some of the hypotheses derived from it. Should states of affairs become known which contradict individual partial hypotheses, this again will not lead to the theory's immediate rejection. As long as the basic hypotheses of the hard core are not affected and the contradictions concern only the hypotheses of the protective belt, attempts will be made to save the theory. There are various ways in which this can be done.

First, *the observations* on which the refutation of the hypothesis was based *can be criticised*. Phenomenological observations, for instance, may be too imprecise or may have been carried out on unsuitable or insufficient objects. Experimental observations may err due to the use of insufficiently sensitive instruments, to disturbances not taken into account, or to incorrect interpretation of the results, etc.

Second, an *ad-hoc hypothesis* may be brought to the rescue, countering the refuting effect of the empirical findings by assuming special circumstances to explain how the observed phenomena came about. In this way they no longer conflict with the application of the theory.

In its early phases the impact theory competed with theories according to which the Barringer Crater in Arizona and the Nördlinger Ries in Southern Germany were created by endogenous terrestrial forces. In both cases the defenders of the impact theory produced empirical evidence designed to refute the theory of an endogenous origin. Their opponents, however, also believed that they could in both cases refute the impact theory through empirical evidence. In order to neutralise the refutation without casting doubts on the empirical findings and their interpretation, the *ad-hoc* hypotheses were introduced that a meteor had hit an already existing volcano in the case of Ries,[39] or that in the case of the Barringer Crater, a meteor had fallen upon the spot where a developing volcanically-induced steam explosion could be triggered by the impact.[40]

Before the hypothesis of continental drift was conceived (now part of the theory of plate tectonics), the theory of the persistence of the oceans and continents was contradicted by empirical evidence of faunal evolution on the continents of South America and Africa. On these continents, now separated by the South Atlantic, the faunas evolved uniformly during the Paleozoic, after which they diverged. An attempt was made to save the theory of the stability of oceans and continents by introducing the *ad-hoc* hypothesis that land or island bridges had once united these continents, only to become later submerged.

Third, *the theory may itself be modified and improved* in such a way that the hypothesis refuted by the empirical findings no longer necessarily follows from the theory, and the empirical findings accord with what should be expected according to the modified theory. While the introduction of an *ad-hoc* hypothesis only saves the theory in its application to isolated phenomena, this change in the theoretical model itself, if successful, must hold for all applications of the theory.

The failure of all attempts at the Barringer Crater (as later in other similar structures) to find the meteorite which created it was at first (see p. 280) regarded as a refutation of the impact theory. Refinements in the theory showed that it is precisely the absence of the meteorite which is to be expected.

A theory can be refuted only if existential statements can be derived from it which are empirically testable. The more testable conclusions can be derived from it the easier it is to refute, or it might also be said, the greater its *empirical content*. Since research into nature and therefore Earth science aims to create theories telling us as much as possible about the world, we must agree with Popper that the goal of research should be to set up theories which can be refuted as easily as possible.[41]

The aim should therefore not be to use all manner of tricks to immunise hypotheses and theories against empirical refutation, since it is above all the boldest and riskiest theories which explain the most and prove the most fruitful for the advancement of inquiry. Following Bunge,[42] we cite the following 'tricks of immunisation' commonly found in the geological literature, encouraged by the false belief that unshakability is a desirable feature of theoretical knowledge.

One means of protecting hypothetical statements from endangerment by empirical testing consists of using adverbial phrases such as 'under certain circumstances', 'under favourable conditions', 'as a rule', or '*mutatis mutandis*'. This produces statements which scarcely anyone can question to good purpose and which therefore have a good chance of surviving empirical tests, but which have correspondingly little explanatory power.

The most effective protection against empirical refutation, however, is displayed by those hypotheses and theories dealing with events or entities whose present or past existence cannot be refuted. The first models of Earth's history posed in the seventeenth and eighteenth centuries contained fantastic events usually of a catastrophic nature not observed today. Geology was founded as a modern science when such hypotheses were discarded and retrodictions became exclusively founded on processes which, in accordance with the principle of uniformitarianism or actualism, can still be observed today, and whose assumed existence at any given time in the past can be refuted since the consequences they must leave behind are known. Even today, hypotheses and theories crop up in the Earth sciences which consist of assertions which are at first non-refutable. Whether they are true or not, they become scientifically fruitful only in so far as developments either in theory or in observational techniques render them refutable. An example of such a non-refutable hypothesis is the assertion that planets inhabited by life forms exist somewhere in space.

A retrodiction which is scientifically unfruitful since it is not refutable is a hypothesis explaining the origin of certain circular depressions on the Franconian Alb and of certain silicification phenomena found in much of Eastern Bavaria[43] as the result of an impact of a comet consisting of silica. The retrodiction may or may not be true; no statements were derived from it through which such an event in the past could be refuted, let alone confirmed. It furthermore lacks any support through external consistency, since siliceous comets are not known and there are no theoretical grounds to suppose that they ever existed.

When the impact theory was used to explain isolated crater structures such as the Nördlinger Ries, its opponents at first considered it a throw-back to the mode of catastrophism, since it could neither be refuted nor confirmed that such a hitherto unobserved event could, like a *deus ex machina*, have brought about the phenomena existing today. The impact theory only became successful when empirically refutable consequences could be derived from the theoretical model.

The history of Earth science shows that the scientific community will decide to reject a theory due to external inconsistency or to empirical conflicts only when another theory becomes available which at least promises to do better at fulfilling the functions of the abandoned theory.[44] As long as no rival hypothesis or theory has appeared, continued attempts will be made to eliminate conflicts with recognised theories or empirical evidence by corrections at the theoretical level, by auxilliary hypotheses, or by accusations of erroneous observations or interpretations. Should none of this succeed, so long as no alternative is available, the conflicts will still be tolerated in silence rather than that the disconfirmed but still usable theory be rejected. This adherence to theories even when they are substantially disconfirmed has its practical reasons. Research, even when purely descriptive, cannot take place in a theoretical vacuum: a disconfirmed theory is preferable to no theory at all.

The rejection of a theory thus occurs as a result of external inconsistency or empirical contradiction, but always *within the context of competing rival theories*.[45] Theories can only compete when they or the hypotheses derived from them behave ambivalently with regard to a certain set of states of affairs, that is, when the subject matter can be explained according to various different hypotheses. Were this not the case, were all the certified states of affairs to point to hypothesis H_1 and against the hypotheses $H_2 \ldots H_n$, then there would be no reason to consider hypotheses $H_2 \ldots H_n$ in the first place.

The various Ries theories, for instance, could only compete because the striations observed on the bedrock around the Ries Basin could be explained equally as well in terms of movements of ice masses (the glacier theory), or in terms of transport of breccia masses, caused either by a steam explosion (the explosion theory), or by collision with a meteor (the impact theory).

In order to decide between several rival theories or hypotheses, those states of affairs must be sought out which can be explained by one hypothesis alone: states of affairs, in other words, which confirm hypothesis H_1 and refute hypotheses $H_2 \ldots H_n$. The search for such *critical states of affairs* S_{cr} makes for an *experimentum crucis*, or crucial experiment, which can only be spoken of when there is a choice between at least two rival hypotheses. Such a possibility does not exist 'for all time', but only relative to a given and generally accepted background of knowledge.[46] Suppose a critical state of affairs S_{cr} permits a decision between one hypothesis H_1 and the rival hypotheses $H_2 \ldots H_n$, such that if S_{cr} is true, H_1 is confirmed and $H_2 \ldots H_n$ are refuted. In that case there is a contradiction between H_1 and $H_2 \ldots H_n$, though it is not always clear how logically to characterise it. It is certain that H_1 and $H_2 \ldots H_n$ are not mutually compatible in regard to S_{cr}. This incompatibility, however, is not always a logical contradiction. Contradictory statements, after all, must satisfy two basic conditions, the rule of the excluded middle and the law of non-contradiction. The law of non-contradiction states that of two contradictory statements, both cannot be true. The rule of the excluded middle states that of two contradictory statements, one is always true.[47]

Some rival hypotheses satisfy these conditions of logical contradiction. These are invariably pairs consisting of a conjunction of a statement and its negation, of the form 'X is a – X is not a', or 'all S are a – not all S are a'. Mutually exclusive contradictory hypotheses usually relate to individual objects. Thus there is a contradictory conflict between the hypothesis H_1 'the Earth has been steadily cooling' and the entire family of hypotheses $H_2 \ldots H_n$, all compatible with the statement 'the Earth has not been steadily cooling.' This latter assertion leaves many possibilities undetermined. The Earth's temperature may have remained constant, it may have been steadily increasing, or it may have fluctuated. None of the hypotheses of the group $H_2 \ldots H_n$ is on its own the contradictory opposite of H_1. Should S_{cr} therefore lead to the refutation of H_1, then H_1 would have to be rejected, and any of the hypotheses $H_2 \ldots H_n$ could possibly be true. Only the confirmation of H_1 would in such a case amount to the unambiguous refutation of $H_2 \ldots H_n$.

Often when explaining isolated phenomena only two mutually incompatible theories or derivative hypotheses are discussed, giving the appearance that the two constitute a logical contradiction. The strategy of producing such an appearance is a seductive one, for if two theories constitute a logical contradiction then the refutation of one simultaneously confirms the other.

When, for example, explaining the diagenetic formation of a solid and dense limestone out of a very porous, unconsolidated calcareous mud deposited in the sea, two hypotheses may be suggested. According to hypothesis H_1, consolidation and compaction take place when unconsolidated calcareous sediment is lifted up out of its marine environment by tectonic processes. This brings it into regions where it can be influenced by CO_2-rich precipitation water, causing recrystallisation which closes up the pore spaces and consolidates it. According to the second hypothesis, H_2, the consolidation does not occur during uplift, but during subsidence, as the calcareous sediment is brought down into greater depths of higher pressures. Here compaction is caused by so-called pressure solution, dissolving the calcareous substance at the pressure-stressed points of contact between the grains and redepositing it again in the pressure-free pore spaces.

In discussing the consolidation of a given limestone assumed to have been deposited as an unconsolidated marine mud, explanations H_1 and H_2 are *incompatible*, but do not form a simple logical contradiction. Should evidence become known against H_1, such as indications that the consolidation took place too quickly after deposition for uplift into precipitational realms to have occurred, then this only supports the statement that this rock was *not* formed in accordance with hypothesis H_1. This does not of itself 'prove' that H_2 is true, since H_1 and H_2 do not form a logical contradiction.

Through this example we see that where two positive hypotheses H_1 and H_2 are juxtaposed, no logical contradiction exists between them, for a contradiction can only obtain between a statement and its negation. In such a case, however, there may in pragmatic terms exist an alternative which behaves like a contradiction, namely if an additional hypothesis establishes that at the time of the explanation of the given state of affairs, the two hypotheses H_1 and H_2 are the only ones which come into question. Only then can the refutation of one hypothesis imply the confirmation of the other.

To explain round crater structures on planetary surfaces, for instance, two hypotheses are available today: the volcanic hypothesis and the impact hypothesis. The additional hypothesis that crater structures can be produced only by one of these two means is generally accepted. Given this presupposition the refutation of one hypothesis simultaneously confirms the other, a conclusion to which, of course, is attached all the uncertainty of the additional hypothesis.

Pairs of logically contradicting hypotheses are not very common, as it is usually very difficult to find secure grounds for the additional hypothesis which rules out all others. While this is true for hypotheses on isolated phenomena, it is particularly true for general hypotheses, and most of all for theories. In discussing rival theories, therefore, the refutation of a given theory does not with logical certainty confirm the other, incompatible one. Opposing parties like to foster the impression that there are only two alternatives, so that the refutation of the opposing view may conveniently serve as 'proof' of one's own views.

7
The growth of geoscientific knowledge

7.1 The internal and external history of science

The character of geoscience as a modern science cannot properly be understood if approached with the medieval conception of science as a summation of knowledge, a system of true knowledge gathered till now on the historic past and future of the planetary bodies, in particular of the Earth and life upon it. Certainly those active in the Earth sciences are conscious that the goal of geoscience is to gather true knowledge of the presently existing world which has undergone a certain past; and certainly the possession of such knowledge, held to be true because of manifold corroboration in practice, accounts for the importance of some Earth sciences in socio-political contexts. Yet what actually takes place in Earth science, i.e. what is actively carried on by persons and institutions under the label 'Earth science' is only to a small degree concerned with ordering, conserving, and passing on certified knowledge, e.g. by teaching or by writing textbooks. More important both according to the internal yardstick of the scientific community and to the external criteria of usefulness to society, is work that brings about growth of geoscientific knowledge. Earth science must be examined in this aspect of *knowledge-accumulating research* to understand its special role within the system of natural science and its usefulness for application in society.

The concept of Earth science as a research activity has been implicit in all the discussions of the previous chapters. To reflect the language and methods used by Earth scientists for everything from the gathering of empirical data to the construction of theoretical systems, it was necessary to arrest Earth science in its development, to reveal it frozen as a four-tiered structure composed of the empirical basis, the level of generalising empirical data, the level of theoretical discovery, and the level of regulative principles. We will now expand this static picture in its dynamic aspect.

Based on the structural insights gained so far, we will examine what constitutes the progress towards which geoscientific research aims, and how it is brought about.[1]

By establishing science as a research activity in the course of the last century the natural sciences subjected themselves to a postulate unknown to previous science. In the centuries before, natural science had been understood to be the knowledge of self-contained reality. 'Geological' world views as envisioned by Descartes, [2] Leibniz,[3] Buffon,[4] Hutton[5], and many others were doctrines, pedagogic descriptions of the Earth's past, as these scholars believed it actually was. The field of science understood as a research activity, on the other hand, constitutes a dialectic unity of objective things and the knowledge of these things. Science as research understands itself, on the one hand, to be an unbounded 'research journey', constantly breaking new ground to discover new realities. At the same time science as research considers itself to be a never-ending process ever approaching its object, never reaching it. Science as research creates its own realms as it is constantly correcting the results of previous research.[6]

Just how research in Earth science visualises and effectuates progress now and in the past is a problem of 'diachronic' philosophy of science, that is, it concerns both history of science and philosophy of science *sensu strictu*. Thorough investigation of the question would have to be founded on the history of the Earth sciences. Here we must limit ourselves in two respects. We can only cast an occasional and cursory glance at the earlier history of geoscientific research and will largely base our investigations on 'current history', that is, on examples of progress in research in the very recent past. At the same time we cannot do total justice to Earth science as a historic phenomenon since progress in research, following a distinction made by Lakatos,[7] has both an 'internal' and an 'external' history. The internal history is our theme here: it focusses on objective growth of knowledge reconstructed and rationally explained according to norms internal to science and using the methods established by philosophy of science. This must be complemented by the empirical sociologically and psychologically founded *external history of science*, which we will keep in mind, but cannot discuss in detail.

From the perspective of *internal history*, progress through research in Earth science can simplistically be described as the proliferation of knowledge in breadth and depth. Knowledge grows in *breadth* through the constant discovery of new, currently observable facts, and through the continued uncovering of conditions and processes in hidden regions and in the geologic past. Knowledge grows in *depth* through the construction of theories of increasing generality and growing explanatory power.

7.2 The proliferation of empirical knowledge

The proliferation of knowledge in scope by observing, experimenting with, and reconstructing hidden and past states of affairs, as we have seen, is guided by theoretical conceptions and expectations. For this reason, an expansion of scope by accumulating knowledge of relevant facts and states of affairs is always closely related to and founded upon expansion in depth by the development of a succession of comprehensive explanatory theories. It is useful, however, to look separately here at each of these two sides of the growth of geoscientific knowledge since they must be evaluated differently when looking back on Earth science's past history, present activities, and future development. We will therefore look at the progress attained by research in Earth science, examining first the accumulation of knowledge that has a direct empirical foundation, and then the dynamics of theories.

The growth of empirically founded knowledge occurs in two differently orientated research contexts. Efforts to *depict the state* of relevant states of affairs are aimed explicitly at gathering information for the empirical basis. One of the most important of these activities is *mapping*. Geological and geomorphological maps, ore deposit maps, gravimetric maps, maps of the magnetic field, etc. are made in order to register facts on the distribution of geologic formations and rock types, surface forms, tectonic structures, types of ore deposits, vectors of the magnetic field or of gravity acceleration, and how these are distributed on the surface of the Earth or other planets. This done with the intention of presenting the phenomena as 'objectively' as possible, avoiding theoretical interpretation, and taking into account that it should be possible to base any conceivable theoretical explanation on the facts presented on the map. We have already seen that this intention cannot fully be realised (p. 131). No map, particularly no geological map, can be made without hypothetical assumptions. Nevertheless it is evident that a hard core of cartographic statement will always remain valid, regardless of changes in the theoretical superstructure.

It will always remain an unquestionable fact, for instance, that the geographic unit of the Canadian Shield north of a certain line and east of a foothill zone of young mountain ranges is, with a few small exceptions, free of non-metamorphic sediments, and consists of igneous and metamorphic rocks of known overall structure and depositional condition. Nor will alterations in theory change the geological maps made over the course of many decades on the distribution of Alpine crystalline and sedimentary rocks. The positions and boundaries of the Mesozoic rocks of the Northern

Calcareous Alps remain stable, quite independent of the truth or falsehood of the geological theory that they originated far to the south and were transported to their present position by horizontal thrusting.

Since the beginning of the nineteenth century when the production of geological maps began in Europe (subsequently on other continents) a significant fund of empirical knowledge has accumulated. This continues to grow because more specialised maps are always being made of known regions, and because there still remain regions such as the Antarctic which are largely unknown.

Other attempts to depict the state of things may be characterised as *explorations of as yet unknown regions*; these are either undertaken without specific expectations, to ascertain whether anything of interest to Earth science may be found, or else they are planned with the specific goal of gathering objects and describing states of affairs. The first kind of exploration includes the planned investigation of the ocean floor carried out in the last few decades by gathering floor samples, cores, and geophysical measurements, or the exploration of the Moon and other inner planets. The latter has involved collecting material samples and enormous masses of empirical data through space research, and has led to the foundation of planetological science. The second kind of exploration, having a specific goal, includes the expeditions sent to the Antarctic in the last few years by the United States and Japan to collect meteorites which have accumulated over long time periods on geologically ancient ice surfaces. More meteorites have been collected within a few years there than had previously existed in all the museums of the world put together.

Knowledge of empirical facts which are largely independent of theory is also supplied by investigations of specialised problems published in monographs, periodicals, or elsewhere. These so-called *original works* are 'original' in that the authors do not simply present new interpretations and conclusions drawn from empirical material, but supply their own, new observational results as an empirical basis for their research. Every original work deserving of the name contains in its empirical basis a core of newly discovered facts contributing to the growth of theory-invariant knowledge. Admittedly, it is more difficult in original work to distinguish between theory-independent and theory-related states of affairs and statements than in the case of geological and geophysical maps. Instead of defining a sharp boundary one does more justice to empirical research by organising its statements into concentric zones showing increasing dependence on theory with distance from a central empirical core of factual statements which depend little or not at all on theory. The closer statements of research results come to the central core of the empirical basis in this concentric diagram, the more they contribute to the growth of theory-invariant knowledge. Some examples of this are the following.

1. 'Sediment dispersal from Vogelsberg basalt, Hessen, West Germany.'[8] This investigates a general question using an isolated local example. The problem involved is the manner in which the volcanic mineral content in sands of rivers draining a volcanic region decrease with distance from the volcanoes, the particular objects investigated being sands from the rivers of the volcanic Vogelsberg. Measurements on sands of different rivers lead to the generalisation that the volcanic mineral content drops from 90% to 5% over a distance of 20 km. This is a theory-invariant statement increasing the knowledge on the local phenomenon 'the Vogelsberg.' The author's additional abductive conclusions attributing the observed phenomenon to conditions of transport and deposition of the sands belong to a more remote zone of the empirically founded knowledge reported in this work. Remoter yet is the hypothesis that this is a regularity that holds for other volcanic regions as well.

2. 'The gravity anomalies along the SW edge of the Bavarian forest and their interpretation.'[9] The problem is to determine which disturbance zones in the Regensburg–Passau region form the deep-lying boundary between the crystalline mass of the Bavarian Forest and its foreland sediments. To answer this question, gravity acceleration values were measured along nine profiles laid at right angles to the boundary between the foreland and the Bavarian Forest. The higher gravity values of the crystalline area relative to the foreland and their variation on the profile constitute the theory-invariant data or core of the empirical basis of the investigation. Using geophysical calculations, a model is abductively created for the position of the crystalline and sedimentary rock masses and for the disturbances which constitute their boundaries; the model assigns these positions to depths which are not directly accessible. This model is so constructed as to explain the acceleration values observed at the Earth's surface; it is then treated as an observable fact contributing to knowledge on the geologic structure of this region: it belongs to one of the outer zones of the empirically founded knowledge transmitted in this report, since it is the result of abductive argument.

3. 'On the age of the Vedrette di Ries (Rieserferner) massif and its geodynamic significance.'[10] Here the question is when the mass of granodiorite-tonalite rocks of the Rieserferner massif in the East Alps originated. The authors report data on the chemical and petrographic

composition of various rock samples from the massif, and their content of rubidium and strontium isotopes. From the isotopic data, they infer an age of 30 million years. The mineralogical–petrographical and chemical data belong to the theory-invariant core of the knowledge conveyed by this investigation. The age calculations, though generally treated in works of this kind as empirical data, derive from the theory of radioactive decay, and would change if, for instance, new physical discoveries were to change the decay constant of strontium. Even more theory-dependent are the further hypotheses of the authors on the origin of the magmas constituting the rocks and on the placement of the igneous event within Alpine history. With certain reservations due to the uncertainties of abductive argument, even these hypotheses could be viewed as additions to the empirically founded knowledge on the history of the Alps, provided the underlying theoretical component is accepted.

4. 'Geochemical trends in the Minette (Jurassic, Luxemburg/Lothringen.'[11] The goal here is to reconstruct the conditions under which the sedimentary iron ores of the so-called Minette from the Jurassic formation were formed; it therefore involves an abductively inferred retrodiction. The authors determined the amount of 21 main and trace components contained in a very large number of samples from several profiles in the Minette. Treated statistically, these data yield regional and stratigraphic distribution trends, which, when plotted in graphs and maps, provide the empirical basis for a hypothesis on the formation of the iron ores in the shallow Jurassic sea. The results of the chemical analyses thus increase the theory-invariant knowledge of the Minette formation. The hypothetical reconstruction of the paleogeographic situation and geochemical and geologic conditions in the Jurassic sea at the time and place when the Minette was formed belong to the outer zones of empirically founded knowledge conveyed by the work.

5. 'On the origin of quartz cement. Estimation of quartz solution in siltstones and sandstones.'[12] Here the author wishes empirically to test hypotheses derived from the theory of sediment diagenesis. He inquires into the origins of quartz cement which, when deposited in the pore spaces of many sands, alters them during diagenesis into solid sandstones. Using many samples obtained from deep cores of sandstones and their accompanying shales of varying geologic age, the author tests the consequences deduced from two mutually non-exclusive hypotheses.

(a) If the silica deposited in the sands as quartz is released by chemical processes in the accompanying shales, then (1) the quartz grains in the shales should be reduced in volume relative to the grains in the neighbouring sandstones, and should be smoothed off parallel to the bedding planes, and/or (2) the SiO_2 content of the concretions (regions of early consolidation, preserving the clay in a pre-diagenetic state) should be higher than in the clay around the concretions.

(b) Sandstones often contain so-called stylolites, solution structures formed by pressure solution during diagenesis. If the quartz deposited in the pore spaces derives from quartz solution along the stylolite margins, then porosity in areas of high stylolite frequency should be particularly sharply reduced.

The results on the size and shape of quartz grains in sandstones and shales, the SiO_2 in and around clay concretions, and the correlation of stylolite frequency and porosity support the prognoses deduced from both hypotheses (a) and (b). Although these results were found in the light of a hypothesis of the theory of diagenesis, they are nevertheless theory-invariant facts which must be accounted for by any other conceivable theory.

Through these examples we wish to show how descriptions and problem-orientated investigations continue as they have in the past to contribute to the growth of empirically-based knowledge of the various branches of Earth science. It is not only the theory-invariant facts observed in natural contexts or produced experimentally which have accumulated: the amount of hypothetically derived insights has grown as well. These are more heavily dependent on theoretical preconceptions, and could therefore be doubted in principle. They are, however, actually treated as facts, that is, as states of affairs which presumably will be modified only insignificantly by any changes in theory. They include all insights made in historical geology concerning the epochs of Earth's history and their divisions in global formations or local stratigraphy, and all results of paleoclimatology and paleobiology regarding the plant and animal species and their communities in the Earth's past.

Since the end of the eighteenth century when geoscientific research began, empirically founded knowledge in Earth science has grown significantly. As far as empirical states of affairs are concerned, knowledge has grown cumulatively in scope at a rate which today is exponential rather than linear. This is partly because the number of people active in Earth science has grown and continues to grow very rapidly,[13] and partly because the gathering of empirical facts has been greatly facilitated by the mechanisation and automation of the equipment and methods used for observations and measurements. This expansion of knowledge should be kept in mind when reconstructing and evaluating older situations and theories of Earth science. We know infinitely more than those researchers who in the so-called 'heroic age of geology'[14] around 1800 argued about the theories of plutonism and neptunism.[15] The limited quantity of scientifically established facts made it easier in the old days to formulate bold theories, but more difficult to refine and test them. Today it is the other way around: thanks to the amount and precision of the data, theories can be formulated more precisely and tested more reliably, but the scarcely over-seeable abundance of relevant facts hampers theoretic innovation. Where a daring attempt is made to modify accepted theories or to set up a new one, a great arsenal of empirical knowledge can be used by opponents for the construction of counter arguments.

7.3 Models of growth in theoretical knowledge

The second, deeper reaching, and historically more striking form of growth in geoscientific knowledge consists of the *changing of theories* which organise the mass of facts within an explanatory scheme of Earth history as understood by natural science. Change occurs as older theories are replaced by newer, more productive ones, a change which involves all the criteria for hypotheses and theories and all modes of empirically testing them discussed in the previous chapter. Having established this methodology for evaluating theories as a foundation, we will now attempt to understand how the development of geoscientific theories can be described as growth of knowledge. In doing this, we understand the development of theoretical knowledge to be not simply something which has taken place in the history of science, but a continuing process in which every active researcher is involved, helping or hindering in word and deed.

Philosophers and historians of science have over the course of time developed different theories of scientific progress[16] giving rise to various models by which the development of theoretical knowledge can be reconstructed. We will examine below their usefulness for understanding Earth science.

7.3.1 *Inductivism*

The model of growth of knowledge through *inductive accumulation* corresponds to the classical justification model of scientific inquiry going back to Bacon.[17] This long formed the undisputed foundation of natural science, and even today underlies the widespread naive conception of scientific research. According to this model, the goal of research is to acquire knowledge in the form of empirically proven statements, and to build these into a structure of 'eternal' truth. This is founded on singular statements of observed facts from which true general statements and true theories are supposed to be inferable; the accumulation of observational statements is thought to parallel the cumulative growth of theoretical knowledge.[18] Given the increase in factual knowledge in the Earth sciences, discussed above, it is clear why progress in knowledge was in hindsight considered to be a data accumulation process of this type. It is frequently still so considered today, and what is less harmful, some researchers active today think they can only contribute to knowledge by following this scheme. Historical surveys especially frequently subscribe to the view that Earth scientists have done and should in future do nothing other than read 'without bias' from the book of nature throughout the world, and that the true constitution and history of the Earth and its inhabitants will then reveal itself. As an example, we cite a passage from *Geschichte der Geologie und Paläontologie*, by K. A. von Zittel (1899).

> The future of our science looks very promising for the next decades . . . In a century, the entire Earth's surface will perhaps be sufficiently well known that current descriptions of details will be replaced by a comparative topographic geology like that which E. Suess has so brilliantly suggested in his *Antlitz der Erde*. Should at the same time, the tectonics of the solid crust have become thoroughly investigated throughout the world, this will provide the history of the physical processes during the evolution of our planet. Complete knowledge of the organic remains and their distribution in the various formations, finally, will also permit more reliable conclusions as to the origin, evolution, and development of organic life on Earth than is possible with our fragmentary knowledge today.[19]

That such a concept of growth in theoretical knowledge is not tenable is evident from what we have learned in the previous chapters. Since the works of Popper[20] at the latest, we know that the inductive method cannot lead to theoretical propositions, and that theories can never once and for all be proven or verified in the strict sense, but at best falsified.

7.3.2 *Conventionalism*

It is not passive observation of facts which led to the theories of the cooling and shrinking Earth, of plate tectonics, of the differentiation of magmas, or of the evolution of plant and animal species; these were produced rather by individual researchers, who through exceptional intellectual efforts dared to create bold explanatory hypotheses. Despite initial difficulties and later resistance, they held fast to their theories, and tirelessly consolidated their initially uncertain assumptions through deductive and abductive application. We may lump together as *conventionalist*[21] all theories of scientific progress which accept that we cannot read the book of nature without some effort on our part, that is, that we must interpret the facts from the outset in the light of theoretical expectations using models which ultimately must be or have been accepted by the scientific community. Following Lakatos and Schäfer, we distinguish between a conservative and a progressive (or in Lakatos, a revolutionary) branch of conventionalist theories of scientific progress.

Conservative conventionalism

Research according to this model is based on the methodological decision to adhere to a long-established theory without permitting it to be rejected in spite of temporarily unexplained facts which seem to contradict the conclusions of individual hypotheses of the theory, or of other inconsistencies. We have already mentioned (p. 282) that even serious disconfirmation of a theory will generally not lead to its immediate rejection, and we have mentioned the feasibility of saving disconfirmed theories or of protecting them against possible attack. Dingler[22] has named this the *exhaustion* strategy, stressing its significance for science. It relies on exhausting all ways in which an established theory may come to terms with new, apparently contradictory facts. Indeed, any theory used in research can be expected to be rich in exhaustive possibilities, and not to be easily refuted by unexpected facts. Should such a conservative strategy successfully defend an old theory against a newer theory which, based on empirical facts, has cast doubts on the established theory, then the older one will emerge from the strife improved and confirmed anew. For conservative conventionalism, growth of knowledge means the expansion and consolidation of long-established theories. Supplementing the previous chapter's discussion on ways of saving a disconfirmed theory, we will describe here one author's attempt[23] to use conservative conventionalistic methods to defend and save the older, 'fixist' theory of the permanence of the oceans and continents against the attack of the newer 'mobilistic' theory of plate tectonics.

The author attempts to resolve inconsistencies between the new facts and the core of the fixistic theory by introducing auxiliary hypotheses and by modifying the older theory. One auxiliary hypothesis explains the lack of older sediments on the ocean floor (something which the older theory could not explain) by assuming that the ocean basins were created at an earlier time by down-warping of portions of the crust; these portions became heavier through 'basification', that is, they were penetrated by heavy, SiO_2-poor magma from the mantle. Such 'reworking' through basification supposedly affected the ocean floors repeatedly, the last time being at the end of the Paleozoic. This is supposed to have destroyed all older sediments.

Modifications to the fixist theory consist of the author's conception of causal links between on the one hand the so-called 'regimes' of geologic events in and on the crust (geosynclines, platforms, graben systems, etc.) and on the other hand the local processes of differentiation in the lower mantle, where heavy materials periodically are supposed to fall into the Earth's core while lighter masses intermittently rise up into the crust. This supposedly produces periodic changes in the intensity of thermal transport from the core outwards, and explains the world-wide phases in mountain building, interspersed with quieter phases.

The author finally claims to find internal inconsistencies which he believes show the theory of plate tectonics to be false. We list three examples. The mobilistic theory claims that the American and European plates are moving apart, yet this supposedly contradicts the geologic situation in the Bering Strait, where the American and Asiatic plates meet without any sign of a subduction zone. Secondly the author cites reports claiming the existence of older sedimentary rocks on the mid-Atlantic ridge where they should not occur according to the mobilistic theory. Finally the author maintains that an 'interplate' tectonics exists on the continents, something not provided for by plate tectonics, since according to this theory mountains should only exist at the plate margins. As we have seen, the exhaustion strategy regarding the accepted fixist theory is here combined with the attempt to show inconsistencies in the competing mobilistic theory.

Progressive conventionalism

We may group under this title those theories of scientific progress which permit and even demand that even long-established hypotheses and theories explaining certain phenomena be replaced by better theories in accordance with certain empirical criteria. Following Lakatos, we distinguish four varieties of the progressive-conventionalistic model of scientific progress: approximating conventionalism (Duhem), methodological falsificationism (Popper), the theory of scientific revolutions (Kuhn) and 'sophisticated' falsificationism (Lakatos).

Approximating conventionalism (Duhem)

This theory[24] is based on the notion that laws and organising or classifying theories are attempts to 'describe' empirical reality. Growth of knowledge as manifested in a change of theoretical system or in the replacement of an obsolete theory by a new and better one is not brought about by compelling logic, but takes place in a field of free decision, i.e.

conventionalistically. Decisions against a theory to be rejected and for one to be accepted are made according to a postulate that research must continually work out 'better' descriptions of empirical reality and approximate that theoretical system which Duhem calls the 'natural classification' of empirically founded laws. A theory shows itself to be 'better' through increased empirical content, that is, through the inclusion of new subject matter and laws permitting more precise deductions and abductions.

Applying Duhem's interpretation of progress to the dynamics of geoscientific theories would mean that the transition from the fixist theory of the Earth's development to the mobilistic one was not logically compelled because certain expectations derived from the fixist theory were disproved and certain of the mobilistic assumptions were confirmed, but simply recommended itself because the new theory developed a better description of old and new facts and transmitted an increased empirical content.

Methodological falsificationism (Popper)

In an earlier chapter we have outlined the basis of Popper's theory of growth of knowledge (p. 274).[25] Its key feature is Popper's criticism of the inductivist illusion: theoretical statements cannot empirically be proved, but at best refuted. From this it follows that the only acceptable theoretical statements are those which correspond to empirically testable basic statements through which the theory may be destroyed. 'A theory is falsifiable when the class of its falsification possibilities is not empty.' (Popper (1934), p. 53). Growth of theoretical knowledge, according to Popper, consists of subjecting existing theories to the strictest possible empirical tests through which they can be destroyed. The decision as to whether the refutation of a basic statement warrants the falsification of the theory presupposes general agreement as to which basic statements are relevant to the theory: this is the conventionalistic component of the Popper argument. The relevance of basic statements is determined by the then accepted techniques of phenomenological and experimental observation, as well as by the total pool of hypotheses and theories regarded as correct, that is, by the background knowledge (see p. 265) necessary to test a hypothesis at all.

The ability, for instance, to test hypothetical statements from the theory of plate tectonics concerning a zone in the Earth's upper mantle of reduced viscosity (the aesthenosphere) on which the plates of the Earth's crust slide laterally, or concerning the structure of the oceanic crust, presupposes a technology for observing and interpreting seismic data, a technology based in turn on physical laws. Should the above cited critic of plate tectonics doubt the seismically-obtained assertions on the structure of the oceanic crust, he must do this on the basis of a different 'background knowledge' and different stipulations as to the relevant basic statements.[26]

In the previous chapter we made clear that the one-sided characterisation of the growth of scientific knowledge which follows from Popper's methodology with its emphasis on negative results, i.e. on refutation, clashes with the actual progress of research. Both the scientific community's recognition of scientific achievement and the intentions of the individual researchers are aimed not at refuting, but at confirming theories (p. 275). Following Lakatos (1970, p. 115) we will, through the following statements, clarify the conflict between the falsificationist model and the actual course of history of science.

According to the falsificationist view:

1 The test of a theory is its confrontation with statements on the results of phenomenological or experimental observations.
2 The only desirable result of such a confrontation is the refutation of the theoretical assertion.

In the history of geoscientific discovery, on the contrary, the case has been that:

1 In empirically testing theoretical assertions, at least two rival theories confront the assertions on the results of observations.
2 In the growth of theoretical knowledge, empirical statements which confirm a given theory are held to be more interesting and more important than those which refute theoretical claims.

As long as philosophy of science conflicts with the history of science and with the self-image of active researchers, there can be no fruitful interaction between them.[27] For this reason, all attempts to bridge the discrepancy are important. Two attempts to form such a bridge have recently attracted much attention, the theories of Kuhn and of Lakatos. We will examine how useful these two models are in reconstructing the dynamics of geoscientific theories.

The theory of scientific revolutions (Kuhn)

In his theory of scientific revolution, Kuhn[28] dispenses with any search for rational grounds motivating an individual researcher and the scientific community to reject an older theory and to explain reality by means of a newer theory. He divides the development of a natural science into longer-lasting phases of so-called 'normal' science and briefer periods of 'extraordinary' science, leading to 'scientific revolutions' after which a new phase of normal science follows. In normal science, research is carried on following established traditions and in accordance with a dominating theoretical structure or *paradigm*. Kuhn wishes the term 'paradigm' to be understood in two ways.[29] 'Paradigm' firstly encompasses the totality of convictions, values, and procedures of the scientific community or of a smaller group of scientists. Kuhn himself calls this meaning of the term 'sociological', since it refers to everything which the members of a group engaged in research have in common. 'Paradigm' secondly refers to a subset of these commonalities, namely the concrete solutions to problems arrived at through 'normal' research which can serve instead of explicit rules as examples or patterns by which to solve still unsettled problems of normal science. By 'paradigm' Kuhn means something more elementary than what is meant by the concept 'theory'. Paradigms are present in the pre-theoretical state; they determine the style of explicitly formulated theories; they form the rationally somewhat inchoate foundation of completed theories and theoretically-guided research activity; they are put into practice as the behavioural patterns of the daily work of research.[30]

The activity of a researcher carrying on normal science ruled by a paradigm is characterised by Kuhn as *puzzle-solving*. Webster's *New World Dictionary* (1970) defines 'puzzle' as follows: '*puzzle*, a toy or problem for testing cleverness, skill or ingenuity.' Kuhn accordingly believes that the aim of puzzle-solving is not to 'produce major novelties, conceptual or phenomenal'[31] but rather to use the equipment and rules of theories canonised by the scientific community to solve such problems as are known to have a solution within the framework of the accepted theory.

> It is no criterion of goodness in a puzzle that its outcome be intrinsically interesting or important. On the contrary, the really pressing problems . . . are often not puzzles at all, largely because they may not have any solution.[32]

What nevertheless spurs a researcher to investigate a question whose solution is neither 'interesting' nor 'important' is, according to Kuhn,

> the conviction that, if only he is skillful enough, he will succeed in solving a puzzle that no one before has solved or solved so well.[33]

The solving of puzzles, according to Kuhn, does not test a hypothesis or theory, but rather the individual researcher:

> for in final analysis it is the individual scientist rather than current theory which is tested.[34]

Should he not succeed in solving the puzzle, then it is only his skill which is cast into doubt, not the accepted theory: 'If it fails the test, only his own ability not the corpus of current science is impugned.'

Most of the works which we have cited as examples of research which expands the empirical basis of individual disciplines in the light of certain theories through theory-invariant information must be grouped under

Kuhn's normal science, for the authors moved within the framework of accepted and corroborated theories, obtained new facts which were important within the context of these theories, and tackled problems which seemed soluble given the instruments of the accepted theories. Original works and descriptions of this kind occupy the bulk of all publications. Researchers active in the Earth sciences are at all times occupied with normal science in the Kuhnian sense, and the individual researcher dedicates most of his strength and life to such work.

Kuhn's interpretation of normal science as puzzle-solving highlights some impulses which undoubtedly characterise the daily conduct of science. The advanced student faced with a specific task in his Masters or Doctoral dissertation and the older researcher as well, invariably regard their work as a puzzle-solving attempt whose success will qualify them as capable scientists. Even modish trends in research and rivalries between schools or individuals can cogently be described as competitions in puzzle-solving ruled by a paradigm which the participants in the game do not question.[35]

Kuhn himself has stressed the sociological character of his paradigm concept.[36] Indeed, the interpretation of normal geoscientific research as puzzle-solving following the rules of a paradigm will be very useful in treating the external history of Earth science, that is, in analysing the sociological and psychological conditions which to an undeniable degree governed and continue to govern the paths of research taken by individuals and groups of scientists. Having restricted ourselves to the internal history of Earth science, however, suffice it to remark that Kuhn's metaphor of puzzle-solving shows the importance of the sociological analysis of science. For the rest, it must be stressed that this metaphor does not cover those functions of normal science addressed by our theme of the internal history of growth of knowledge. If we look at the works filling geoscientific journals which fall into the category of normal science, and study their contribution to the internal development of science we will see that Kuhn's metaphor does not suffice. Every successful solution of a problem, even when done completely within the framework of accepted theories and traditional methods is essentially different from the solving of a cross-word puzzle whose result is already pre-determined. The examples cited at the outset of this chapter show that each such work produces something new and previously unknown, and hence brings about a growth in knowledge. This can involve the following.

1 The discovery of new facts as theory-invariant components of the empirical basis.
2 Additions to the hypothetically inferred states of affairs of the geologic past or of inaccessible regions.

3 Improvements, refinements, or additions to hypotheses and theories.
4 Opening upon new aspects, problems, or questions (many works close with a plea for further investigation).
5 Every specialised investigation is based on expectations, conjectures, or prognoses which belong to the framework of accepted theories. Whether confirmatory or not, the results influence the evaluation of this framework, even when this is not a stated goal or result of the investigation.

The phase of 'extraordinary' science which, according to Kuhn, follows a long period of normal science, is characterised by an accumulation of empirical findings which becomes increasingly difficult to explain according to the dominant paradigm. A crisis of confidence in the accepted paradigm sets in. Some researchers abandon conservative conventionalist attempts to correct anomalies in the dominant theory; they turn to a new paradigm and demand the rejection of the old one. Kuhn calls the changeover from the hitherto accepted paradigm dominating research to the new one a 'scientific revolution'. Here again a metaphor appears, the significance of which is not always clear when analysing the dynamics of theories and which has therefore become the subject of much discussion.[37] According to linguistic usage in human history, a 'revolution' is understood to be a fundamental reordering in which the former conditions are reversed or 'stood on their heads.'[38] By 'revolution' Kuhn means that the change undertaken by the scientific community (or by the majority of its members) from one theory to the other primarily involves a change in paradigm which cannot be comprehended rationally. The old and the new theories are supposedly not only incompatible but also incommensurable. As Watkins[39] has shown this is a logical impossibility, since incommensurable theories cannot be incompatible. Nevertheless, Kuhn uses this as the basis for his argument that in the course of a scientific revolution communication breaks down and the superiority of the new theory over the old cannot be shown in debate. Instead, according to Kuhn, the one party can only use 'persuasion' to 'convert' the other party to the new paradigm.[40]

It has in fact occurred in Earth science that after a conflict between 'traditionalists' and 'modernists', the majority of the scientific community abandoned the previously accepted explanations in favour of theories so new as to warrant speaking of a 'revolutionary' change in views. Such was, for instance, the acceptance of Darwin's theory of evolution which introduced a new era in paleontology in the 1860s. A revolutionary phase took place in the development of tectonics between the years 1903 and 1910 when it became accepted that mountain ranges of the Alpine type were not created by vertical crustal movements, as previously assumed, but by widespread horizontal thrusting of rock units (the nappe theory). The acceptance of the impact theory in the 1960s to explain structures on the Earth and Moon also carried the hallmarks of a revolutionary change, since extra-terrestrial masses and energies were then first acknowledged to influence the structure of the Earth's surface. Above all, however, the 'victory' of Wegener's hypothesis of continental drift and the theory of plate tectonics over fixist theories of the Earth's evolution have been expressly interpreted as a revolution.[41] In the controversy during such change-overs, phenomena were indeed evident corresponding to those which Kuhn sees as typical for the break-down of communication. Between 1912 and 1960 most researchers did not take Wegener's hypothesis of the lateral drifting of continents seriously. So strong was the adherence to the traditional, corroborated paradigm according to which the ocean basins were of great age and the continents had remained in their respective positions since ancient times, that in the United States an open commitment to Wegener's mobilistic hypothesis endangered one's chance of a professorship. For older researchers such a commitment was tantamount to isolation within the scientific community.[42] One symptom of the endangered state of communication in the debates between the modernists and the traditionalists of the time was the appearance of arguments *ad hominem*. These, instead of substantive reasons, were supposed to bring about the downfall of the new theory. As an example we reproduce here three such attacks on the person of Alfred Wegener (after Hallam (1973)).

> Whatever Wegener's own attitude may have been originally, in his book he is not seeking truth; he is advocating a cause, and is blind to every fact and argument that tells against it. Much of his evidence is superficial. Nevertheless he is a skilful advocate and presents an interesting case (P. Lake).

> Wegener's method, in my opinion, is not scientific, but takes the familiar course of an initial idea, a selective search through the literature for corroborative evidence, ignoring most of the facts that are opposed to the idea, and ending in a state of auto-intoxication in which the subjective idea comes to be considered as an objective fact. (E. W. Berry).

> Wegener's hypothesis in general is of the foot-loose type, in that it takes considerable liberty with our globe, and is less bound by restrictions or tied down by awkward, ugly facts than most of its rival theories. Its appeal seems to lie in the fact that it plays a game in which there are few restrictive rules and no sharply drawn code of conduct (R. T. Chamberlin).[43]

The personal antipathy directed by geologists towards Alfred Wegener was also due in large part to the fact that he was not an accredited member of the geological community, but a meteorologist, and therefore carried no authority among the geologists.[44] Similar arguments questioning the credibility, competence, integrity, and scientific conduct of authors of newer theories are also known from the debates of the 1960s about the impact theory.

Kuhn's interpretation of change in theory as a scientific revolution does justice to these and other threats to rational communication during the debates for and against new theories. Kuhn's theory describes with equal accuracy the experience of the individual researcher for whom it must be difficult to surrender mastered and long-practiced forms of thought, theoretical models, and concepts on the course of Earth history and geologic processes, or even non-empirical, tacitly recognised regulative principles of research; he must perhaps face the charge of having worked for long years under false premises, and now faces the laborious task of making a new beginning. This was expressed by the American geologist Chamberlin in 1928 as follows (after Hallam (1973):

> If we are to believe Wegener's hypothesis we must forget everything which has been learned in the last 70 years and start all over again.[45]

It is these interruptions in rational discourse between the defenders of the old and the champions of the new theories to which Kuhn refers in his concept of scientific revolution. This term, along with the concepts 'persuasion', 'conversion', 'paradigm' and the metaphor of 'puzzle-solving' belong to the sociologically and psychologically-determined external history of science. For our purpose of understanding the dynamics of geoscientific theories as a growth of knowledge we must bear in mind that rational communication should be possible not only among the adherents of one theory, but also between defenders of alternative theories, even when such possibilities are forgotten in the heat of battle, and are only recognised in retrospect. For our rational reconstruction of the internal history of theoretical inquiry in geoscience, the threat to communication during changes in theory stressed by Kuhn is not essential. It is a contingent phenomenon in history of science, which one must understand as a historian, but oppose as a researcher and philosopher.

'Sophisticated' falsificationism (Lakatos)

Of all the models of the more recent philosophy of science, Lakatos' model[46] of 'sophisticated falsificationism' developed out of Popper's insights provides the most useful guideline for rationally reconstructing the growth of knowledge in Earth science. According to Popper's methodological falsificationism, theories are changed in the progress of inquiry when one theory is falsified due to contradictions with empirical facts and is replaced by a new theory which is acceptable as soon as the existential statements deducible from it can be empirically tested (i.e. falsified), and as long as it withstands all attempts at refutation. For sophisticated falsificationism, these conditions for the rejection of the old and acceptance of the new theory are neither necessary nor sufficient. According to Lakatos, a theory is never rejected simply on the basis of confrontation with isolated contradictory facts. Any theory can survive the existence of (as yet) unexplained anomalies, or can eliminate these through suitable *ad hoc* hypotheses. No theory was even given up as 'falsified', despite many anomalies, until such time as a new theory was available which proved itself better by performing as well as the rejected theory, by explaining those facts which had seemed anomalous in light of the earlier theory, and by providing an excess of empirical content, that is, by predicting the existence of new and previously unexpected facts. Sophisticated falsificationism holds as true the following model of progress from a theory T (or several theories T_1, T_2 . . .) to the new theory T′.

1 Compared with T, T′ should contain *excess empirical content*, that is, empirically testable statements should be deduced directly or indirectly from T′ concerning states of affairs which could not have been expected in light of T, or were even ruled out by it. Whether or not this condition is fulfilled is shown by logical analysis of T previous to all experience.
2 T′ should explain the previous success of T, that is, all unrefuted empirical statements derived from this theory should belong to the empirical content of T′.
3 Some of the statements belonging to the excess empirical content of T′ should be confirmed. To what extent this condition is fulfilled will become evident in the course of continued phenomenological and experimental observations, that is, through a process having no definite end.

Growth of theoretical knowledge according to sophisticated falsificationism, does not consist of a sequence of theories irregularly interrupted by empirical refutation (falsificationism, after Popper), or by revolutionary changes in paradigm (after Kuhn). It is rather a constant process fed by competition between rival theories which are judged in light of statements of the empirical basis accepted by the scientific community. Progress, according to Lakatos' model, differs in two respects from all other models: in the recognition that theories do not fail simply due to empirical facts, but are only given up in favour of rival theories, and in the notion that the growth of theoretical knowledge can be presented as a progressive, constant, and rationally reconstructible process.

To permit that rivalry between theories which nurtures progress, a proliferation of theories is not only desirable according to Lakatos,[47] but necessary. Alternative theories must be present in outline before a previously-accepted theory can be given up. A new theory is not simply created after an old theory has been falsified by empirical findings, or, as according to Kuhn, after research operating by the old theory has wound up in an unsupportable crisis of confidence.

Continuity of progress is evident in the history of science as a series of connected and mutually replacing systems of theories which Lakatos calls *research programmes*. Continuity within a research programme is guaranteed by adhering to methodological rules determining which paths research must avoid (negative heuristics) and which paths research should pursue (positive heuristics).

Every research programme is characterised by a 'hard core' of basic ideas which are often not stated explicitly. *Negative heuristics* prohibits any questioning of this hard core. Growth in knowledge through a series of mutually replacing theories takes place by building up a 'protective belt' of partial hypotheses around the hard core. This belt is developed, modified, or radically changed through constant confrontation with the results of phenomenological and experimental observation. As long as this is possible the hard core and thus the research programme are consolidated by corroboration. A programme's sequence of theories is linked by their shared hard core of basic hypotheses, while they differ in the partial hypotheses forming their protective belt. The hard core of a research programme may be compared with Kuhn's paradigm.

According to the type of basic ideas in the hard core, research programmes can be distinguished in the history of science according to varying degrees of generality. Among the basic ideas of natural science as a whole are 'metaphysical' principles, such as the assumption that physical and chemical laws always hold in unlimited time and space; the regulative principles characterising geoscientific research programmes will be discussed in the next chapter, including the principle of uniformitarianism and actualism, and the principle of the endogenous causes of geologic changes. More specialised basic ideas forming particular research programmes are assumptions such as that of organic evolution, the cooling of the Earth, the permanence of the oceans and continents, or the shaping of the Earth's surface through impacts out of planetary space. As the basic idea of the hard core becomes less generalised, the concept of a research programme grades into the concept of an isolated theory.

A research programme remains valid as long as the changes and developments of the theories within its framework are *progressive*, that is, as long as the hard core is retained and new empirical findings can be dealt with by creating new theories which, compared with their predecessors, have excess empirical content: theories, that is, that open up scientifically uncharted territory. The research programme *degenerates* and becomes obsolete when it can be maintained only by defensive strategies, that is, by *ad-hoc* hypotheses which resolve contradictions and anomalies without helping to discover new facts or relationships. Degenerating research programmes are defeated in competition with new, progressive programmes. When the scientific community rejects an old and adopts a new research programme, it undertakes a change in the direction and methodology of research which best corresponds to the paradigm change which Kuhn characterises as a scientific revolution.

Positive heuristics encompass all the work carried out by researchers in the course of a research programme. These investigations do not concern the hard core, but involve the theories, hypotheses, generalisations, and singular statements which can be tested, changed, refined, and completed within the framework of the research programme. For every theory there is always a vast number of unsolved problems and unexplained anomalies. These should not be tackled in random order by arbitrary selection as one would Kuhn's puzzles, whose solution, since predictable, would contribute nothing significant to the growth of knowledge. Rather, as long as the programme is in a progressive phase, research should ignore some unexplained anomalies to aim selectively at those problems and phenomena by whose explanation hypotheses and theories will be improved or replaced. Thus according to the model of sophisticated falsificationism, all research work is governed by the postulate of positive heuristics: the protective belt of basically debateable hypotheses and theories surrounding the hard core of a research programme should be expanded and consolidated for as long as the programme can be carried out in a progressive manner.

In conclusion we will attempt a sketchy reconstruction of the basic development of the theory of plate tectonics (discussed in its static aspect in a previous chapter, see p. 240), according to Lakatos' model of sophisticated falsificationism.[48]

The history of this theory, involving a period of around fifty years, was marked by two phases of particularly dramatic controversy. These were the period of disputes over Alfred Wegener's hypothesis of continental drift proposed in 1912, and the early 1960s, when the hypothesis of oceanic spreading was conceived and united with Wegener's hypothesis. Of the basic ideas which went generally unquestioned around 1912, two assumptions deserve emphasis. Firstly, scientists were convinced of the steady cooling of the Earth and believed that all major and minor geotectonic changes of the Earth's crust could be attributed to contraction associated with cooling. Secondly, as a corollary of the principle of actualism it was assumed that the ancient rocks of the continental shields and the deep ocean basins were created in a very early period of Earth's history and had since undergone no changes in their respective positions. These basic ideas belonged to the hard core around which the various disciplinary hypotheses and theories formed a protective belt. It was necessary, for instance, to explain the paleontological evidence in the southern hemisphere indicating the identity or close relationship of the Paleozoic and Mesozoic faunas and floras of the areas of South Africa, Madagascar, India, Australia, South America, and Antarctica, areas separated by oceans today. Since the principle of the permanence of the oceans could not be violated, European geologists supported the hypothesis that land bridges had existed as late as the Mesozoic between the now-separated continents and islands, and that these land bridges had later sunk. Since geophysical gravity measurements gave no indication of the existence of sunken continental blocks in the sea floor, American geologists later assumed that island chains had once existed across which the exchange of plants and animals had occurred.

Alfred Wegener published his hypothesis for the first time in 1912 in two journal articles,[49] and in 1915 in a book which by 1929 had gone through three further editions, each larger than the last.[50] It cannot be said that Wegener suggested his hypothesis, contradicting both the idea of the contracting Earth and of the permanence of the oceans and continents (therefore questioning the hard core of geological research at the time), at a time when geological science was in a state of crisis. To be sure there existed, as always, a number of unsolved problems, but there was no cause to think that the unanswered questions could not be solved in accordance with the research programme carried out up till then.

Wegener's new basic idea, though sketchy at first, consisted of the bold assumption that the continents of the Earth's crust are not anchored to the internal mass of the Earth, but that, like icebergs on the sea, they swim on the heavier material of the Earth's mantle, and have moved laterally. As in the development of other geoscientific hypotheses, Wegener did not do precisely that which he should have done according to methodological falsificationism, making only peripheral attempts to 'falsify'[51] the dominant theory and its basic ideas; with regard to his own ideas he matched Popper's image of the conscientious researcher just as little. He wanted to be completely right, and endeavoured to defend his own hypothesis, making no attempt to refute it. Wegener and his first followers undertook positive heuristics in accordance with sophisticated falsificationism: by creating hypotheses in the most varied of fields, he began to construct a protective belt around the hard core of his basic assumptions. These empirically founded hypotheses with testable consequences included geophysical and petrographical assertions as to the structure of the Earth's oceanic and continental crust, the geographic position and direction of mountain chains, morphologic, tectonic, and structural similarities between land areas now separated by oceans, paleoclimatology, and faunal and floral evolution. This laid down a new research programme which Wegener supported since he considered it to be a progressive one. The individual hypotheses not only thought to explain all phenomena explained by the earlier theory; they also explained findings which had hitherto not been understood, and contained excess empirical content, for they opened new paths for research, and promised the discovery of new facts.

Wegener's hypothesis was rejected by the majority of researchers in the Earth sciences, since they did not at all feel themselves to be in a crisis forcing them to surrender the basic assumptions which they had never held in doubt. The following reasons may have had particular importance.

1 The accepted geoscientific hypotheses and theories had not been falsified by Wegener in a sufficiently detailed and thorough manner. It was thought possible to eliminate the anomalies and inconsistencies pointed out by Wegener by corrections and auxiliary hypotheses in the tradition of conservative conventionalism.
2 The main evidence produced to support the 'theory of continental drift' as Wegener called it, could all be explained by the prevailing theories, though not all of it fully satisfactorily. The excess empirical content of Wegener's theory in the first publications was evident more in promise than in substance.
3 The hypothesis of the lateral mobility of the great continental masses was not consistent with the existing geophysical theories on the mechanical properties of the Earth's crust and mantle.

The period between 1915 and the 1950s was marked by the rejection of Wegener's ideas by the majority of influential geologists. Although Wegener and a few other researchers worked at gathering new arguments and applying the new theories,[52] the above-mentioned socio-scientific phenomenon of break-down in rational communication came about between the adherents of the 'fixistic' and 'mobilistic' research programmes. In Germany the school of Hans Stille was particularly influential, building up a large-scale synthesis of mountain building based on the theory of contraction.

The reversal which has been interpreted as a 'scientific revolution' in Kuhn's sense was triggered in the 1950s by the discovery of the ocean floor phenomena described elsewhere (p. 246): the young age of the oceanic sediments, the ridge-and-rift system of the oceans with their basaltic volcanism, the symmetrical pattern with respect to these ridges of magnetisation in the submarine basalts. The hypothesis of ocean spreading was successfully worked out in accordance with Wegener's ideas. This not only explained the observations on the ocean floor, but united the hypothesis first sketched out by Wegener of the mobility of parts of the Earth's crust with the theory of plate tectonics. Wegener's hypothesis had in the mean time been further expanded and had become physically plausible in light of new ideas regarding the upper mantle. The unification of plate tectonics with Wegener's theory produced the mobilistic research programme, which has since been accepted by nearly all researchers in the Earth sciences.[53]

The following features can be recognised in the further development of the programme up to the present. Only a few isolated attempts have been made in recent times to rehabilitate the older fixistic programme or to falsify the mobilistic programme. These include the above-treated attempted rescue using conservative-conventionalistic methods (p. 299). Researchers working within the new programme have not attempted any more than Wegener to refute the older theories in detail, nor have they tried in the spirit of Popper to refute their own theories. What has occurred in the plate tectonics research programme since the 1960s consists entirely of positive heuristics. The new theory has encountered extensive application in the most varied areas, as in orogenies, oceanic evolution, volcanism, magmatism of the depths, formation of ore deposits, earthquakes, or the evolution of faunas and floras. The individual hypotheses and thus the theory as an entire programme have been and continue to be in a progressive phase. To be sure one can cite any number of unsolved problems, e.g. regarding the explanation of individual tectonic structures through lateral plate thrusting, or regarding the mechanisms and causes of convectional streaming in the upper mantle which drives the plates; as a whole, however, the hard core of the programme is being corroborated, in that the hypotheses deduced from it not only explain that which is known, but continue even today to lead to the discovery of new facts.[54]

The change in theory initiated by Wegener and the researchers of the 1950s may be called a revolution, if by this nothing more is meant than the fact that within the space of a few years, hypotheses, theories, and regulative principles were dropped which had not been doubted for decades. The term 'conversion' may also be used when an individual researcher suddenly decides to solve his problems in the light of the new theory. This, however, all occurred and occurs in the course of the growth of theoretical knowledge, which can be rationally understood and reconstructed in accordance with the model of sophisticated falsificationism.[55]

8

Regulative principles of geoscientific research

8.1 Introduction

The background from which geoscientific hypotheses and theories are produced contains, as we have seen (p. 262) both objectively valid factors (that empirically founded knowledge recognised by the scientific community at the time) and subjectively valid factors (so designated because they include convictions, preconceptions, and premises which are not or cannot be subjected to scientific testing). These include specialised factors pertaining to individual researchers or small research groups as well as more general factors functioning as a framework for research as a whole. The latter comprise above all the principles which underlie the hard core of research programmes and which remain unquestioned within their context. The oft-noted longevity of such principles and the tenacity with which they are adhered to are due to various reasons. Firstly, they are not inferred inductively from experience, but are predeterminations which alone make possible experiences of a certain kind; secondly they have a heuristic value for research, serving as a beacon to guide researchers along the path considered to be the right one and to save them from aimlessly speculating and probing in the dark. Thirdly, they often help integrate geoscientific theories into larger contexts, particularly into that of a universal science of nature. Because they provide direction for empirical research and for constructing theories, we call them *regulative principles* of geoscientific research.

There is a hierarchy of regulative principles. The highest level consists of general principles of natural science, such as the principles of causality or the principle of homogeneity of space and time. We will not discuss these here, however, but will limit ourselves to looking at the lower, more specific regulative principles which have played and continue to play a role in Earth science in particular. These are considered by researchers to precede experience and are used as a foundation for hypotheses and principles; but insofar as further discourse must decide whether and to what extent theories based on such principles are validated by experience, regulative principles also mediate between the most general 'metaphysical' principles of natural science and objectivised, empirically testable theoretical assumptions.[1]

A study of the role played by regulative principles in geoscientific research is even more dependent on historiography of science than is the study of the dynamics of theories, for it is only by looking back over long periods of development that we can recognise these principles, often not explicitly named, and appreciate their functions and heuristic advantages. In the following discussions we will therefore not limit ourselves as in the pevious chapters to the present state of science and research, but will draw examples from the past history of the Earth sciences.

Of the various principles used by Earth scientists to guide them when creating theories, we stress several typical of the historical character which we have emphasised as being characteristic of Earth science. In investigating Earth's past and prognosticating future events, the Earth sciences have had to deal with problems of real time to a degree not met with in any other natural science except cosmology. For this reason the regulative principles specific to geology provide, first and foremost, rules to deal with the difficult task of using observations of present states of affairs to build up a science of the non-observable past and future.

8.2 Uniformitarianism and actualism

A principle of great importance for the development of the Earth sciences from the end of the eighteenth century till today has been a regulative principle called *Aktualismus* ('actualism') in German, and 'uniformitarianism' in the English speaking world.[2] This principle presupposes *temporal uniformity* of geologic processes and their causes; its general form can be expressed more precisely in the following three statements.[3]

(1) The physical and chemical laws of nature have temporally unlimited validity.

(2) The 'geologic forces', those causes, founded on laws of nature, of the changes in and on the Earth's body, were qualitatively the same in the past as in the present, and will not change qualitatively in the future.

(3) The geologic forces were the same in the past as in the present, and will stay the same in the future not only qualitatively, but also quantitatively, that is, with respect to their energy or work performed.

Statement (1) has not been questioned since Earth science came into existence; it is moreover not a geoscientific principle, but is inherent in the idea of physical and chemical laws of nature and is implicit in Earth science when laws of nature are used to explain phenomena. The principle of uniformity has been held in various forms with respect to statements (2) and (3), however.

Uniformitarianism in the strict sense assumes the validity of statements (1), (2), and (3). This strict *uniformitarianism* was the guiding ideal of James Hutton,[4] celebrated as the 'founder of modern geology' because his ideas committed geological method to the observation of that which can currently be observed upon the Earth. From presently observed events it should be possible to infer timelessly valid laws underlying all geologic processes of the past, present, and future. Retrodictions and prognoses can be carried out with their help, especially as not only the laws, but all initial conditions should be both qualitatively and quantitatively the same throughout the past, present, and future. The subtitle of Hutton's first publication (1785) proclaims the research programme in condensed form: *An investigation of the laws observable in the composition, dissolution, and restauration of land upon the globe*. Through this principle Hutton eliminated the real course of time from the Earth sciences. For him there is no evolution and above all, no beginning and no end; everything is perpetual present; he found in the rocks 'no vestige of a beginning, no prospect of an end.' The strict principle of uniformitarianism created an 'ahistoric' Earth science, which in its pure form was intended to be a science of law patterned after physics and chemistry.

Two advantages obviously attach to this uniformist principle. By restricting himself to the events which can be observed today in various regions of the Earth and to the regularities which can be inductively inferred from them, the researcher gains the heuristic advantage that he need not reckon with the uncertainties of unobserved processes and unusual energies to explain phenomena and retrodictively uncover the geologic past. The future, too, can be anticipated as an extrapolation from the present. An ahistorical geology moreover fits more smoothly with physics and chemistry than historical geology, and achieves there the desired recognition, even if only as an 'application' of the more fundamental sciences of natural law.[5]

Hutton's uniformitarianism consisted of the assertion, based at his time on only a few observations, that the present distribution of land and sea, mountains, rocks, and minerals, were all created by forces and processes which, though generally very slow, are still at work today. Some of these processes take place at the Earth's surface, such as the destruction of rocks and mountains through water and weathering and the deposition of clays and sand in the sea, while other processes take place invisibly at depth, such as the consolidation by subterranean heat of sediments accumulated on the sea floor. In the years after Hutton, researchers who agreed with his principles found that the rapid accumulation of geological discoveries (expansion of the empirical basis) presented them with the task of making the general principle concrete, that is, of interpreting and applying it in the form of hypotheses to explain individual phenomena. This, it turned out, could not be done without restricting the principle in ways which have affected developments right up to the present.

The stoutest defender of strict uniformitarianism was Charles Lyell.[6] His work, which appeared in 12 editions between 1830 and 1874 had great influence in many countries. Its subtitle announced the uniformitarian programme: *Being an attempt to explain the former changes of the Earth's surface by reference to causes now in operation.* In the spirit of this programme Lyell rejected the hypothesis of the constant cooling of the Earth, as well as, until the second to the last edition, the evolution of life from lower forms to higher, worldwide climatic changes, and the catastrophic upheaval of mountain ranges. To explain the individual empirical findings, he created *ad hoc* hypotheses conforming to and protecting the programme from rejection: a protective belt around the hard core. He believed the temperature of the Earth's interior to have remained constant within the geologic past; he admitted that changes had taken place in the world of organisms: old forms died out and new forms took their place, but he believed that there was not sufficient evidence for continuous or for directed evolution.

The *principle of actualism* is a modification of uniformitarianism. This was initiated by K. E. A. von Hoff,[7] and supported by various German geologists at the beginning of the nineteenth century. It is based only on statements (1) and (2), admitting therefore that the geologic forces, while unchanged in kind, have changed or may change in intensity (i.e. in energy). von Hoff believed that certain traces of past volcanic activity can only be explained by the hypothesis that volcanic energies were formerly much greater than today, that volcanic events took place more frequently, and that the volcanic activity of the Earth as a whole has decreased with time. While uniformitarian theories state that igneous rocks such as granites were formed at all times and continue to be formed today, von Hoff believed that the granites of the mountain cores were formed in ancient times when magmas were produced and solidified in far vaster quantities.

A necessary corollary to the uniformitarian and actualistic principles is the regulative *principle of cycles* which can be found even in Hutton. Experience teaches that under the influence of geologic forces the Earth's surface undergoes constant destructive change so that morphologic structures are planified and qualitative differences rendered homogeneous:

mountains are degraded, rivers carry gravels, sands, and clays into basins, seas, and lakes and fill these with homogeneous, mineral-poor sediments created by weathering from the various 'primary' rocks and minerals. The constancy demanded by uniformitarianism and actualism can only be maintained if destruction is from time to time compensated for by construction. Cyclical processes must therefore be assumed, repeating themselves in endless sequences in various regions of the Earth. Out of the material deposited in the sea from the degraded mountains and continents, new land and mountain masses must be built up; from the homogeneous sediments, the qualitative variety of the 'primary' rocks must be produced again. Downwarping must alternate with uplift, and the planification of the Earth's surface with differentiation at depth in order for the uniformitarian or actualistic principle to agree with our experience of constant change. Within any one cycle there are directed sequences through time, but since the cycles are repeated constantly as typical processes, real time is again eliminated overall, and the perpetual present, the real referent of a uniformitarian, ahistorical geoscience, is preserved. The goal of this ahistorical uniformitarianism is to discover the timelessly valid laws at this level within the cyclically repeated processes.

Many hypotheses and theories both of older and of most recent Earth science agree with the principle of cycles, such as the theory of cyclical mountain-building out of thick sedimentary packets derived from the destruction of older mountain ranges and deposited in deep, narrow marine basins (geosynclines); the theory proposed by H. Stille of worldwide 'phases of mountain building' periodically repeating themselves in the same way over and over again; or the geochemical cycles by which chemical elements are thought to travel through the atmosphere, hydrosphere, and lithosphere.

The regulative principle of uniformity, particularly in the form of *actualism*, has profoundly influenced the development of the Earth sciences, and is regarded even today as the most important basis for geoscientific research.[8] Given a phenomenon to be explained, the principle of actualism limits the number of admissible hypotheses. This is heuristically advantageous, even though it often makes it very difficult for the researcher to explain extraordinary and unusual phenomena actualistically; in this case it is often simpler to have recourse to non-actualistic hypotheses. The geologists' unwritten code however, still regards as facetious any attempt, save for very weighty reasons to fall back on a non-actualistic hypothesis as a *deus ex machina*.

The basis for actualistic explanations, hypotheses, and theories is created by all disciplines which investigate current conditions and processes, including soil science, geomorphology, hydrology, limnology, oceanography, sedimentology, volcanology, and, bringing together all important actualistic geologic knowledge, general geology. The importance of these fields in research and teaching reflects not only their practical value, but also the significance of actualism as a fundamental regulative principle of geoscientific research.

Dogmatic uniformitarianism came under critical fire very early. The mineralogist William Whewell[9] accused Lyell of arbitrarily regarding the period in which we live as the standard for all other periods. Without denying the heuristic value of the principle of uniformity, he rejected Lyell's ontological interpretation, stating that it could not be ruled out that catastrophes and cataclysms might have taken place of a kind not recorded in human history.

Since the beginning of the nineteenth century, in fact, Earth scientists have repeatedly disregarded the principles of uniformitarianism and actualism when setting up specialised hypotheses and general theories. They also presupposed the temporal unchangeability of laws of nature, but allowed that the geologic forces may have differed in the past both in kind and in intensity from what they are today. Geoscientific theories have since developed in a field of dialectic tension, with ahistorical actualism and uniformitarianism at the one end, and historical non-actualism and non-uniformitarianism at the other.

In actualistic and uniformitarian theories the Earth's past is a perpetual present. Their aim is to remove the superficially foreign appearance of objects surviving from the past (the rocks, minerals, fossils, and structures) to reveal the present in them and to reconstruct from them the 'past present'. Isolated past events and conditions are not relevant as such to the actualist or uniformitarian, since for him everything geologic occurs and exists in the present as well. The important things, viewed actualistically or uniformitarianly, are the laws which are manifested equally in that which is present as in that which is past.

In non-actualistic and non-uniformitarian theories the Earth's past is not a perpetual present, but the history of 'other times' and events 'foreign to the present'. Along with the laws which affect present, past, and future equally, the main theme of such research concerns the singular nature of events, conditions, and epochs[10] and their sequence through time. It is theories of this type which remove Earth science furthest from physics and chemistry, making it a historical science resembling human history, which is more interested in the specific events and novelties that appear through time than in general laws.[11]

The historical character of non-actualistic and non-uniformitarian theories is manifested in *two notions of the structure of geologic time*. According to one notion, *catastrophes* recur through time: These are brief, sudden, and rare happenings whose intense effects distinguish them from the preceding and succeeding quiet flow of events; they have never been observed due to their rarity relative to the short time period available for human observation. According to the other notion, events are to be understood as a *directed development* of large or small spatial, temporal, or object fields and ultimately of Earth and planetary history as a whole. Catastrophism and development are the two extremes of non-actualistic and non-uniformitarian theories between which there are various intermediate degrees, just as elements of actualism and uniformitarianism appear in predominantly non-actualistic and non-uniformitarian theories.

8.3 Catastrophism

We may lump under the title of *catastrophism*[12] all theories and research programmes whose regulative principles include two elements: the rejection of any strict actualism relating to all events, and the notion that important changes to the Earth's surface, fauna, and flora have been brought about not only by slow, 'actualistic' factors, but also by catastrophic and rare events. The various catastrophic theories proposed since the beginning of the nineteenth century share several features.

Both the catastrophic theories which are non-uniformitarian in assuming the existence of geologic forces which do not appear today, and those which are only non-actualistic in assuming a massive intensification of geologic forces at work today face enormous resistance among geologists both now and in the past, a resistance fed by a *horror miraculi*,[13] for catastrophes appear to break miraculously through the 'normal' course of nature. Geologists are afraid to admit the existence of such events, since they must then give up the security of an actualistic or uniformitarian methodology based on empirical facts. They therefore resist the dangers of boundless speculation lacking adequate empirical foundation. Catastrophic theories moreover create models for past events to which observations of nature as it is today cannot testify. Their plausibility, that is, the certification that they are at all possible, must be based on other things, such as on extrapolations from observations in nature, on small-scale model experiments, or on thought experiments and calculations conforming to general laws of nature. Like all theories, catastrophic theories must be corroborated through empirical confirmation of the deductions derived from them and they are accepted despite initial resistance only to the extent that they are progressive, that is, that they lead to the discovery of new connections and states of affairs.

George Cuvier's *Discours sur les Révolutions de la surface du Globe*[14] is an early example of a catastrophic theory. In the Tertiary sediments around Paris Cuvier observed repeated changes in the lithologic composition of the superimposed strata. Investigating the fossils within the strata, he found that the sudden change in lithologic composition at a stratigraphic boundary was invariably accompanied by an abrupt change in fossil fauna. Such stratigraphic breaks with abrupt changes in fossil content and lithologic composition can be observed in other areas as well. Cuvier believed that they could only be explained by the hypothesis of repeated wide-spread catastrophic floods of a power inconceivable today. These destroyed preexisting life forms and created new conditions in which new life forms arose and different rocks were formed.

The term 'revolution' was also used by L. Élie de Beaumont,[15] who, following Cuvier, proposed the catastrophic theory of mountain building. During the course of Earth's history, the steady decrease in the size and thermal energy of the Earth is supposed to have led from time to time to cataclysmic upheavals of the major mountain chains. These brief episodes of orogenic revolutions which radically changed the face of the Earth are supposed to alternate with longer-lasting quiescent periods, in which, as in the present, earthquakes prove that the cause of orogenies are only 'asleep' as Beaumont put it, not wholly dead.

The catastrophic theories of Cuvier and Beaumont were based on observed discontinuities which they believed could not be explained by continuous and steady processes in the Earth's history. Cuvier based his theories on observations of lithologic and paleontologic breaks along stratigraphic boundaries, Beaumont based his on discontinuities displayed by the upended, folded, and distorted strata raised high in the mountains over their original formational milieu. Both interpreted these discontinuities manifested in space as temporal discontinuities, as sudden outbreaks of otherwise sleeping energies.

A catastrophe theory of more recent origin is the impact theory which we have discussed in earlier chapters (see p. 278). This had to face strong opposition, since it violates to a particularly marked degree uniformitarian and actualistic principles. Moreover, structures of the Earth's surface which cannot have been formed exogenously (through geologic factors active on the surface) are generally attributed to energies and forces within the Earth's interior. This basic rule, regarded as so trivial that it is not explicitly stated, can be called the regulative *principle of endogenous causes*. The fact that this principle is violated by the impact theory has hindered and continues to hinder recognition of the latter's application to individual structures. Where a structure on the Earth's surface could not be explained exogenously, it has been customary to look for local and regional geologic evidence on the surface and subsurface which might distinguish the location of a given structure from other places and form the basis for a hypothesis explaining the structure through the activity of processes and forces of the Earth's interior, and explaining why the event occurred at that particular time and place. Thus the opponents of the theory that the Ries was formed by an impact noted that the Ries Basin lies at the very intersection of two important tectonic trends, that the basement is particularly near the surface here, and that the Ries lies on the extension of a line connecting the volcanic regions of Urach and the Hegau with the Steinheim Basin. All these elements, according to the principle of endogenous causes, mark the site where the Ries formed as geologically distinct from the surrounding area. Similar arguments are presented today with respect to the Vredefort Dome in South Africa, which some authors consider to be a deeply eroded impact crater. Opponents of this hypothesis point out its special geological situation and stress that the structure is aligned with the Bushfield Complex and the Great Dike, both of which are known for great igneous-tectonic activity.

All these arguments are irrelevant to explanations of such structures according to the impact theory. A collision with an extra-terrestrial mass must be viewed as a random event which can happen at any time and any place on the Earth's surface. An explanation by the impact theory implies discarding the principle of endogeous causes and consequently sacrificing the 'geological' method, which consists in searching for clues to an event's cause in peculiarities of the locality's geologic structure as it existed before the event.[16]

8.4 The principle of evolutionism

The developmental laws, particularly the goal-orientated, that is, teleological laws (see p. 184) correspond to the regulative principle of evolutionism. This principle requires as a methodological postulate that a search be made for *a fronte* explanations of events instead of or in addition to causal *a tergo* explanations (see p. 183). Among the infinite number of temporally successive occurrences, those conditions and events must be sought out which are interrelated. This should involve more than simply establishing individual links of 'cause' and 'effect' explainable by causal laws, but rather discovering long-term processes which can be distinguished within the confusion of everything else that ever happened by defining the concept of evolutionary 'development'. Evolutionary developments are processes taking place upon a substrate of continuous or discontinuous changes and characterised by a certain direction or tendency, so that beginning and end stages as well as many intermediary stages may be differentiated. A developmental process is irreversible since every stage embodies a unique situation which never recurs during the course of the development. Each stage is qualitatively new with respect to the preceding one, and in turn influences the new one produced from it. The unity of such a process is expressed as a developmental law.

Before the historical nature of the geologic past as a research problem was diminished in importance by the triumph of uniformitarianism and actualism, a self-evident goal of all hypotheses and theories about the Earth was to reconstruct from the documents of the rocks *geohistory as a directed development*. The regulative principle of evolutionism was particularly prominent in early hypotheses of the seventeenth and eighteenth centuries. According to Descartes[17] and Leibniz,[18] the molten Earth formed a hardened crust initially covered entirely by a paleo-ocean. As the water cover receded through the course of time, continents and islands emerged. During the eighteenth century people began to look more closely at the sedimentary rocks in particular, and many researchers attempted to explain the new empirical facts as evidence of a more or less steady development whose direction and irreversibility were supposed to have been governed largely by steadily falling sea levels. J. L. Leclerc de Buffon[19] drew up a division of past history in seven epochs of terrestrial evolution. Beginning with total coverage by a paleo-ocean the level of the sea was supposed to have dropped constantly through these epochs, while at the same time the internal heat of the Earth, which according to Buffon controlled the climate, decreased constantly, a development which would inevitably terminate in general death from cold.

The classic early geological theory based on the principle of evolutionism was Abraham Gottlob Werner's system,[20] widely disseminated by his students. This clashed with the uniformitarians and actualists in the famous 'plutonist' versus 'neptunist' controversy,[21] but was nevertheless very influential far into the nineteenth century. Following his observations on the stratification in rocks of Saxony, Werner distinguished successively formed 'formations'[22] whose lithologic composition was thought to reflect the evolutionary change of conditions in the seas as they regressed through time.

Concurrent with and despite the ahistorical conceptions of uniformitarianism and actualism, theories following the principle of evolutionism continued to be put forward and the tendency to practice geoscience as a historical science remained alive throughout the nineteenth century and into present times. The evolutionary theories of paleontology may be cited here above all. At the beginning of the nineteenth century, J. B. de Lamarck[23] and E. G. Saint-Hilaire set up the hypothesis, vigorously attacked by Cuvier, of the continuous evolution of living things, supposedly connected to and caused by a corresponding evolution in the abiotic environment. The hypothesis of biological evolution and the discarding of belief in the immutability of species held by uniformitarian biologists and geologists could only take place after Darwin[24] succeeded in ingeniously reconciling the regulative principles of uniformitarianism and evolution through his theory of natural selection. Darwin gave up the dogma of the immutability of species, a dogma which is neither provable nor disprovable. He explained how evolution comes about through a uniformitarian mechanism however, that is, through a mechanism which functions the same way now and in the past as it will in the future. The starting point was provided by the experience shared by all breeders that heritable mutations arise spontaneously in many species. The human breeder can thus bring about new 'races' by deliberate selection. In nature the motor for evolution was and is, according to Darwin, the natural selection of those mutants capable of living and reproducing to a greater degree under given spatially and temporally varying environmental conditions. The ever-operative 'natural selection' generates evolution.[25] Not all paleontologists accept the ahistorical explanation of evolution through natural selection, as is clear from the continued attempts right up to the present to form teleological hypotheses which would account for a direction in evolution as Darwin does not. We have discussed several such teleological hypotheses elsewhere (p. 189).

An example from nineteenth century geology is Élie de Beaumont's theory[26] of the continuous cooling and contraction of the Earth causing sporadic paroxysmal episodes of mountain building. He proposed this theory as early as 1830 and defended it against Lyell's uniformitarianism. The cooling theory in its Beaumontian and later forms is an evolutionary theory, and has been described as such by several authors.[27] It proved quite fruitful in its later forms until replaced in the 1960s by the theory of plate tectonics; over a period of more than 100 years, then, it remained the undisputed foundation of geological theories, especially those dealing with the formation of mountains. We have described elsewhere (p. 283) the tenacity with which Earth scientists clung to the theory of cooling, even though weighty counter arguments from physics were never refuted. The cooling theory was maintained not only because no other mechanism was available to explain crustal movements, but also because Earth scientists did not wish to give up the principle of evolutionism and therewith the conception of a unified earth history.

Beaumont's theory is also an example of a theory which agrees both with principles of evolutionism and of catastrophism. Evolution can take place either continuously or discontinuously through time, but the individual discontinuous events or catastrophes must follow the evolutionary trend of movement, in Beaumont's case, the progressive contraction of the Earth's crust.

Since the middle of the last century the words 'development' and 'evolution' have come to be used with increasing frequency in all the Earth sciences, not only in paleontology. One speaks today of the evolution or development of soils and landforms, of individual mountains, of certain crustal structures, of entire continents, and of the igneous activity of certain areas; of the evolution of the Earth's mantle, of the oceans and atmosphere, and finally of the evolution of the Earth as a whole, of the Moon, and of all the solar planets.

From this increased usage of these words and from the number of evolutionary hypotheses and theories it can be inferred that the regulative principle of evolutionism confers a significant heuristic advantage on geoscientific research. This advantage consists of the fact that this principle makes it possible to review and comprehend conditions and events in their chronological sequence and to set up phenomenological laws without having to provide piecemeal explanations for complex causal relations.

8.5 The dialectics of regulative principles and Earth science as a whole

We have not treated all of the regulative principles which have governed research in the Earth sciences[28] and which have determined the creation of geoscientific hypotheses, theories, and research programmes in particular. As stated at the outset, we have not dealt with general principles of scientific research because we were concerned with the specific features involved in forming geoscientific theories. Thus the principle of causality underlies both actualistic and non-actualistic explanations, and may thus still apply even when sequences of events are regarded evolutionistically *a fronte*. Nor have we discussed principles with more specialised application, as these can be discovered by analysing the reasons for which the theories of Earth science were developed. A particularly typical instance of this for older geology is the principle of the local homogeneity of geologic events, according to which that which was observed or inferred within a limited region was considered to be indicative of world-wide events.

Uniformitarianism, actualism, catastrophism, and evolutionism are without doubt the most important regulative principles in the formation of theories in all the Earth sciences. In discussing how these principles guide the conception of theories, most of our examples have been drawn from the past, for the principles in question are revealed more clearly and openly in older theories than in those of our time. This is not because they have lost their force or importance in current theories: uniformitarianism and actualism, catastrophism, and the concept of evolutionism continue to determine the style and methods of theory and research, but in a different way than before. We will identify this difference more precisely in order to discover a characteristic feature of geoscience as a current research activity in the diachronic context of the history of science, and to further derive perspectives for placing Earth science in the synchronic context of natural science.

We have stressed the heuristic advantage to research of giving isolated principles a dominant role. But the decisiveness with which authors formerly prescribed a given principle, a decisiveness that often appears strange to us now, arose from different causes. The principles from which theoretical models and methods were developed tended to be regarded as statements on the 'true' and 'real' nature of the world. Such principles were regulative in function, but ontological in intention.[29] They allowed the construction of homogeneous geological 'world views', each principle guaranteeing universal research methods and subjects both within individual disciplines and in Earth science as a whole. Thus for Lyell and all uniformitarians the geologic universe was a cohesive whole due to the constant and unvarying activity of the geologic forces. Beaumont described the history of the Earth as a uniform process of cooling relieved by catastrophes of mountain formation. Cuvier's concept of revolutions was for him not simply a heuristically fruitful hypothesis for explaining stratigraphic boundaries with abrupt faunal changes; his world picture consisted of a succession of catastrophes which he compared to human history 'shaken by the ravages of war or the oppression of the mighty'.[30] Lamarck, on the other hand, saw in fossils a document of constant change in living forms: for him the whole of life's past consisted of the reality of the evolution of forms.

Ontologically understood principles with a claim to absolute validity must lead to mutually contradictory theories. This is particularly evident in the conflict between theories based on the principles of actualism and historical theories based on the principle of evolutionism. Stubborn polemical conflicts have often taken place between the adherents of rival theories, such as the conflict between the historical Neptunists of Werner's school and the ahistorical plutonist followers of Hutton and Playfair, Cuvier's polemics against Lamarck and Geoffroy Saint-Hilaire, or Lyell's rejection of Beaumont's views. These and other controversies were based on antimonies which were unresolvable so long as the parties involved insisted on the absolute validity of their principles and respective world views.

Geoscience today embraces the entire heritage of regulative principles developed during its history, but total domination by any one principle has been replaced by a plurality of regulative principles. We have stressed the importance of actualism for research in many disciplines. Yet the principle of actualism which provides Earth science's closest link to the so-called 'exact' sciences is no longer the sole guideline in constructing hypotheses and theories. Phenomena are being explained with increasing frequency by means of theories which would be inadmissible according to strict uniformitarian or actualistic views. Thus in paleontology the principle of evolutionism has long governed the interpretation of phenomena as documenting a history of life, yet paleoecology (the reconstruction of past environments) operates primarily along the lines of actualism. In reconstructing the abiotic history of Earth the present trend follows Whewell's criticisms, quoted above, rather than an actualistic dogma despite the great success of actualistic methods, hypotheses and theories. The actualistic principle is still fruitful and binding for many hypotheses and theories in general geology, petrogenesis, and geochemistry, yet in abductive inference one also treads the uncertain grounds of historical hypotheses and theories which assume that conditions in past epochs may have differed from those of the present, and that events and developmental processes which have never been observed many have been effective.

Certain volcanic rocks, for example, previously interpreted as solidified lavas, are now called 'ignimbrites' and interpreted as deposits of enormous eruptions of glowing clouds never observed in Recent volcanoes; thick sedimentary packets of clastic rocks occurring in many folded mountains are interpreted as 'turbidites', deposits of sudden mudflows along submarine continental margins, although such catastrophic events have not yet been observed in oceans today; structures and breccias on the Earth and other planets are interpreted as the effects of impact catastrophes the like of which no human has yet experienced.

Now that research has reached back into the earliest epochs of Earth's history, a flood of hypotheses and theories have been proposed contradicting the principle of actualism. There is no doubt, for instance, that in the long periods when no plants existed yet upon the continents, the same physical and chemical laws governed transport and deposition as now, but it is reckoned that these geologic processes may have differed from today in kind and intensity since the protective blanket of vegetation was lacking and the sea water and atmosphere had a different composition. In each case one must decide how to explain something in accordance with the principle of actualism, and under which conditions to abandon the path prescribed by actualism. When, for instance, conglomerate strata containing clasts of pyrite and uraninite are found in Precambrian sediments, the physical processes of transport, rounding and deposition of the clasts will be explained using actualistic principles patterned after processes taking place today along the sea shore or in rivers. It would be difficult, however, to explain the chemical composition of the clasts actualistically, for pyrite in contact with the atmosphere would now be oxidised to iron oxide and sulphate, while uraninite (an oxide of U^4) would be oxidised to compounds of U^6. If absolute allegiance to actualism is given up, the hypothesis becomes possible that the Earth's atmosphere contained no free oxygen at that early time. This hypothesis, of course, cannot be left as an *ad hoc* assumption but must be supported by theoretical considerations, such as the gradual production of atmospheric oxygen out of carbon dioxide by the photosynthesis of the developing plant world, or observations on the mineralogic and chemical composition of rocks of early formational epochs. The most extreme non-actualistic theories are those of planetology, based on research of the Moon, Mars, and Mercury concerning the evolution of all the inner planets. Events which have never been observed figure prominently in these, such as a vigorous bombardment from outer space experienced in very early epochs by the Earth and other planets.

Decisions for or against the principle of actualism, the evolutionary view of processes, or the role of cataclysmic events may be based upon, but are not necessarily dictated by empirical discoveries. This is evident, for instance, in the controversies which break out even today when hypotheses are suggested which either contradict the actualistic principle canonised since Lyell's time, like the idea of the mobility of parts of the Earth's crust, or as in the case of the impact theory, revive the generally suspect theory of catastrophism. Regulative principles still serve as guidelines today, opening up specific methodological and argumentational pathways to the researcher. Even the methods of treating a phenomenon to form an empirical basis are determined by preconceptions which can be traced back to regulative principles. This is true both for selecting phenomenological observations and for designing experiments. Phenomena which may be ignored as anomalous in the context of permanently valid regularities or laws may turn out to be of the greatest relevance in terms of the possibility of catastrophic events or of geohistoric changes in important conditions. Regulative principles used as a basis become further integrated in hypothetical explanations, theories, and research programmes and determine their direction or tendency,[31] but they do not guarantee their 'correctness'. Unlike before, when such principles were understood ontologically, the scientific community now decides according to the state of inquiry and the degree to which hypotheses and theories have been corroborated whether it is justified to interpret a certain class of phenomena by means of one of the regulative principles.

The function of regulative principles in geoscientific research corresponds to that of Kant's transcendental ideas,[32] which 'determine the use of the understanding within the whole realm of experience, according to principles' (A 321). Operating by the rules of logic they direct the understanding to fulfill the needs of reason, that is, to systematise the multiplicity of phenomena into an unconditioned whole, such as an evolutionary series unfolding through time, or a cyclically renewed and therefore permanent configuration. A transcendental idea, however, is not 'a *constitutive principle* of reason, enabling us to extend the concept of the world of sense beyond all possible experience, but it is merely a principle of the greatest possible continuation and extension of our experience, allowing no empirical limit to be taken as an absolute limit. It is therefore a principle of reason, which, as a *rule*, postulates what we ought to do in the regressus, but does *not anticipate* what may be given in the object, before the regressus. I therefore call it a regulative principle of reason' (A 509).

Understood in this light, what regulative principles can and cannot accomplish in research becomes clear. Regulative principles, of which we have named the most important, though perhaps not all, are principles of the use of understanding. They point the way towards the goals of hypotheses, theories, and research programmes even when these goals can never absolutely be achieved, and they determine the problems with which the understanding must grapple in research. Since as transcendental concepts they say nothing about the world itself, but relate solely to our attempts to organise empirical phenomena within a totality of explanatory contexts, any contradictions between such principles are wholly illusory. Thus, a contradiction appears to exist between theories of historical and ahistorical character: in one case an empirical state of affairs is interpreted wholly as a result of perpetual and omnipresent forces and conditions, while in the other case it is interpreted wholly as a developmental stage. However the two are not *de facto* mutually exclusive; a given piece of rock can be explained on the basis of its chemical and mineralogic composition and fabric within the context of all geochemical petrological laws of rock formation, or on the basis of specific features as a product of certain unique conditions of a past epoch, i.e. by being incorporated into a different frame of reference of geohistorical development. The one explanation can belong to the arguments of an actualistic theory of rock formation, the other to arguments of a geohistorical theory of the Earth's evolution, each of which can be confirmed or disconfirmed in the course of scientific inquiry.

Geoscientific theories, regardless of how certain they may seem, can no more constitute a geological world view as a certified representation of reality in time and space than can their underlying regulative principles.

Theories transcend empirical experience; for this reason they must always remain provisional and subject to correction; they are neither 'true' nor 'false', but are regarded as valid or invalid to the degree to which they have been corroborated or disconfirmed. Yet the quest for truth remains a legitimate impulse of research. Every scientist desires to know 'how it actually happened' (Ranke). Even if he recognises that no absolutely certain knowledge can ever be obtained through science, still he hopes to contribute to the process whereby in the succession of changing theories, the truth is approached.[33] Such a hope should be based on two conditions.

A growth of knowledge that approximates truth, as Peirce recognised,[34] can not take place through solitary inspirations but only within a community of scientists, whose affirmation or rejection of theories through discourse decides whether theories are corroborated or refuted. Science and research are therefore, as we stated in the beginning, necessarily a form of collective practice.

Secondly nothing hinders the growth of knowledge so much as the prejudice created by the tyranny of a single regulative principle. It was formerly believed that the integrity and autonomy of Earth science could only be guaranteed by the primacy of a single principle, particularly that of actualism. In geoscientific research at present, however, there are a multiplicity of regulative principles, some of which partially contradict each other. What nevertheless unites the various disciplines into a whole of *one* science is not only the material basis of a common subject, but the very dialectic of regulative principles, particularly that between actualism and non-actualism. Research in all disciplines is characterised by this tension. All phenomena can be interpreted on the one hand through endlessly repetitive sequences of timelessly valid laws of nature ('like the dead striking of the clock's pendulum' – Schiller), and on the other hand as the result of unique events and situations which will never recur in the irreversible course of Nature's history.

This double aspect determines Earth science's special position among the natural sciences. Earth science not only, like all natural sciences, transcends that which can be experienced here and now through the search for laws which are valid for all times and places, but also transcends the experiential world of today through hypotheses and theories concerning the irreversible, non-recurring past of nature and the events yet to occur in nature's future.

NOTES

Chapter 1: The structure of communications in Earth science

1 Compare Kuhn (1969: 210) 'Scientific knowledge is intrinsically the common property of a group or else nothing at all. To understand it we shall need to know the special characteristics of the groups that create and use it.' To do this in detail is the task of the sociology of science, as presented in Weingart's review (1976).

2 Compare Ravetz (1971: 88ff). 'Although tools are auxiliary to the advancement of scientific knowledge, their influence on the directions of work done is important and frequently decisive. New tools make possible the production of entirely new sorts of data and information' p. 90.

3 In his 'postscript' to his influential book *The structure of scientific revolutions*, Kuhn emphasises that an investigation of the communications structure of a science is one of the most important tasks of research into science. 'If this book were being rewritten, it would therefore open with a discussion of the community structure of science.' Kuhn (1969: 176). We follow Kuhn's advice here, using, however, not only sociology of science as a basis, but also work on linguistics and philosophy of language, as these provide the most appropriate systematic framework for our analysis.

4 The term goes back to Peirce, who used it, however, as a deliberate idealisation. As a category in sociology of science, it corresponds to Fleck's 'thought collective'. Fleck's treatise *Genesis and Development of a Scientific Fact* which appeared in 1935 as *Introduction to the doctrine of style of thought and thought collectives* exercised great influence on Kuhn, and has since been republished. On the significance of Fleck, see the introduction by L. Schäfer and T. Schnelle in the new edition of the work.

5 On the 'fine structures' of the scientific community, see Hagstrom (1965).

6 An introduction to these is furnished in Weingart (1976), and in both of Weingart's collected essays on the sociology of science (1972).

7 'Discourse' is a key concept in contemporary theory of knowledge and philosophy of language. Compare Apel (1973), Habermas (1973 & 1976) and Schnädelbach (1977).

8 Wunderlich (1974).

9 *Proceedings of the lunar and planetary science conferences* (1970–80).

10 Merton has labelled this the 'Matthew effect', referring to the evangelist Matthew, who states 'whosoever hath, to him shall be given' (Merton, 1968).

11 Solla Price (1963: 80ff).

12 On the concept of 'identity-orientated communication', see Schlieben-Lange (1975: 98f).

13 Jakobson (1960) The basic functions of object-related, speaker-related, and hearer-related linguistic use were developed in the linguistic theory of K. Bühler (1934) as the trio of 'representation' *(Darstellung)*, 'expression' *(Ausdruck)*, and 'appeal' *(Appell)*.

14 When we speak here and subsequently of 'the' author, we mean this to include a group of people in cases where more than one person takes credit for a text.

15 Kant distinguishes 'problematic', 'assertoric', and 'apodictic' modes of judgement. 'The problematic are accompanied by the awareness of the mere possibility, the assertoric by the awareness of the reality, and finally the apodictic by the awareness of the necessity for judgement', *(Logik* A 169). In place of Kant's 'problematic', we use the term 'hypothetical'.

16 In the assertoric mode verbs such as 'describe', 'observe', 'determine', 'record', or 'identify' are used.

17 On the distinction between 'states of affairs' *(Sachverhalte)* and 'facts', see Stegmüller (1969–73 vol. I: 82 and 250 ff).

18 Carnap (1942).

19 We therefore deliberately use a 'pragmatic' concept of explication here, in the sense of a context and group-specific 'elucidation' of the meaning of expressions, not as a logically founded synonymy postulate as in Carnap (1947). For a criticism of this stricter concept of explication, see Quine (1960: 258 ff).

20 On types of definition, see p. 46.

21 Lavoisier provided a classic formulation of this in his development of a new chemical terminology. 'And thus, while I thought I was dealing solely with the nomenclature of chemistry, and that my efforts were directed towards perfecting chemical language, my work changed imperceptably in my hands into an elementary tract on chemistry. The impossibility of differentiating the two emerges from the fact that all natural sciences necessarily deal with three things: with the sequence of facts which is its actual subject, with the ideas with which we recall these facts, and with the words through which we designate them. The words are intended to recall the ideas, the ideas should reflect the facts. These form three prints from one and the same plate, and since it is the words which capture and transmit the ideas, it follows that we cannot perfect language without science, nor science without language.' (After Cassirer, 1907: 440ff.)

22 Some examples of negative expressive statements are: 'The mania of plate tectonics has spread so monstrously rapidly and uncritically as to give an eerie feeling.' 'Overlooking doctrinary scientific philosophers, whose publications are based less on a knowledge of natural science than on an arbitrary interpretation of the gaps in our knowledge and in the texts of natural philosophy, . . . we find . . . a period of time in the Precambrian from which sediments have been preserved which . . . contain structures that . . . can be interpreted as microorganisms.' 'The interpretation of the Ries as a meteoritic crater has been repeatedly attempted, but only by people who know the Ries but scantily.'

23 In expressive statements of the positive kind, the author uses terms such as 'expect', 'hope', 'intend'; he notes that 'there can be no doubt that . . . ' 'it is therefore clear that . . .', 'all arguments and observations seem to me to indicate that . . .', he is 'fully conscious of the significance of the new observations', he considers his conclusions to be right with 'a probability bordering on certainty', believes that certain observations 'are particularly

important', and that it is his 'agreeable duty to thank his colleagues for valuable suggestions and encouraging discussions.' A special case of the expressive use of language is found in statements of aesthetic evaluation, e.g. 'the lavas of this phase can be recognised . . . by the abundance of beautiful plagioclase, hypersthene and biotite phenocrysts.'

24 This must be especially emphasised here, as it is still a widely-held opinion that 'value judgements' do not or should not occur in the natural sciences. Against this view Ravetz (1971) rightly stresses that evaluations are needed at every stage of a scientific investigation. 'Judgements of value are commonplace in the informal, conversational communication among scientists' (p. 162); they are, however, also a normal component of the argument within a scientific text. Ravetz distinguishes three types of value criteria involved in weighing-up a scientific problem. (1) Criteria of 'internal' worth: 'To what extent the solved problem will, or does, advance knowledge of the objects of inquiry of the field, either directly or through suggested descendent-problems' (p. 162). (2) Criteria of 'external' value: 'The contribution that the completed project makes to the solution of problems or the accomplishment of tasks, outside the given field' (p. 163), and (3) 'subjective purposes of the individual scientist in undertaking a particular project' (p. 165). Even judging the 'internal' and 'external' value, however, implies an act of personal evaluation, even if it is only an evaluation according to the guidelines laid down by the 'recognised' authorities.

25 The term 'persuade' is ambiguous: it can mean 'talk into' or 'convince'. We use it in the former meaning here. The 'positive' alternative aimed at rationally founded conviction through critically reviewable arguments has recently been put forward, particularly by Kopperschmidt (1973).

26 Jakobson (1960: 355f).

27 Appeals to the reader's attention are usually made implicitly through expressive or evaluative statements, using formulae such as 'the high values of . . . are remarkable', or 'of particular importance are . . .'. Direct appeals are rarer, such as 'it must be kept in mind . . .', or 'let us turn now to other authorities'. Rhetorical questions seem to be more common.

28 Kuhn (1962: 35ff) sees 'puzzle-solving' as the most normal trait of 'normal' science. There is no reason, however, why the problem-solving aspect should be restricted to routine questions of daily scientific activity: a 'revolutionary' theoretical hypothesis must also be amenable to statement in the form of a problem.

29 Compare Ravetz (1971: 157ff).

30 Tondl (1973: 126ff).

31 Tondl (1973: 129).

32 An *experimentum crucis* can, of course, only be carried out relative to a certain stage of research. A crucial experiment 'for all time' is impossible. Compare Schäfer (1974: 153ff).

33 For an analysis of argumentation see Toulmin (1958), Naess (1975), Perelman (1977), and Kopperschmidt (1980).

34 Compare Böhme (1975).

35 See Böhme (1975).

36 An elaborate and most influential presentation of this is found in Quintilian, *Institutio oratoria*, 1.4–6.

37 On the role of the 'material' provided in the scientific process of argumentation, see Ravetz (1971: 141ff).

38 Compare Ravetz (1971: 191ff).

39 According to Ravetz (1971: 136ff), the components of a scientific investigation manifest in the outline of a paper can be understood according to the model of

the four Aristotelian causes. The material cause (*causa materialis*) is the natural raw material of the investigation. The efficient cause (*causa efficiens*) is the worker, who, by observation and experiment creates empirical data or extracts possible states of affairs from the literature. The formal cause (*causa formalis*) is the shape of the argument, which, though first embodied during the course of the work, already existed previously as an idea. The final cause (*causa finalis*) is the goal of the work, the intended solution of the problem.

40 Compare Ravetz (1971: 194).
41 'Diagenetische Entwicklung und faziesabhängige Na-Verteilung in Karbonatgesteinen Sloveniens', Ogorelec & Rothe (1979).
42 Geochemische Aspekte der Diagenese von marinen Ton- und Karbonatgesteinen', Wedepohl (1979).
43 Die Sandsteindiagenese im Spiegel der neueren Literatur', Füchtbauer (1979).
44 On the function of scientific papers and their 'life-span' relative to the frequency of their citation, see Solla Price (1963: 74ff).
45 This is a more rapid reduction than the one calculated by Solla Price (1963) for citations from the year 1926 in a scientific journal. He calculates a fall by a factor of two every 15 years for the space of time between 1926 and the present.

Chapter 2: The language of geoscience

1 On the analysis of scientific languages, see Janich, Kambartel & Mittelstrass (1974: 41ff), Ströker (1974), Bar-Hillel (1972), von Weizsäcker (1971: 39ff), Fluck (1976), and Böhme (1978).
2 Compare '*Entstehung und Entwicklung der Fachsprachen im Überblick*' in Fluck (1976: 27ff).
3 On the language of physics, see von Weizsäcker (1971: 61ff); for the language of chemistry, see Wolff (1971).
4 von Weizsäcker (1971: 71).
5 We thus recommend differentiating between 'terminology' and 'nomenclature', terms which are often used synonymously today.
6 See '*Fachsprachliche Normung*' in Fluck (1976: 110ff).
7 A survey of these terms is found in Jakoby (1980) and in Denis and Atwater (1974).
8 See Lüschen (1968).
9 On the semantics of scientific languages, see Carnap (1947), Hempel (1952), Stegmüller (1957), Stegmüller (1969–73 vol. I), Quine (1960), Kamlah & Lorenzen (1967), Lorenz (1970), Lorenzen & Schwemmer (1973), and Janich, Kambartel & Mittelstrass (1974).
10 Stegmüller (1969–73 vol. I: 251).
11 Of the very large number of recent studies on this dichotomy, we may mention the paper by Schäfer (1976).
12 Eco, (1968: 140ff).
13 See Quine's influential essay (1951), where the analytic–synthetic distinction is relativised from the pragmatic point of view.
14 On the characterisation of proper names in the philosophy of language, see Quine's discussion of 'singular terms', Quine (1960: 90ff. and 114ff).
15 Another form of simple subject expression used for the identification of singular objects can be omitted here, as it fulfills a function only in certain contexts of oral communication. This is the demonstrative 'this', 'that', 'these' or 'those', which refer deictically as verbal substitutes for pointing to an object.
16 Compare Quine (1960: 90ff).

17 Quine (1960: 91).
18 See the chapter '*Junktoren und Quantoren*' in Janich, Kambartel & Mittelstrass (1974: 60ff).
19 This distinction has become customary in philosophy of science largely due to Carnap (1947).
20 See Zimmermann (1973: 188ff).
21 Lorenzen & Schwemmer (1973: 431ff).
22 'Sense' (*Sinn*) and 'meaning' (*Bedeutung*) are used in very different ways in the philosophy of language and in linguistics. We cannot go into this tangled and often confusing discussion here. A quick survey may be found under the key words '*Semantik*' and '*Bedeutung*' in the *Handbuch wissenschaftstheoretischer Grundbegriffe*, Speck, ed., 1980. Compare also Bunge (1974).
23 Coseriu (1973: 55).
24 On the linguistic characterisation of these terms, see Lyons (1968: 456ff).
25 On the theory of definitions, see Hempel (1952: 2ff), Essler (1970), von Savigny (1970), and Pawlowski (1980).
26 This stands at the core of constructivistic attempts at establishing a rational and critical foundation for scientific argument; see the essay on '*Erfahrung und Begründung*' in Mittelstrass (1974: 56ff).
27 See, for instance, Kant, *Logik*, A 150.
28 Here we follow the distinctions laid down by Lyons (1968: 463ff).
29 On the concept of semantic opposition, see Coseriu (1973: 11ff).
30 This threefold distinction has been fairly long established in philosophy of science; see Hempel (1952: 54ff) and Stegmüller (1969–73 vol. II: 19ff).
31 On operationalism as a position in philosophy of science, see Klüver (1971).
32 Compare Essler (1970 vol. 1: 41).
33 Modern semiotics was founded by Peirce, the most detailed presentation being provided in his letter exchange with Lady Welby, edited by Hardwick (1977). (On Peirce's semiotics, compare Zimmermann (1973) and Oehler (1979). A complete survey of Peirce's philosophy, with special emphasis on its relevance to philosophy of science, is found in Apel's treatment (1975) which contains the introduction to the German selection of Peirce's writings.) The distinction between the syntactic, semantic, and pragmatic dimensions of sign goes back to Morris (1938 and 1946). The introductory presentations by Barthes (1964) and Eco (1968 and 1973) follow the structuralistic tradition. We adhere largely to Bertin (1967), since we are dealing specifically with the analysis of graphic representational forms within scientific texts, as well as with the analysis of maps.
34 Peirce (1970 vol. II: 324).
35 On degrees of 'iconicity' compare Morris (1964: 68).
36 Eco (1968: 200).
37 Bertin (1967: 10).
38 In remote sensing of the surface of the Earth, Moon and planets, pictures are produced not only directly through photography, but also by pointwise measurements of light intensity in various sections of the visible and non-visible spectrum. Through suitable blending of digital data (to suit the spectral sensitivity of the eye) pictures can be formed reproducing the object as the eye would see it. The data can also be blended in other ways (changing the relative intensity of the spectral area, adding to or reducing the spectral intensity, etc.) to create pictures in which the object represented would not readily be recognised by the viewer. Such pictures may convey important information on the morphology or composition of the surface represented. They are used in research as icons, i.e. as representations which 'resemble' the designated object,

even though their iconicity cannot be measured relative to the visually perceived picture.

39 In Bertin's terminology (1967) these include diagrams, charts, and maps.

40 Bertin (1967: 27). 'A graphic representation only makes sense when the invariant and the components are known. A drawing is equipped with a title so that the invariant and components can be recognised without ambiguity.'

41 More on this is found in Bertin (1967: 42ff).

42 Bertin (1967: 147ff).

43 Bertin (1967: 168ff).

Chapter 3: The foundations of geoscientific research

1 For a more detailed account, see Stegmüller (1969–73 vol. II: 293ff).

2 Spinner (1974: 131) calls this the 'hierarchical model of inquiry' (*Stufenmodell der Erkenntnis*), and criticises it in the light of the theoretical preconceptions which underly all scientific research. We consider it here only as a structural scheme useful for an overview of the various levels of geoscientific research. The dependence of what may be called the 'empirical basis' on theoretical preconceptions will be stressed by us in another place (see p. 98).

3 '*Sedimenttransport im Mündungsgebiet des Alpenrheins*', Müller & Förstner (1968).

4 Althaus (1967).

5 '*Prüfung einer theoretisch abgeleiteten Gesetzmässigkeit über den Transport von Schwebstoffen in einem Fluss*', von Engelhardt (1973: 81).

6 Kröner (1981).

7 Compare this to the types of questions discussed in connection with the structure of argument in geoscientific texts, p. 21.

8 From the abundance of literature on the concept of explanations, we particularly recommend Nagel (1961), Hempel (1965), Hempel (1966), Stegmüller (1969–73 vol. I), Essler (1970–79 vol. IV), Rescher (1971), Weingartner (1971), Kutschera (1972 vol. I) and Ströker (1973).

9 Compare here Stegmüller (1969–73 vol. I: 72ff).

10 This model was first developed in the essay 'Studies in the logic of explanation' by C. G. Hempel and P. Oppenheim (1948) (reprinted in Hempel, 1965: 3ff). For more detail, see Stegmüller (1969–73 vol. I: 75ff).

11 Aristotle, *Analytica Hystera*, I, 2; Nagel (1961: 42ff); Stegmüller (1969–73 vol. I: 131ff).

12 On analogy models in science, compare Hesse (1963), Achinstein (1964), Harré (1970: 33ff), Stegmüller (1969–73 vol. I: 133ff) and Weingartner (1976).

13 Peirce, *Schriften* (vol. I: 373ff, vol. II: 357ff); *Collected Papers* (2.508ff, 2.619ff, 2.636ff, 1.65ff, 7.97ff, 7.183ff). In some of Peirce's texts, inference by analogy is presented as a fourth mode of inference (see, e.g. CP 1.69); his explication (e.g. 2.513), however, appears to present it as an operation resulting from a combination of other modes of inference, so that we may omit it here.

14 Within philosophy of science several variants of inductive logic are distinguished which we will not enter into here. See Essler (1970), Kutschera (1972 vol. I) and Stegmüller (1969–73 vol. IV).

15 The term was introduced by Peirce into more recent philosophy of science but has so far found little use. In his early texts, Peirce still speaks of 'hypothetical' inference in contrast to deductive or inductive inference. The term 'retrodiction' is also found (e.g. CP 1.68). Hanson above all (1958) refers to Peirce and Peirce's emphasis on the function of abductive inference; Achinstein (1968) in turn focusses on Hanson. In the logic of Ajdukiewicz (1958) inferences from

the resulting state of affairs to the controlling state of affairs appear under the entry 'reduction': 'Given the facts of reflection, refraction, interference, and polarisation of light, one performed a reductive inference by reasoning from the effects to the cause in stating that, given those facts, light consists of a wave' (p. 161).

16 On the role of deduction in scientific explanation, see the literature cited for note 8, as well as Ajdukiewicz (1958 §19).

17 This is also true of deductive inference which, when used prognosticatively, for example, can refer to future events, that is, to events which cannot be established at the time the inference is made.

18 Peirce: *Schriften* (vol. I: 392).

19 Modern teachings on categories usually combine ontological and semantic considerations since 'things', 'properties', etc. can only be spoken of relative to an existing linguistic structural scheme. See above all Quine (1960), Quine (1969) and Waismann (1965). Texts on categories in the narrower sense can be found in Goodman (1951), Strawson (1959), Körner (1966), and Körner (1974).

20 Körner (1966).

21 On the category of the 'thing', see Körner (1966), Goodman (1951: 93ff), Strawson (1959: 17ff) and Stegmüller (1969–73 vol. II: 191ff).

22 On the principles of scientific classification compare Hempel (1952: 50ff), Apostel *et al.* (1963), and Bunge (1967 vol. I).

23 The literature cited does not mention this category as it can be analytically reconstructed as a complex of properties. Given the preconceptions of the scientist, however, we feel justified in distinguishing categorically between 'configurations' and 'properties'.

24 See Jammer's book (1960) about changes in the concept of space in philosophy and in natural science.

25 On the categorical specification of the field concept see both Körner (1966) and Hölling (1968).

26 For an analysis of time concepts see Reichenbach (1928), Böhme (1966), Grünbaum (1968), and Janich (1969); on geologic time compare Wagenbreth (1966).

27 For the following compare Harland (1978).

28 On the specification of states of physical systems see Stegmüller (1969–73 vol. I: 208ff).

29 Popper (1934: 55ff) distinguishes 'process' and 'event' in a manner which diverges considerably from the colloquial usage. 'We define a "process (P)" as the class of all events P_k, P_l. . . which differ *only* through the variance of the individuals (the space–time positions). Thus, of the sentence, "here and now a glass of water will be spilled" we would say that it is an element of the process "spilling of a glass of water".' We suggest, on the other hand, that the criterion of differentiation (which of course cannot always be unambiguously clear in practice) be the length of the time interval.

30 More on evolution and evolutionary laws will be found in Chapter 6.

31 On the problem of a 'natural' classification, see Duhem (1906: 20ff), Quine (1969), and Schäfer (1974: 179ff).

32 Hubaux (1970, 1973).

Chapter 4: Problems at the empirical basis

1 '*Une expérience scientifique est alors une expérience qui contredit l'expérience commune*'. (Bachelard, 1977: 10).

2 Compare Spinner (1974: 131ff).

3 Carnap (1966: 226).
4 Carnap's Two-level model of scientific language has since been widely criticised. For a survey of this discussion and in particular of the objections raised by Hempel, Kuhn and Feyerabend, see Stegmüller (1969–73 vol. II: 189ff and 293ff).
5 Schäfer (1974: 135ff).
6 Duhem (1906: 189f).
7 Duhem (1906: 192).
8 Popper (1934: 31).
9 Compare Schäfer (1974: 197ff).
10 '*Die homogenen Bausteine der Erdrinde nennt man Mineralien. Die wissenschaftliche Forschung muss versuchen, die Mineralindividuen zu Arten zusammenzufassen*'. Niggli (1941: 1).
11 '*Mineralien sind im physikalischen und chemischen Sinne homogene Naturkörper, die uns fast ausschliesslich in Form von Kristallen oder wenigstens in feinst-bis grobkörnigen kristallinen Aggregaten entgegentreten.*' Machatschki (1953: 1).
12 '*Mineralien sind Kristallarten, die in der Natur ohne Zutun der Organismen entstanden sind, bzw. entstehen können . . . Die Natur, in der wir leben, gliedert sich in das Reich der Mineralien, das Reich der Pflanzen und das Reich der Tiere.*' Strunz (1966: 3).
13 Gary *et al.* (1977: 455).
14 On the criteria for applying substance names or 'mass-terms' see Quine (1960: 91ff and 97ff).
15 Lüschen (1968).
16 That igneous rocks are classified only according to their content of quartz, alkali feldspars, plagioclase, and feldspar replacers ignoring their content of iron and magnesium minerals (mica, amphibole, pyroxenes, etc), is purely a convention founded theoretically on the assumption that the leucocratic components (Q, A, P, F) indicate the petrogenetic formational conditions. Recent research indicates this is only true at a first approximation. Ongoing refinement of petrogenetic theory necessitate taking the iron and magnesium minerals into account and nomenclatorial refinements will necessarily ensue.
17 The history of rock names was discussed briefly on p. 33.
18 The boundaries here are fluid. Fossils, too, may have the character of indices as 'indications' of past living or preservational conditions. A collection of certain fossils ('fossil bonanzas') can, for instance, indicate conditions of a particular environment or the conditions to which the remains of dead organisms were subjected. Trace fossils can be interpreted iconically: the tracks of a dinosaur, for instance, show the shape of its feet.
19 On problems of classification in biology compare Ruse (1969), Pratt (1972), and Oeser (1974).
20 Maull (1958); preceding quotations also from here.
21 Tricart (1965).
22 Machatschek *et al.* (1973).
23 Machatschek *et al.* (1973).
24 Tricart (1965).
25 Popper (1972: 342). On the role of observations in natural science, see also the collection of essays edited by von Körner (1957).
26 '*Aut enim libera est natura et cursu consueto se explicans ut in coelis, animalibus, plantis, et universo naturae apparatu . . . aut ab arte et opera humana constringitur et fingitur, et tamquam novatur ut in artificialibus*' (nature is either free, proceeding in her ordinary course as in the heavens, the animals, plants, and the entire apparatus of the natural universe . . . or it is bound and wrought upon by human means, for the production of things artificial). '*Quemadmodum*

enim ingenium alicuis haud bene noris aut probaris, nisi eum irritaveris . . . similiter etiam natura arte irritata et vexata se clarius prodit, quam cum sibi libera permittitur' (for as man's temper is never well known until he is crossed; in like manner the turns and changes of nature cannot appear so fully, when she is left at her liberty, as in the trials and tortures of art). Francis Bacon, *De dignitate et augmentis scientiarum* (vol. II: 2) quoted from English transl (1853).

27 We have called this simply 'observation' above.

28 On the role of experimentation in the natural sciences, compare Duhem (1906), Dingler (1952), Campbell (1957), Kolb (1963), Hempel (1966: 19ff) and Wartofsky (1968: 181ff).

29 Hempel (1966: 21).

30 On the theory of experimental models' 'resemblence' to conditions in nature, see Hubbert (1973).

31 We will not go into more detail here on the development and problems of cartography in general. An overview of these can be found under the entry '*Kartographie*' in Fochler-Hauke (1968) with bibliographic references.

32 Wagenbreth (1958: 9).

33 On the following compare Harrison (1963).

34 On the dangers of an uncritical use of simplicity as a criterion in science and philosophy of science, see Bunge (1963).

Chapter 5: Problems of inductive organisation of empirical data

1 Compare Schäfer (1974: 132).

2 This is then a law of determinate property. Compare p. 168.

3 See p. 66.

4 E.g. Haseloff & Hoffmann (1970: 105).

5 Reiner (1949: 107).

6 Chappel (1974).

7 'Rules' in contrast to 'laws' have a fundamentally pragmatic character and status. They can often hardly be justified as anything more than 'rules of thumb' which have proved useful in certain contexts but which lack any adequate theoretical or empirical foundation. A 'rule' becomes a 'regularity' when it is empirically sufficiently well founded and when heuristically at least, it suggests theoretical perspectives.

8 Compare here Dence, Grieve, & Robertson (1977).

9 For the following compare Stegmüller (1969–73 vol. IV, Part A: 104ff and vol. IV, Part D: 15ff), von Kutschera (1972 vol. I: 45ff) and Essler (1970–79 vol. III).

10 Carnap (1950); Carnap & Stegmüller (1959).

11 Scientists somewhat naively suppose that the statistical probabilities which they calculate are objectively founded and 'in the nature of things'. Essler on the other hand stresses that no one has yet been able to state 'when the given relative frequencies permit an inference as to certain objective probabilities'. As a rule, any testing of an objective probability occurs, according to him, *without* knowledge of the actual relative frequencies of a certain property in the *entire* domain, and thus implies a generalising inference as to that domain, i.e. assumptions based on *subjective* probability judgements. That there can be no 'immediate' access to objective probability is also due, among other things, to the fact that a statistically assessed sample can never be regarded as wholly representative and invariably valid. (Essler, in Speck, ed., 1980: 695.)

12 Compare Essler: 'We are interested in the relative frequencies with which we must deal *now* (or in the *immediate future*) and we must *evaluate* these on the basis of the data available to us.' (Essler, in Speck, ed., 1980: 694).

Chapter 6: Problems of theoretical knowledge

Laws

1 Stegmüller (1969–73 vol. 1: 273ff)
2 For the following compare Nagel (1961), Stegmüller (1969–73 vol. I: 273ff), Hempel (1966 Chapter 4), Achinstein (1971), and Rescher (1970).
3 von Engelhardt (1973: 230ff).
4 An analysis of this problem, which has since become 'classic' is found in Goodman (1955).
5 Compare here Stegmüller (1969–73 vol. I: 283ff).
6 Nagel (1961: 64).
7 See the discussion above, p. 157.
8 Rescher (1970: 105ff).
9 On the importance of conventions in the formation of laws compare Schäfer (1974: 134ff).
10 Compare Nagel (1961: 75ff).
11 For example by Stegmüller (1969–73 vol. I: 433ff).
12 von Wright (1971: 72).
13 Kant, *Kritik der reinen Vernunft* (B XIV) cited from English edition, p. 503.
14 von Engelhardt (1969: 32).
15 This example of a thought experiment functions heuristically in showing that the imagined event of an impact would, on the basis of physical causal laws, produce the phenomena observed in basins interpreted as impact craters. Not less important are thought experiments with a critical function, which serve to reduce hypotheses *ad absurdum*. Thus Kelvin (p. 283) showed that the hypothesis that the Earth gradually cooled from a molten condition, would on the basis of physical causal laws allow too short a time for the geologic past, providing neither enough time for the slow evolution of organisms, nor for the sedimentation of the great masses of sedimentary rock.

On the use and misuse of thought experiments employed apologetically in defense, or critically for the rejection of hypotheses, see Popper (1934: 397ff). (New appendix, section XI from 1959).
16 Nagel (1961: 74ff).
17 Stegmüller (1969–73 vol. I: 452ff).
18 In Lorenzen and Schwemmer's 'constructive theory of science', for example, the geosciences are dealt with as follows in the chapter 'Theory of historical knowledge': 'History includes . . . not only the origin of culture . . . but also the history of life (with the evolutionary history of plants, animals, and *Homo sapiens*) and the element history of the heavens and Earth (the history of the non-living elements). Element history and life history, as "natural history" can be considered the counterparts of "cultural history"'. (Lorenzen & Schwemmer, 1973: 197.)
19 The self-reliance of the social sciences which claim to rely on hermeneutic methodology is contested by a theory of science orientated towards a deductive nomological ideal of explanation. von Wright, on the other hand, stresses the procedural differences very convincingly. In the social sciences, causal, and therefore deductive-nomological explanations are not relevant *directly*; only teleological explanations are relevant, since these are appropriate to the intentional nature of historic and social subject matter, and therefore necessarily involve understanding and interpreting as specific hermeneutic procedures (von Wright, 1971: 122ff). On the relationship between hermeneutics and philosophy of science, compare Apel (1973 vol. II: 96ff) and Zimmerman (1975).

20 Kant, *Kritik der Urteilskraft* (§§72 and 75, B 320, 337) from English translation, pp. 237, 247.
21 Compare Mach (1883: 359ff).
22 See also Rescher (1970) and Stegmüller (1969–73 vol. I: 510ff).
23 Stegmüller (1969–73), speaks in this case of 'material teleology'.
24 Compare von Wright (1971: 83ff).
25 The most clear-cut contemporary representative of teleological metaphysics in nature in Ernst Bloch. He attributes a goal-pursuing character not to a Creator God, but to matter itself, understanding nature to be directed not towards an ultimate goal, but towards an open future. (Bloch, 1959: 148ff).
26 Compare Viallet (1958).
27 Public lecture at the university of Tübingen, winter semester 1933/34. Quoted after von Engelhardt & Hölder (1977: 224).
28 Kant, *Kritik der Urteilskraft* (§68, B 308) from English translation, p. 230.
29 Stegmüller (1969–73 vol. I: 522).
30 von Wright calls such explanations 'quasi-teleological': these are explanations which 'may be couched in teleological terminology but nevertheless depend for their validity on the truth of nomological connections. Explanations of this kind more frequently answer questions as to *how* something is or became *possible* . . . than questions as to *why* something happened *necessarily*' von Wright (1971: 84).
31 Schindewolf (1950: 198).
32 This should be supplemented by mentioning that teleological explanations of processes pose a considerable problem in contemporary biology. True, it is agreed that evolution in the sense of Darwinian selection is not a teleological process: 'natural selection rewards the events of the past, namely the creation of successful recombinations of genes, but in no way creates plans for the future' (Mayr, 1971: 96). On the other hand, teleological processes have been recognised in organisms and organs of living nature, for which the term 'teleonomic' has been suggested (Pittendrigh, 1958). According to Mayr (1971), goal-directedness in teleonomic processes or behaviours is based on a goal in the action of a program. (A program here is understood as coded and previously established information controlling given processes in such a way that internal or external disturbances are compensated for by a back-linking mechanism, so that a given goal is attained). The program of teleonomic processes are either definitively encoded in the DNA of the genotype in closed programs, or in open programs, encoded in such a way as to enable additional information to be incorporated (e.g. through learning). Teleonomic processes are thus goal-orientated, but still consistent with causal explanations, since the program is a material object of DNA, and existed before the beginning of the process. This corresponds to case M_2–T_2 of our scheme (p. 183).

    Mayr (1971) considers teleonomic processes to be the characteristic feature of the world of living organisms, and calls the goal-directed processes in the inorganic world 'teleomatic'. Whether there are teleonomic processes in ecosystems consisting of several living and inorganic components with their homeostatic mechanisms, and whether changes in some inorganic systems can be interpreted teleonomically, are questions which can only be raised here.
33 Erben (1975).
34 Schindewolf (1950: 228ff).
35 Erben (1975: 222).
36 Schindewolf (1950: 255).
37 Erben (1975: 203).
38 Erben (1975: 229).

39 Erben (1975: 250).
40 Schindewolf (1950: 338).
41 Schindewolf (1950: 430f).
42 Erben (1975: 221, 199, 223 and 282).
43 The description of a probabilistic law in the philosophy of science is closely connected to the analysis of the concept of statistical probability (see the discussion in Section 5.2). Hempel characterises the connection as follows.

'A law of universal form is basically a statement to the effect that in *all* cases where conditions of kind *F* are realised, conditions of kind *G* are realised as well; a law of probabilistic form asserts, basically, that under certain conditions, constituting the performance of a random experiment *R*, a certain kind of outcome will occur in a specified percentage of cases', Hempel (1966: 66).

Chapter 6: Problems of theoretical knowledge

Abduction and deduction

1 Peirce, *Collected Papers*, 7.219.
2 Peirce, *Collected Papers*, 5.171. Since Peirce's discussions of philosophy of science have scarcely been appreciated by mainstream philosophers of science, we reference here the most important sections of his *Collected Papers* (CP) dealing with the various types of inference and their uses within scientific inquiry. (The summarising titles were given by the editors). CP 1.82 (Generalization and abstraction); CP 1.92ff (Reasoning from samples); CP 1.99ff (Observation); CP 1.116ff (The paucity of scientific knowledge); CP 1.120ff (The uncertainty of scientific results); CP 2.508ff (Induction and hypothesis); CP 2.619ff (Deduction, induction and hypothesis); CP 2.636ff (Empirical formulae and theories); CP 5.574ff (Methods for attaining truth); CP 5.590ff (On selecting hypothesis); CP 7.97 (Kinds of reasoning); CP 7.110ff (Kinds of induction); CP 7.162ff (Regularity and explanation); CP 7.202ff (Abduction, induction, deduction); CP 7.208ff (Three kinds of induction); CP 7.218ff (Abduction); CP 7.223ff (The logic of history); CP 7.280ff (Measurement); CP 7.327ff (Observation and reasoning). On the significance of the 'logic of abduction' specifically, see Habermaas (1968: 143ff) and Apel (1975: 297ff). One of the few more recent works in philosophy of science inspired by Peirce's concept of abduction is Hanson (1958). An up-to-date survey of Peirce's theory of science will be found in Rescher (1978).
3 Ajdukiewicz (1958: 90). He calls this hypothetical form of inference 'reduction'. Many standard works in philosophy of science do not discuss the basic difference between deductive and abductive reasoning. This can lead to serious misunderstandings, particularly in evaluating the methods of Earth science. Thus Carnap defines 'explanation' sweepingly as 'prediction' of an 'unknown fact' with the aid of 'facts already known', and remarks that the 'unknown fact' may even lie 'in the past'.

A geologist may infer from striations on boulders that at one time in the past a region must have been covered by a glacier. I use the term 'prediction' for all these examples because in every case we have the same logical schema and the same knowledge situation – a known fact and a known law from which an unknown fact is derived. (Carnap, 1966: 17).

That this does *not* involve either the same logical scheme or knowledge-situation will be made clear in the following discussions. Peirce, on the other hand, was aware that geology above all is a prime example of a discipline in which abductive reasoning is employed, and that therefore the conditions controlling its theories differ from those which control the deductively-operating theories of physics. The theoretical uncertainties of abduction, he stresses, must make geology an 'arena of controversy'. (Peirce, *Collected Papers*, 5.578).

4 The following scheme is based on an attempted systematisation by Stegmüller (1969 vol. I: 153ff). We ignore unnecessary complications here, and examine only those types of explanation which are actually used in scientific research.

5 Carnap (1966: 14) uses the term 'prediction' for every 'deduction' from an explanatory scheme, thereby stressing that in every explanation the conclusion is offered 'later' than the premises, and that the state of affairs to be inferred therefore becomes known 'later' than the initial state of affairs given at the outset. This leads to a misleading reduction of the various types of explanation, as we have explained in Note 3. That the term 'prediction' can have two meanings with regard to the actual temporal relations is rightly noted by Stegmüller (1969 vol. I: 204). 'One can predict a natural event which will only take place in the future, but one can also predict the result of a future human act (e.g. an observation), relating to an already existing objective state of affairs (e.g. the height of a mast)'. In order to avoid such confusions, we will distinguish not only between predictive, retrodictive, and codictive inferences, but also single out 'prognoses' within the realm of prediction, where a prognosis is a prediction relating only to the 'real' future. Bunge (1967 vol. I: 69ff), differentiates between the 'inferred' and the 'predicted' by the terms 'projection' and 'prediction'.

6 The concept of 'prediction' is used analagously in Tondl (1973).

7 On the structure of retrodiction, compare Stegmüller (1969 vol. I: 163ff and 221ff).

8 This rarely used term is found in Tondl (1973: 236); '*codiction*, if i and j are simultaneous'.

9 Compare Stegmüller's definition (1969 vol. I: 154): a prognosis is involved 'when *antecedent conditions* and *laws* are given at the outset and the *explanandum* deduced from them, *before* event e described by E has taken place'.

10 Peirce, *Collected Papers*, 5.172

11 Peirce, *Collected Papers*, 5.173.

12 Popper (1934: 6).

13 Hanson (1958: 90).

14 Peirce, *Collected Papers*, 5.581.

15 Satir (1976).

16 Klootwijk (1976).

17 Haack (1976).

18 Gall, Müller & Stöffler (1975).

19 On the concept of 'chains of explanation' see Stegmüller (1969 vol. I: 120).

20 Schubert & Valastro (1974).

21 On the distinction between abductive and deductive codiction, see p. 222.

22 Pohl & Angenheister (1969).

23 Scrudato & Estes (1976).

24 Kukla, Matthews & Mitchell (1972).

25 Kukla & Matthews (1972).

26 Schwarzbach (1974: 314ff).

27 Stegmüller (1969 vol. I: 165).

28 Hubbert (1977).

29 Klitsch, Sonntag, & Weistorfer (1976).

30 In this context we refer again to Peirce, in whose argument the interplay of abduction, deduction, and induction plays a central role.

31 Compare the case cited by Stegmüller (1969 vol. I: 166) of a 'retrospective rationalisation of an irrational prediction'.

Chapter 6: Problems of theoretical knowledge
Hypotheses and theories

1 We have at various times stressed the basic primacy of theoretical knowledge and it has become widely accepted in philosophy of science. Spinner (1974: 109) summarises it as follows: 'Scientific knowledge is theoretical knowledge *raised to a higher power*. Theories, even covert, implicitly theoretical principles and perspectives rule the process of discovery from beginning to end, in all its phases and dimensions . . . Theories guide even the choice of problems, shape the entire problem-solving situation, and determine the relevance, both of the initial information provided by the problem's history, and of all additional peripheral information (auxilliary hypotheses and 'data' of all kinds)'.

2 On the history of the Ries hypotheses, see von Engelhardt (1982b).

3 See Bunge in particular (1967 vol. I: 222ff). There are detailed discussions on the formation of scientific hypotheses in Nagel (1961), Hempel (1966), Stegmüller (1969–73 vol. II) and Schäfer (1974).

4 On the 'theory' of theory, see the literature in note 3 above, as well as the contributions in Suppe (1974) and the concentrated presentation on '*Theorien und Metatheorien*' in Spinner (1974: 109ff).

5 Bowen (1928: 63–4).

6 A brief summary will be found in Hallam (1973).

7 Compare Bunge (1967 vol. I: 385ff).

8 von Engelhardt (1973: 173).

9 Compare Bunge (1967 vol. I: 391ff).

10 In contrast to a strictly unified theory of scientific concepts (generally drawn from mathematical physics), Rapaport (1958) defends judging theories by a plurality of criteria relative to their specific scientific domain. This is particularly true, of course, for theorising in the humanities, but also to a lesser extent for various kinds of theories in the natural sciences.

11 Here Stegmüller (1969–73) vol. IIA: 124) remarks 'A theory as a *whole* has a factual content. Every theory, however, is simultaneously based on descriptions. It may . . . turn out *that the precise place where we hit upon convention cannot be identified.* Although the theory as an entirety, containing factual content, certainly cannot be transformed into a system of stipulations, still, *individual* statements of the theory can be considered mutually interchangeable either as conventions or as empirical propositions. What in one interpretation is an empirical proposition, can in another interpretation be a stipulation; empirical hypotheses keep reoccurring, however, only in different places.'

12 The most rigorous recent analysis of 'mature' physical theories may be found in Sneed (1971).

13 Compare Bunge (1967 vol. I: 400ff).

14 Carey (1976), Jordan (1966).

15 Van Bemmelen (1976).

16 Meyerhoff (1970).

17 We refer here to the discussion of this problem in the chapter on the empirical basis, p. 98.

18 After Bunge (1967 vol. I: 248ff).

19 Bunge (1967 vol. I: 455).
20 This problem area is generally called 'heuristics' today. Spinner (1974: 174ff) stresses that the influence of Carnap and Popper on philosophy of science had led to a tendency to exclude the 'context of discovery' from analysis, as a result of which the heuristics of theory formation belong to the least studied areas in philosophy of science. Suitable material for such studies could best be found in case studies from the sociology of science and history of science.
21 Compare here Polanyi (1964).
22 Hanson (1958: 90). Hanson is one of the few philosophers of science who turned to problems of heuristics of theory formation relatively early.
23 Spinner (1974: 177).
24 On the problem of theory reduction, compare Nagel (1961) Chapter 11, and Hempel (1966), Chapter 7.
25 As an area of research becomes theoretical, discourse and argument become necessary components of the research process itself. 'The various works of scientists are therefore no longer isolated pieces of information which accumulate to form an explanation of an area or subject; rather they are all inter-related and evaluate each other through mutual support, presentation of problems, or contradiction; together they create new problems and thus cooperatively contribute to the progress of inquiry' (Böhme 1975: 249f).
26 This has been emphasised with particular energy by Quine. According to him, all of science, as an ensemble of theories is 'sub-determined' by experience; only the 'periphery' of science must be kept 'in accordance' with experience (Quine 1951: 42ff).
27 This logical argument underlies Poppers' now-famous refutation of verificationism, verificationism being the demand that scientific hypotheses must be definitely shown as 'true' through empirical confirmation. (Popper 1934: 34ff).
28 In this context Stegmüller stresses that there is not just *one* correct use of the predicate 'observable', but rather a 'whole continuum of possible usages of this term.' 'The boundary between those things to be designated as observable *must* be drawn somewhere if an intersubjectively comprehensible scientific language is to be arrived at. *Where* the boundary is drawn is a matter of *convention*. This naturally does not mean that it should be drawn arbitrarily. The ultimate decision will rest on many theoretical and practical considerations; nor does it need to be made once and for all, but can vary and be adapted to the given scientific purpose' (Stegmüller 1969 vol. II: 190).
29 Bhattacharyya (1974).
30 On this criticism, see Popper (1934), Albert (1968), Chapter 1, and Spinner (1974: 135ff).
31 Popper (1972: 47).
32 It is remarkable that one of the founders of conventionalism in the scientific formation of theories explicitly sticks to the concept of the desire to find the 'true' theory. The more (physical) theory becomes perfected, 'the more we dimly perceive that the logical order in which it describes the facts of experience is a reflection of an ontological order. The more we suspect that the relationships it creates between observational results correspond to the relationships between things, the more we can prophecy that they approach a natural classification. The physicist could not justify this conviction. The methods he uses are limited to the results of observations, and therefore cannot prove that the organisation of the experimental laws is a reflection of an order which transcends experience, equally little can his methods suggest the nature of the real relationships which correspond to the relationships set up by the

theory. But while it is not possible for the physicist to verify his convictions, it is equally impossible for him to deny them. . . . In an intuition which Pascal would have seen as one of the judgements of the heart 'which knows no rationality', he stresses his faith in a real order, of which his theories present a daily clearer and more faithful picture' (Duhem 1906: 30f).

33 Popper (1971: 223).
34 Lakatos (1970).
35 Gilbert (1896).
36 Shoemaker (1963).
37 Shoemaker & Chao (1961).
38 On Kelvin's criticisms, see Hubbert (1967).
39 Weiskirchner (1962).
40 Upham (1894).
41 Popper (1934: 77ff).
42 Bunge (1967 vol. I: 262ff).
43 Rutte (1974).
44 Philosophy of science has lately come to recognise this by developing the 'pluralistic' model of theory testing. See Spinner (1974: 164ff).
45 Feyerabend above all considers such 'competition' to be a constitutive feature of scientific progress. 'The best criticism is provided by those theories which can replace the rivals they have removed.' (Feyerabend 1965: 227).
46 This is shown by historic examples such as the controversy between the adherents of the corpuscular and wave theory of light. Certain experiments which were regarded as 'crucial' were later shown to have been quite rash decisions. See Duhem (1906: 249ff).
47 See Ajdukiewicz (1958: 79).

Chapter 7: The growth of geoscientific knowledge

1 This adoption of a dynamic approach is particularly characteristic for discussions of philosophy of science in recent years. It has led to closer links between philosophy and history of science, both in the sense of mutual *feed-back*, (*Rückkoppelung* (Stegmüller 1969–73 vol. II–2: 11), and in the sense that history of science has even become a 'corrective for philosophy of science' (Schäfer 1974: 216). Today philosophy of science may, *vice versa*, be considered a precondition for a 'rational reconstruction' of historical developments in science. According to Spinner (1974: 180–2), the adoption of a dynamic approach has meant that 'insights by historians of science into the changeability of methodological criteria, standards, and rules, have pulled the rug out from under old and new *sub-specie-aeternitatis* methodologies.' The trend, he says, is towards a '*selective synthesis* differentiating according to typical problem situations ascertained by the history of science; a selective synthesis, that is, of several methodological paradigms within the framework of complex scientific concepts where monomethodological concepts seem inadequate to deal with actual methodological problems of research practice with their changing, very diverse problem situations'.
2 Descartes (1644).
3 Leibniz (1693).
4 Buffon (1749).
5 Hutton (1785).
6 On the character of modern natural science as a research activity, see Schulz (1972: 91ff).
7 Lakatos (1971).

8 Blatt (1978).
9 'Die anomalien der Schwere am Südwestrand des Bayerischen Waldes und ihre Interpretation', Führer (1978).
10 Borsi *et al.* (1978).
11 'Geochemische Trends in der Minette (Jura, Luxemburg/Lothringen)', Siel & Thein (1978).
12 'Zur Herkunft des Quarzzements. Abschätzung der Quarzauflösung in Silt- und Sandsteinen', Füchtbauer (1978).
13 On the growth of scientific 'manpower' in the natural sciences, see Solla Price (1963: 13ff).
14 von Zittel (1899).
15 von Engelhardt (1982a).
16 A systematic overview of the various theories of scientific progress will be found in Spinner (1974: 158ff and 182ff).
17 Compare Schäfer's discussion (1974: 29ff) of Bacon's 'atheoretical model of experience'.
18 Newer inductivism was heavily influenced by the theory of cognition of the physicist, Ernst Mach, who defined scientific research as 'the adaptation of ideas to facts.' Our practical and intellectual demands are satisfied 'as soon as our ideas are capable of completely representing the perceived facts'. The most precise form of such representation is 'quantitative investigation', which culminates in the formulation of functional laws describing the 'permanence of the connections' between the various perceived facts. Mach (1905: 239, 323, 252).
19 von Zittel (1899: 226).
20 Popper (1934: 12ff) also (1963).
21 For the following see Lakatos (1970 and 1971), as well as Schäfer (1974).
22 Dingler (1938).
23 Beloussov (1979).
24 Duhem (1906): a detailed presentation will be found in Schäfer (1974: 107ff).
25 Popper (1934).
26 On the pragmatic and relativistic evaluation of hypotheses and theories and their validity, see Hempel (1966: 45) where he stresses that 'the credibility of a hypothesis at a given time depends, strictly speaking, on the relevant parts of the total scientific knowledge at that time, including all the evidence relevant to the hypothesis and all the hypotheses and theories then accepted that have any bearing upon it; . . . strictly, therefore, we should speak of the credibility of a hypothesis relative to a given body of knowledge'.
27 It should be pointed out that Stegmüller (1969–73 vol. II–2: 11) expressly considers that in addition to the 'feedback between philosophy of science and history of science', 'feedback between philosophy of science and the individual sciences' is also indispensable. This includes, among other things, taking into consideration the scientist's self image as shown in his actual procedures in developing and testing theories.
28 Kuhn (1962 and 1969).
29 Kuhn (1969: 175).
30 The concept of paradigm corresponds more or less to the concept of 'style of thought' (*Denkstil*) in Fleck, whose *Introduction to the doctrine of style of thought and thought collective* in Kuhn's own judgement was a strong influence on Kuhn. 'We may . . . define style of thought as an oriented apprehension, with corresponding mental and material processing of that which is apprehended. It is characterised by the shared features of those problems of interest to the thought collective, by the judgements which the thought

collective regard as obvious, and by the methods which the thought collective uses to make discoveries'. (Fleck 1935: 130).

31 Kuhn (1969: 35).

32 Kuhn (1969: 36ff).

33 Kuhn (1969: 38).

34 Kuhn (1970: 5).

35 We discussed such aspects of the scientific enterprise in Section 1.3, under the heading 'participation in discourse'. See also the discussion in Fleck (1935: 135ff) on the scientific 'thought collective'.

36 Kuhn (1969: 175).

37 See for instance, Lakatos and Musgrave (1970).

38 Fleck (1935: 124) attempts to forge a direct link between upheavals in 'style of thought' and historical upheavals. An empirical justification capable of being generalised for this thesis, however, is hardly possible. The undoubtedly 'revolutionary' development of Darwin's theory of evolution, for instance, fell in an epoch which historians do not regard as a time of significant upheaval, but rather as a period of restoration.

39 Watkins (1970).

40 Kuhn (1969: 198ff).

41 Wilson (1968) and Hallam (1973).

42 In such cases Fleck even uses the term 'thought compulsion' (*Denkzwang*) exercised by a certain thought-collective and its 'style of thought'. 'The organically closed nature of every thought community is paralleled by a stylistic restriction of the problems it allows. . . . From this there arises a specific evaluation and characteristic intolerance: communal features of every closed community. . . . Thought compulsion, thought habit (*Denkgewohnheit*), or at least a decided distaste for foreign styles of thought are what ensure harmony between the thought style and application'. (Fleck 1935: 137).

43 Hallam (1973: 25).

44 Compare Fleck (1935: 142f) 'The greater the difference between two thought styles, the less will be their exchange of thought. . . . The principles of a foreign collective are found, when noticed at all, to be arbitrary, their possible legitimisation to be *petitio principii.* The foreign thought style seems mysterious, the questions it rejects often the very ones which are most important, its explanations inconclusive or irrelevant, its problems unimportant or trifling. Facts and concepts, according to how closely the collectives are related, are either regarded as mere inventions to be simply ignored, . . . or in less divergent collectives they are interpreted differently, that is, they are translated and incorporated into a different thought speech (*Denksprache*)'.

45 Hallam (1973: 113).

46 Lakatos (1970, 1971).

47 The original formulation of this principle goes back to Feyerabend (1965): 'Invent and elaborate theories which are inconsistent with the accepted point of view, even if the latter should happen to be highly confirmed and generally accepted'. See here Schäfer (1974: 74ff).

48 On the history of the theory of plate tectonics, see Hallam (1973), Marvin (1973), Frankel (1979 and 1980), and Laudan (1980). For interpretations according to the Lakatos model, see von Engelhardt (1977), Frankel (1979), and Hallam (1980).

49 Wegener (1912a and b).

50 Wegener (1915–29).

51 Wegener devoted only five pages of the first edition of his book to criticisms of the contraction theory and the supposed sunken land bridges.

52 Successful applications of Wegener's ideas during the early period include in particular the interpretations of Alpine tectonics by R. Staub (1924) and E. Argand (1922), and of the relationships between orogenies on both sides of the North Atlantic (Holmes, 1929), the explanation of the similarities in the Paleozoic and Mesozoic geologic structures of South Africa and South America by A. L. du Toit (1927, 1937), and A. Holmes' attempt (1929) to explain continental drift physically by convection streams in the substrata of the Earth's crust, set in motion by radioactive heat.

53 According to a survey carried out by Nitecki *et al.* (1978) of 215 members of the Geological Society of America and the American Association of Petroleum Geologists, the theory of plate tectonics (or of continental drift) had been accepted by 7% previous to 1940, 22% previous to 1960, and 87% in 1977. Of the latter 87%, 40% considered the theory 'essentially established', and 47% 'fairly well established'. The remaining 11% considered the theory to be still 'inadequately proven'. None of those questioned considered it to be 'erroneous'.

54 The problems brought up by the theory of plate tectonics are so wide-ranging in nature and involve so many disciplines that they cannot be tackled without mutual agreement and planned cooperation between the researchers of all geoscientific disciplines and countries. For this reason, international projects have been set up which were planned, coordinated, and controlled by the committees of international unions (the International Union of Geodesy and Geophysics, and the International Union of Geological Sciences) and financed by the individual countries, something completely new in the history of geoscientific research. The 'Upper Mantle Project' ran from 1965 to 1970, and the 'Geodynamics Project' ran from 1969 to 1979. The following two paragraphs are quoted from the programme for the 'Dynamics and Evolution of the Lithosphere' project, started in 1981 and laid out for ten years. The first paragraph shows that research here is planned entirely in accordance with a research programme in Lakatos' sense, without dogmatically asserting the 'truth' of the theory. The second paragraph illustrates the two main concerns of geoscientific research: the exploration of the conditions and processes in the lithosphere and deeper zones of the Earth's sphere through abductive codiction, and the reconstruction of the history of the Earth's crust through retrodiction.

1. 'The major scientific objectives of the programme are formulated primarily as tests of both the postulates and the consequences of the plate tectonics theory of global geological processes. This has been done, not because the Steering Committee accepts uncritically the 'truth' of the current versions of the plate tectonics theory, but because this theory has had great success in unifying and accounting for vast amounts of independent data and much research will inevitably be directed to searching for fallacies in the theory or developing modifications (including the possibility of radical revisions) needed to bring the theory into accord with all observations . . .'

2. 'The proposed research falls naturally into two major groups of studies, those of the current state and ongoing processes in the lithosphere and subjacent parts of the Earth and those devoted to the reconstruction of the history of the lithosphere . . .' (Proposal of the Steering Committee, International Union of Geodesy and Geophysics, International Union of Geological Sciences, April, 1980).

55 We must here forego an analysis of the term 'rational', although we are aware that it is very difficult to find general consensus on the criteria of rationality. Feyerabend stresses this difficulty, at first admitting that Lakatos comes 'much

closer' to science with his conception of sophisticated falsificationism than preceding accounts of scientific progress. '. . . The methods of revision involve history in an essential way and thus close the gap between the *theory* of knowledge and the material . . . that is *actually* being assembled'. By qualifying a scientific procedure 'in accordance' with that conception as 'rational', however, Feyerabend accuses Lakatos of playing with the ambiguity inherent in this concept: 'In his arguments against naive falsificationism he emphasises the new 'rationalism' of his standards which permits science to survive. In his arguments against Kuhn and against anarchism he emphasizes the entirely different 'rationality' of common sense but without informing his audience of the switch, and so he can have his cake – have more liberal standards – and eat it too – have them used conservatively, and he can even expect to be regarded as a rationalist in both cases'. (Fayerabend 1975: 213 and 199ff). Feyerabend therefore sees adherence to or rejection of theories based on 'progressive' or degenerative problem development as a strategy that has only *pragmatic* justification, since in principle the possibility can never be ruled out that a degenerative research programme might yet recover (Feyerabend 1975: 257ff). This argument is no longer compelling, however, if the criteria of rationality are not looked at *sub specie aeternitatis*, but as a consequence of a collective learning process, which depends naturally on certain historical and social conditions. Exactly when the adherence to a degenerating research programme becomes 'irrational' in the eyes of competent members of the scientific community can only be decided by an exact analysis of the given research situation.

Chapter 8: Regulative principles of geoscientific research

1 The specification of the relationship between metaphysics and science has changed with the determination of thought in theory of science. The invention of alternative theories of scientific progress may certainly initially display traits of a 'new metaphysics' (Feyerabend 1962: 27ff). Schäfer (1974: 81ff) distinguishes between a 'generative', a 'critical', and an 'explicative' function of metaphysics relative to the development of science. For a thorough discussion of this problem see Agassi (1975).
2 These concepts are usually used synonymously, but differed in their original meanings, as we will show.
3 A detailed presentation of the concepts and history of uniformitarianism and actualism is given in Hooykaas (1963, 1970), which we follow here.
4 Hutton (1785). On Hutton, see Bayley (1967), and Eyles (1970).
5 This was doubtless one of the reasons why geology was accepted as a natural science around 1800, and the relationship of Earth science to physics and chemistry is still often seen in this light even today. C. F. von Weizsäcker, for example, cites physics and chemistry as central to all inorganic disciplines, continuing: 'The numerous peripheral disciplines, . . . which, however, deal predominantly with areas of practical importance (geology and paleontology, meteorology, oceanography, etc), can largely be regarded as applications of physics' (1971: 23).
6 Lyell (1830–3).
7 von Hoff (1822–34).
8 Thus a newer textbook of geology states: 'With the formation and general acceptance of the theory of actualism 150 years ago, geology finally entered the circle of the natural sciences. For only when the yardstick of currently valid judgements can be applied to the past do scientifically reliable statements become possible' (Brinkmann 1964: 40).

9 Whewell (1837).
10 The designation of segments of geologic time as 'epochs' contradicts the uniformitarian and actualistic principle, for this term, borrowed from human history, refers to segments of a historical process in which events occur which are new relative to the past and which prepare in turn for the new events of the future.
11 'History is characterised by its interest in actual, singular, or specific events rather than in laws or generalisations' (Popper 1957: 143).
12 The history of catastrophism in geology has been described by Hooykaas (1970).
13 Hooykaas (1963: 162).
14 Cuvier (1825).
15 Beaumont (1829, 1830).
16 The difficulties of such sacrifice are shown by the following statement made by a geologist at the time of the most heated debates on the impact origin of the Ries: 'A meteoritic impact . . . is a slap in the face for research into the Earth's history, since historical geology attempts to show the terrestrial-historic preconditions necessary for the occurrence of a geologic event, to form a special image of a piece of Earth's history' (Hölder (1962)).
17 Descartes (1644) *Pars Quarta: De Terra.*
18 Leibniz, *Protogäa* (1693). Leibniz wished to use this treatise as an introduction for the history he had planned to write about the Welfian rulership and the Welfian lands: Earth history was intended to lead into human history. At the conclusion of the work, Leibniz commented: 'Thus the nature of things shows us the changes in history' (*Ita rerum natura praestat nobis historiae vicem*).
19 Buffon (1778).
20 Werner (1787).
21 von Engelhardt (1982a).
22 In the order of their formation: *Urgebirgsarten* ('paleo-mountain types': granite, gneiss, glimmerschiefer, marble, porphyry, basalt), *Flözformationen* ('Flöz formations: greywacke, clay shales, sandstones, limestones), *Aufgeschwemmte Gebirgsarten*: ('sedimentary mountain types': gravel, sand, marl, clay), and *Vulkanische Gebirgsarten* ('volcanic formations': lava, pumice, ash).
23 Lamarck (1809).
24 Darwin (1859).
25 The degree to which Darwin succeeded in making the evolution of life forms, a conspicuously non-uniformitarian state of affairs, acceptable to the uniformitarians and actualists is shown by the fact that Charles Lyell, the most decisive representative of uniformitarian geology, who had for thirty years defended the immutability of species, immediately agreed with Darwin's theory, and incorporated it into the tenth edition of his book.
26 See note 15 above.
27 As early as 1850, Bernhard von Cotta spoke of an evolutionary law of the Earth (*Entwicklungs-Gesetz der Erde*). In his *Geologie der Gegenwart* (third edition, 1872), he defined this more precisely as the increasing differentiation of the substances and structures of the Earth due to its continued cooling and to the cumulative effects of the successive processes.
28 Probably the most famous 'classic' attempt philosophically to justify the most general principles of modern natural science is Kant's treatise (1786) on the *Metaphysical foundations of natural science*. Since Kant assumed that Newtonian physics formed a universal and eternally valid paradigm, many of the principles which he considered to be 'transcendental' have in historical perspective proved subject to revision. For an evaluation of Kant's foundational efforts, see Schäfer (1974: 46ff). Schäfer argues that the

'transcendental' question has not disappeared, even though its content was by necessity relativised, but that it has only shifted. 'It is no longer aimed at the conditions for the possibility of all experience, but rather at the conditions necessary for the possibility of having new experiences which may be so new that different types of categorical frameworks must be invoked'.

29 James Hutton gives two justifications for the uniformitarian principle. On the one hand he is convinced of the uniformity of nature, since this corresponds to God's plan to create a 'habitable' world; this is an ontological interpretation of the principle. Elsewhere he appears to understand nature's uniformity in a Kantian sense as a transcendental form of our capacity for knowing. He states that 'It is not given to man to know what things are in themselves but only what those things are in his thought' (Hutton 1788: 297). Lyell later adopted a decidedly ontological stance when he called himself 'a staunch advocate for absolute uniformity in the order of nature' (letter to Fleming, 3 February, 1810, according to Hooykaas (1963: 28).

30 '*troublé par les ravages de la guerre ou par l'oppression des hommes en pouvoir*' (Cuvier 1825: 6f).

31 Some large-scale theories characteristically contain hypotheses patterned after several principles. The theory of plate tectonics is non-actualistic insofar as it has given up the concept of the persistence of the oceans and continents and allows for the evolution of the great oceans. On the other hand it contains cyclic hypotheses such as that of the migration of basaltic materials rising in the oceanic ridges and sinking again into the Earth's mantle in the subduction zones. Furthermore the recently proposed possibility that oceanic spreading and continental drift took place even in earlier epochs implies that this process has lost its unique character and may instead involve cyclic recurrences.

32 We follow Schäfer here (1974). 'Finally Kant, in his theory of the regulative functions of ideas, may have handed us a key to solving current methodological problems, for finite reason includes the concepts of an unconditioned whole without which our knowledge cannot take on the form of a system. The function of the idea is therefore to systematise our knowledge, i.e. to develop it to the fullest completion. It has a methodological rank. Although metaphysical ideas are involved, they can function heuristically at the experiential level. Ideas provide the directional guidelines for empirical research, much as do research programmes' (p. 54f). The quotations given are from Kant's *Kritik der reinen Vernunft* (1787), English translation, pp. 206, 310–11.

33 Our formulation here naturally implies a normative demand in the context of research ethics. Prevailing practice in science can deviate far from this ideal, as is shown by a statement of Feyerabend's which at the same time provides a contrast to the developments which we consider desirable.

As opposed to its immediate predecessor, late twentieth century science has given up all philosophical pretensions and has become a powerful *business* that shapes the mentality of its practitioners. Good payment, good standing with the boss and the colleagues in their 'unit' are the chief aims of these human ants who excel in the solution of tiny problems but who cannot make sense of anything transcending their domain of competence. Humanitarian considerations are at a minimum and so is any progressiveness that goes beyond local improvements (Feyerabend 1975: 188).

34 'The real, then, is that which, sooner or later, information and reasoning would finally result in, and which is therefore independent of the vagaries of me and

you. Thus, the very origin of the conception of reality involves the notion of a *community*, without definite limits, and capable of a definite increase of knowledge'. (Peirce, *Collected Papers* 5.311).

'On the other hand, all the followers of science are animated by a cheerful hope that the processes of investigation, if only pushed far enough, will give one certain solution to each question to which they apply it . . . This great hope is embodied in the conception of truth, and reality. The opinion which is fated to be ultimately agreed to by all who investigate, is what we mean by the truth, and the object represented in this opinion is the real. That is the way I would explain reality' (Peirce, *Collected Papers* 5.407).

BIBLIOGRAPHY

Achinstein, P. (1964). Models, analogies, and theories. *Philosophy of Science*, **31**, 328ff.

Achinstein, P. (1968). *Concepts of science*. Baltimore: Johns Hopkins.

Achinstein, P. (1971). *Law and explanation*. Oxford: Clarendon Press.

Agassi, J. (1975). *Science in flux*. Dordrecht: Reidel.

Ajdukiewicz, K. (1958). *Abriss der Logik*. Berlin: Aufbau Verlag.

Albert, H. (1968). *Traktat über kritische Vernunft*. Tübingen: J. C. B. Mohr.

Althaus, E. (1967). The triple point andalusite–sillimanite–kyanite. *Contributions to Mineralogy and Petrology*, **16**, 29ff.

Apel, K.-O. (1973). *Transformation der Philosophie*, **I & II**. Frankfurt: Suhrkamp. Transl. (1979): *Towards a transformation of philosophy*. London: Routledge & Kegan Paul.

Apel, K.-O. (1975). *Der Denkweg von Charles S. Peirce: Eine Einführung in den amerikanischen Pragmatismus*. Frankfurt: Suhrkamp/KNO.

Apel, K.-O. (1976). ed., *Sprachpragmatik und Philosophie*. Frankfurt: Suhrkamp/KNO.

Apostel, L., *et al.* (1963). *La classification dans les sciences*. Gembloux: Duculot.

Argand, E. (1922). La tectonique de l'Asie. Comptes Rendu 13, *Congrés Géologique International*. Liège, 171ff.

Aristotle, *Analytica hystera*.

Bachelard, G. (1977). *La formation de l'esprit scientifique*. Paris: J. Vrin.

Bacon, F. (1623). *De dignitate et augmentis scientiarum*. Cited from Joseph Devey, transl. (1853). *The Physical and Metaphysical Works of Lord Bacon*. London: Henry G. Bohn.

Bar-Hillel, Y. (1972). Language. In *Scientific thought*, ed. Bar-Hillel *et al.*, 107ff. Den Haag: Mouton.

Barth, T. (1969). *Feldspars*. London: Wiley.

Barthes, R. (1964). *Éléments de sémiologie*. Paris: Gonthier. Transl. A. Lavers & C. Smith (1967). *Elements of semiology*. London: Cape.

Battacharyya, D. S. (1974). Criticism on the paper: Regional metamorphism in the Sini-Sonapet quadrangle. In *Geologische Rundschau*, **63**, 558ff.

Bayerisches Geologisches Landesamt (1977). *Geologische Karte des Rieses 1:50 000*. Munich.

Bayley, E. G. (1967). *James Hutton – the founder of modern geology*. Amsterdam: Elsevier.

Beaumont, L. Élie de (1829 & 1830). Extrait d'une série de recherches sur quelqu'unes des révolutions de la surface du globe, présentant différents examples de coincidence entre le redressement des couches de certains systémes de montagnes et les changements soudains des terrains de sédiment. *Annales des Sciences Naturelles* **18 & 19**.

Beloussov, V. V. (1979). Intercontinental structural ties and mobilistic reconstructions. *Geologische Rundschau*, **68**, 393ff.

Beloussov, V. V. (1979). Why I do not accept plate tectonics. EOS *Transactions of the American Geophysical Society*, 207ff.
Bertin, J. (1967). *Sémiologie graphique*. Paris: Mouton Gauttier-Villars. Transl. W. J. Berg (1983). *Semiology of graphics: diagrams, networks, maps*. Madison: University of Wisconsin Press.
Blatt, H. (1978), Sediment dispersal from Vogelsberg basalt, Hessen, West-Germany. *Geologische Rundschau*, **67**, 1009ff.
Bloch, E. (1959). *Das Prinzip Hoffnung*, **I** & **II**. Frankfurt: Suhrkamp.
Böhme, G. (1966). *Über die Zeitmodi*. Göttingen: Vandenhoeck & Ruprecht.
Böhme, G. (1975). Die Ausdifferenzierung wissenschaftlicher Diskurse. In *Wissenschaftssoziologie*, ed. N. Stehr & R. König, Sonderheft 18 of the *Kölner Zeitschrift für Soziologie und Sozialpsychologie*, 231ff.
Böhme, G. (1978). Wissenschaftssprachen und die Verwissenschaftlichung der Erfahrung. In *Sprache und Welterfahrung*, ed. J. Zimmermann, 89ff. Munich: W. Fink.
Borsi, S., Del Moro, A., Sassi, F. P. & Zirpoli, G. (1978). On the age of the Vedrette di Ries (Rieserferner) massif and its geodynamic significance. *Geologische Rundschau*, **68**, 41ff.
Bowen, N. L. (1928). *The evolution of igneous rocks*. Princeton: Princeton University Press.
Boyd, F. R. & England, J. L. (1960). The quartz–coesite transition. *Journal of Geophysical Research*, 65, 749ff.
Brinkmann, R. (1964). *Lehrbuch der Allgemeinen Geologie*. Stuttgart: Ferdinand Enke.
Brockamp, O., Goulart, E., Harder, H. & Heydemann, A. (1978). Amorphous Copper and Zinc in the metalliferous sediments of the Red Sea. In *Contributions to Mineralogy and Petrology*, **68**, 85ff.
Bühler, K. (1934). *Sprachtheorie*. Jena: Fischer.
Buffon, G. L. Leclerc de (1749). *Théorie de la Terre*. Paris.
Buffon, G. L. Leclerc de (1778). *Histoire naturélle des époques de la Nature*. Paris.
Bunge, M. (1963). *The myth of simplicity. Problems of scientific philosophy*. Engelwood Cliffs: Prentice-Hall.
Bunge M. (1967). *Scientific research*, **I** & **II**. Berlin: Springer.
Bunge M. (1974). *Treatise on basic philosophy: Semantics*, **I** & **II**. Dordrecht: Reidel.
Campbell, N. R. (1957). *The foundation of science. The philosophy of theory and experiment*. New York: Dover Publications.
Carey, S. W. (1976). *The expanding Earth*. Amsterdam: Elsevier.
Carnap, R. (1942). *Introduction to semantics*. Cambridge (Mass.): Harvard University Press.
Carnap, R. (1947). *Meaning and necessity*. Chicago: University of Chicago Press.
Carnap, R. (1950). *Logical foundations of probability*. Chicago: University of Chicago Press.
Carnap, R. (1966). *Philosophical foundations of physics*. London: Basic Books.
Carnap, R., & Stegmüller, W. (1959). *Induktive Logik und Wahrscheinlichkeit*. Vienna: Springer.
Cassirer, E. (1907). *Das Erkenntnisproblem in der Philosophie und Wissenschaft der neueren Zeit*, **I**. Cited from new edition (1974), Darmstadt: Olms. Transl. Woglom, W., & Hendel, C. (1950). *The problem of knowledge: philosophy, science and history since Hegel*. New Haven: Yale University Press.
Chamberlin, T. C. (1899). Lord Kelvin's address on the age of the earth as an abode fitted for life. *Science* **9**, 10ff.
Chappell, J. (1974). The geomorphology and evolution of small valleys in dated coral reef terraces, New Guinea. *Journal of Geology*, **82**, 795ff.
Correns, C. W. (1968). *Einführung in die Mineralogie*. Berlin: Springer Verlag. Transl. Johns, W. D. (1969). London: Allen & Unwin.

Coseriu, E. (1973). *Probleme der strukturellen Semantik*. Tübingen: Tübinger Beiträge zur Linguistik.

Cotta, B. von (1872). *Geologie der Gegenwart*, 3rd ed. Leipzig.

Cuvier, G. (1825). *Discours sur les révolutions de la surface du globe, et sur les changements qu'elles ont produits dans le règne animal*. Paris: G. Dufour et E. d'Ocagne.

Darwin, C. (1859). *On the origin of species by means of natural selection, or the preservation of favoured races in the struggle of life*. London: John Murray.

Dence, M. R., Grieve, R. A. F. & Robertson, B. P. (1977). Terrestrial impact structures. In *Impact and explosion cratering*, ed. D. J. Roddy, R. O. Pepin & R. B. Merrill, 247ff. New York: Pergamon Press.

Dennis, J. G. & Atwater, T. M. (1974). Terminology of geodynamics. *American Association of Petroleum Geologists Bulletin*, **58**, 1030f.

Descartes, R. (1644). *Principia Philosophiae*. Amsterdam: Ludovicum Elzevirium.

Dingler, H. (1938). *Die Methode der Physik*. Munich: Reinhardt.

Dingler, H. (1952). *Über die Geschichte und das Wesen des Experiments*. Munich: Eidos Verlag.

Duhem, P. (1906). *La théorie physique, son objet et sa structure*. Cited after the German edition, *Ziel und Struktur physikalischer Theorien*, (1978). Hamburg: Meiner.

Eco, U. (1968). *La Struttura assente*. Milan: Bompani. Cited after the German edition, (1972). *Einführung in die Semiotik*. Munich: Fink.

Eco, U. (1973). *Segno*. Milan: Mondadori.

Engelhardt, W. von (1969). *Was heisst und zu welchem Ende treibt man Naturforschung?* Frankfurt: Suhrkamp.

Engelhardt, W. von (1973). *Die Bildung von Sedimenten und Sedimentgesteinen*. Stuttgart: Schweizerbart.

Engelhardt, W. von (1977). Das Erdmodell der Plattentektonik – ein Beispiel für Theorienwandel in der neueren Geowissenschaft. In *Die Struktur wissenschaftlicher Revolutionen in der Geschichte der Wissenschaften*, ed. A. Diemer, 91ff. Meisenheim: Hain.

Engelhardt, W. von (1982a). Plutonismus und Neptunismus. *Fortschritte der Mineralogie*, **60**.

Engelhardt, W. von (1982b). Hypotheses on the origin of the Ries Basin, Germany, 1792–1960. *Geologische Rundschau*, **71**.

Engelhardt, W. von., Arndt, J., Müller, W. F., & Stöffler, D. (1976). Shock metamorphism and origin of the regolith of the Apollo 11 and 12 landing sites. In *Proceedings of the Second Lunar Science Conference Houston*, **I**, 373ff.

Engelhardt, W. von., & Hölder, H. (1977). *Mineralogie, Geologie und Paläontologie an der Universität Tübingen von den Anfängen bis zur Gegenwart*. Tübingen: Mohr.

Engelhardt, W. von, Hurrle, H., & Luft, E. (1976). Microimpact-induced changes of textural parameters and modal composition of the lunar regolith. In *Proceedings of the 7th Lunar Science Conference Houston*, **I**, 373ff.

Erben, H. K. (1975). *Die Entwicklung der Lebewesen*. Zürich.

Essler, W. K. (1970). *Induktive Logik*. Freiburg: K. Alber.

Essler, W. K. (1970–79). *Wissenschaftstheorie*, **I–IV**. Freiburg: K. Alber.

Eyles, V. A. (1970). *James Hutton's System of the Earth (1785), Theory of the Earth (1788), Observations on granite (1794)*. Darien (Conn.): Hafner Publishing Company.

Feyerabend, P. (1962). How to be a good empiricist. A plea for tolerance in matters epistemological. In *Philosophy of Science. The Delaware Seminar*, **II**, ed. B. H. Baumrin, 302ff. New York: Wiley.

Feyerabend, P. (1965). Reply to criticism. In *Boston Studies in the Philosophy of Science*, **II**, ed. R. S. Cohen & M. W. Wartofsky, 223ff.

Feyerabend, P. (1975). *Against method. Outline of an anarchistic theory of knowledge*. London: NLB.

Fleck, L. (1935). *Entstehung und Entwicklung einer wissenschaftlichen Tatsache. Einführung in die Lehre vom Denkstil und Denkkollektiv.* Basel: Benno Schwabe. Cited after new edition (1980), ed. L. Schäfer and T. Schnelle. Frankfurt: Suhrkamp. English transl. *Introduction to the doctrine of style of thought and thought collectives.* Chicago: University of Chicago Press.

Fluck, H.-R. (1976). *Fachsprachen.* Munich: UTB/BRD/Francke.

Fochler-Hauke, G., ed. (1968). *Geographie.* Frankfurt: Fischer-Taschenbuch-Verlag.

Förstner, R. & Müller, G. (1974). *Schwermetalle in Flüssen und Seen.* Berlin: Springer.

Frankel, H. (1979). The reception and acceptance of continental drift theory as a rational episode in the history of science. In *The reception of unconventional science*, ed. S. H. Manskopf, 51ff. Washington, D.C.: American Association for the Advancement of Science.

Frankel, H. (1980). Hess's development of his sea-floor spreading hypothesis. In *Scientific discovery: Case studies*, ed. T. Nickels, 345ff. *Boston Studies in the Philosophy of Science.*

Freiesleben, J. C. (1817). *Abraham Gottlob Werners letztes Mineralsystem.* Freiberg.

Füchtbauer, H. (1978). Zur Herkunft des Quarzzements. *Geologische Rundschau*, **67**, 991ff.

Füchtbauer, H. (1979). Die Sandsteindiagenese im Spiegel der neueren Literatur. *Geologische Rundschau*, **68**, 1125ff.

Füchtbauer, H., & Müller, G. (1970). *Sedimente und Sedimentgesteine.* Stuttgart: Schweizerbart.

Führer, F. X. (1968). Die Anomalien der Schwere am Südwestrand des Bayerischen Waldes und ihre Repräsentation. *Geologische Rundschau*, **67**, 1087.

Fürbringer, W. (1977). Zur Sedimentologie eines arktischen Deltas. *Geologische Rundschau*, **66**, 577ff.

Gall, H. (1971). *Geologische Karte von Bayern 1:25 000, Blatt Nr. 7328 (Wittislingen).* Munich.

Gall, H., Hüttner, R. & Müller, D. (1977). Erläuterungen zur Geologischen Karte des Rieses 1:50 000. *Geologica Bavarica,* **76**.

Gall, H., Müller, L. & Stöffler, D. (1975). Verteilung, Eigenschaften und Entstehung der Auswurfmassen des Impaktkraters Nördlinger Ries. *Geologische Rundschau*, **64**, 915ff.

Gary, M., McAfee, R., & Wolf, C. L. (1977). *Glossary of geology.* Washington: American Geological Institute.

Gilbert, G. K. (1896). The origin of hypotheses, illustrated by the discussion of a topographic problem. *Science*, **3**, 1ff.

Goodman, N. (1951). *The structure of appearance.* Cambridge (Mass.): Harvard University Press.

Goodman, N. (1955). *Fact, fiction, and forecast.* Cambridge (Mass.): Bobbs-Merrill.

Grieve, R. A. (1980). Impact bombardment and its role in protocontinental growth on the early Earth. *Precambrian Research*, **10**, 217ff.

Grünbaum, A. (1968). *Geometry and chronometry in philosophical perspective.* Minneapolis: University of Minnesota Press.

Gwinner, M. P. (1971). *Geologie der Alpen.* Stuttgart: Schweizerbart'sche Verlagshandlung.

Haack, U. (1976). Rekonstruktion der Abkühlungsgeschichte des Damara-Orogens in Südwest-Afrika mit Hilfe von Spaltspuren-Altern. *Geologische Rundschau*, **65**, 967ff.

Habermas, J. (1973). Wahrheitstheorien. In *Wirklichkeit und Reflexion*, ed. H. Fahrenbach, 211. Pfullingen: Neske.

Habermas, J. (1976). Was heisst Universalpragmatik? In *Sprachpragmatik und Philosophie*, ed. K.-O. Apel, 174ff. Frankfurt: Suhrkamp/KNO.

Hagstrom, W. O. (1965). *The scientific community.* New York: Basic Books.

Hallam, A. (1973). *A revolution in the Earth sciences: From continental drift to plate tectonics.* Oxford: Clarendon Press.

Hallam, A. (1980). How secure is plate tectonics? In *Oceanography: The Past*, ed. M. Sears & D. Merriman, 622ff. New York: Springer.

Hansen, E. (1971). *Strain facies*. Berlin: Springer Verlag.

Hanson, J. F. (1958). *Patterns of discovery. An inquiry into the conceptual foundations of science*. Cambridge: Cambridge University Press.

Hardwick, C. S., ed. (1977). *Semiotics and significs. The correspondence between Charles S. Peirce and Victoria Lady Welby*. Bloomington: Indiana University Press.

Harland, W. B. (1978). Geochronological scales. Contributions to the geologic time scale. In: *American Association of Petroleum Geologists*.

Harré, R. (1970). *The principles of scientific thinking*. London: Macmillan.

Harrison, J. M. (1963). Nature and significance of geological maps. In *The fabric of geology*, ed. C. C. Albritton, 225ff. Stanford: Addison-Wesley.

Haseloff, O. W., & Hoffmann, H. J. (1970). *Kleines Lehrbuch der Statistik*. Berlin: de Gruyter.

Heier, K. S. (1969–78). Rubidium. In *Handbook of geochemistry*, ed. K. H. Wedepohl. Berlin: Springer Verlag.

Hempel, C. G. (1952). Fundamentals of concept formation in the empirical sciences, *International Encyclopedia of Unified Science*, **II**, no. 7. Chicago: University of Chicago Press.

Hempel C. G. (1965). *Aspect of scientific explanation*. New York: Free Press.

Hempel, C. G. (1966). *Philosophy of natural science*. Englewood Cliffs (N.J.): Prentice-Hall.

Hempel, C. G. & Oppenheim, P. (1948). Studies in the logic of explanation. *Philosophy of Science*, **15**, 135ff.

Herrmann, A. G. (1979). Geowissenschaftliche Probleme bei der Endlagerung radioaktiver Substanzen in Salzdiapiren Norddeutschlands. *Geologische Rundschau*, **68**, 1076ff.

Herrmann, A. G., Siebrasse, G. & Könnecke, K. (1978). Computerprogramme zur Berechnung von Mineral- und Gesteinsumwandlungen bei der Einwirkung von Lösungen auf Kali- und Steinsalzlagerstätten (Losungsmetamorphose). *Kali- und Steinsalz*, **7**, 288ff.

Hesse, M. B., (1963). *Models and analogies in science*. London: University of Notre Dame Press.

Hölder, H. (1962). Zur Geschichte der Riesforschung. *Jahreshefte des Vereins für Vaterländische Naturkunde in Württemberg*, **117**, 10ff.

Hölling, J. (1968). Zur Kategorialanalyse des physikalischen Feldbegriffs. *Philosophia Naturalis*, **10**, 343ff.

Hoff, K. E. A. von (1822–34). *Geschichte der durch Überlieferung nachgewiesenen natürlichen Veränderungen der Erdoberfläche*. Gothenburg: Perthes.

Holmes, A. (1929). Radioactivity and earth movements. *Transactions of the Geological Society of Glasgow*, **18**, 559ff.

Hooykaas, R. (1963). *The principle of uniformity in geology, biology, and theology*. Leiden.

Hooykaas, R. (1970). Catastrophism in geology. *Medeling koninklijke Nederlandse Akademie van Wetenshappen* N. R. **33**, no. 7.

Hubaux, A. (1970). Description of geological objects. *Mathematical Geology*, **2**, 89ff.

Hubaux, A. (1973). A new geological tool – the data. *Mathematical Geology*, **9**, pp. 159ff.

Hubbert, M. K. (1937). Theory of scale models as applied to the study of geologic structures. *Bulletin of the Geological Society of America*, **48**, 1459ff.

Hubbert, M. K. (1967). Critique of the principle of uniformity. *Geological Society of America, Special Paper*, **89**, 3 ff.

Hubbert, M. K. (1977). Role of geology in transition to a mature society. *Geologische Rundschau*, **66**, 654.

Hüttner, R. (1956). *Geologische Untersuchungen im SW Vorries auf Blatt Neresheim und Wittlingen*. Dissertation, University of Tübingen.

Hutton, J. (1785). *Theory of the Earth*. Edinburgh: Royal Society of Edinburgh.
Jacoby, W. (1980). Zur Nomenklatur für ein neues geowissenschaftliches Konzept: die Plattentektonik. *Zeitschrift der Deutschen geologischen Gesellschaft*, **131**, 579ff.
Jakobson, R. (1960). Linguistics and poetics. In *Style in language*, ed. A. Sebeok. Cambridge (Mass.): MIT Press.
Jammer, M. (1960). *Der Begriff des Raumes*. Darmstadt: Wissenschaftliche Buchgesellschaft. Transl. (1954) Cambridge Mass.: Harvard University Press.
Janich, P. (1969). *Die Protophysik der Zeit*. Mannheim: Suhrkamp.
Janich, P., Kambartel, F. & Mittelstrass, J. (1974). *Wissenschaftstheorie als Wissenschaftskritik*. Frankfurt: Campus.
Jordan, P. (1966). *Die Expansion der Erde*. Braunschweig: Friedrich Viehweg. Transl. A. Beer (1971). *The expanding Earth*. Oxford: Pergamon Press.
Kamlah, W. & Lorenzen, P. (1967). *Logische Propädeutik*. Mannheim: Bibliographisches Institut. Transl. H. Robinson (1984). *Logical propaedeutic: pre-school of reasonable discourse*. Lanham: University Press of America.
Kant, I. (1787). *Kritik der reinen Vernunft*. Riga: Hartknoch. Cited after F. M. Müller, transl. (1961). *Critique of pure Reason*. New York: Doubleday & Co.
Kant, I. (1786). *Metaphysische Anfangsgründe der naturwissenschaft*. Riga: Hartknoch.
Kant, I. (1790). *Kritik der Urteilskraft*. Berlin: Lagarde. Cited after J. H. Bernard, transl. (1968). *Critique of Judgement*. New York: Hafner Publishing Co.
Kant, I. (1800). *Logik*. Königsberg: G. B. Täsche.
Kimberley, M. K. (1979). Origin of oolite iron formations. *Journal of Sedimentary Petrology*, **49**, 111ff.
Klitzsch, E., Sonning, C., & Weistorfer, K. (1976). Grundwasser der Zentralsahara. *Geologische Rundschau*, **65**, 264ff.
Klootwijk, C. T. (1976). The drift of the Indian subcontinent; an interpretation of recent paleomagnetic data. *Geologische Rundschau*, **65**, 885ff.
Klüver, J. (1971). *Operationalismus*. Stuttgart: Frommann-Holzboog.
Kolb, J. (1963). Erfahrung im Experiment und in der Theorie der Physik. In *Experiment und Erfahrung in Wissenschaft und Kunst*, ed. W. Strolz. Freiburg: Alber.
Kopperschmidt, J. (1973). *Allgemeine Rhetorik*. Stuttgart: Kohlhammer.
Kopperschmidt, J. (1980). *Argumentation*. Stuttgart: Kohlhammer.
Körner, S., ed. (1957). *Observation and interpretation. A symposium of philosophers and physicists*. London: Academic Press.
Körner, S., ed. (1966). *Experience and theory*. London: Routledge & Paul Kegan.
Körner, S., ed. (1974). *Categorical frameworks*. Oxford: Basil Blackwell.
Köster, E. (1964). *Granulometrische und morphometrische Messmethoden*. Stuttgart: Ferdinand Enke.
Kröner, A. (1981). Precambrian crustal evolution and continental drift. *Geologische Rundschau*, **70**, 412ff.
Krumbein, W. C., & Graybill, F. A. (1965). *An introduction to statistical models in geology*. New York: McGraw Hill.
Kuhn, T. S. (1962). *The structure of scientific revolutions*. Chicago: University of Chicago Press.
Kuhn, T. S. (1969). Postscript to *The structure of scientific revolutions*, 2nd. ed. (1970), 174ff. Chicago: University of Chicago Press.
Kuhn, T. S. (1970). Logic of discovery or psychology of research? In *Criticism and the growth of knowledge*, ed. I. Lakatos and A. Musgrave, 1ff. Cambridge: Cambridge University Press.
Kukla, G. J. & Matthews, R. K. (1972). When will the present interglacial end? *Science*, **178**, 190ff.
Kukla, G. J., Matthews, R. K. & Mitchell, J. M. (1972). The end of the present interglacial. *Quaternary Research*, **2**, 261ff.
Kutschera, F. von (1972). *Wissenschaftstheorie*, **I & II**. Munich: UTB/BRO/Fink.

Lakatos, I. (1970). Falsification and the methodology of scientific research programs. In *Criticism and the growth of knowledge*, ed. I. Lakatos & A. Musgrave, 91ff. Cambridge: Cambridge University Press.
Lakatos, I. (1971). History of science and its rational reconstruction. In *Boston Studies in the Philosophy of Science*, **VIII**, 91ff.
Lamarck, J. B. de (1809). *Philosophie zoologique*. Paris.
Langheinrich, G. (1977). Zur Terminologie von Schieferungen. *Geologische Rundschau*, **66**, 336ff.
Laudan, R. (1980). The method of multiple working hypotheses and the development of plate tectonic theory. In *Scientific discovery. Case studies, Boston studies in philosophy of science*, ed. T. Nickels, 331ff.
Lavoisier, A. L. (1864). *Traité élémentaire de Chimie. Oevres de Lavoisier*, **I**. Paris.
Leibniz, G. W. (1693). *Protogäa*. Posthumously published (1748). Göttingen: C. L. Scheid. New edition with German translation by W. von Engelhardt (1949). Stuttgart: Kohlhammer.
Lippmann, F. (1973). *Sedimentary carbonate minerals*. Berlin: Springer.
Lorenz, K. (1970). *Elemente der Sprachkritik*. Frankfurt: Suhrkamp.
Lorenzen, P. & Schwemmer, O. (1973). *Konstruktive Logik, Ethik und Wissenschaftstheorie*. Mannheim.
Lüschen, J. (1968). *Die Namen der Steine*. Thun: Ott.
Lyell, C. (1830–3). *Principles of geology*. London: John Murray.
Lyons, J. (1968). *Introduction to Theoretical Linguistics*. Cambridge: Cambridge University Press.
Mach, E. (1883). *Die Mechanik in Ihrer Entwicklung*. Leipzig: Internationale Wissenschaftliche Bibliothek.
Mach, E. (1905). *Erkenntnis und Irrtum. Skizzen zur Psychologie der Forschung*. Cited from 2nd edition (1906). Leipzig: J. F. Barth.
Machatschek, F., Graul, H. & Rathjens, C. (1973). *Geomorphologie*. Stuttgart: Teubner.
Machatschki, F. (1953). *Spezielle Mineralogie auf geochemischer Grundlage*. Vienna: Springer.
Marvin, U. B. (1973). *Continental drift: the evolution of a concept*. Washington, D.C.: Smithsonian Institute Press.
Maull, O. (1958). *Handbuch der Geomorphologie*. Vienna: Deuticke.
Mayr, E. (1971). Teleological and teleonomic. A new analyses. In *Boston Studies in the Philosophy of Science*, **14**, 91ff.
Merton, R. I. (1968), The Matthew effect in science. *Science*, **159**.
Meyerhoff, A. A. (1970). Continental drift: implications of paleomagnetic studies, meteorology, physical oceanography and climatology. *Journal of Geology*, **78**, 1ff.
Mittelstrass, J. (1974). *Die Möglichkeit von Wissenschaft*. Frankfurt: Suhrkamp/KNO.
Morris, C. W. (1938). *Foundations of the theory of signs*. Chicago: University of Chicago Press.
Morris, C. W. (1946). *Signs, language, and behaviour*. New York: Braziller.
Morris, C. W. (1964). *Signification and significance*. Cambridge (Mass.): MIT Press.
Müller, A. H. (1963). *Lehrbuch der Paläontologie*. Jena: Gustav Fischer Verlag.
Müller, G. & Förstner, U. (1968). Sedimenttransport im Mündungsgebiet des Alpenrheins. *Geologische Rundschau*, **58**, 229ff.
Naess, A. (1975). *Kommunikation und Argumentation*. Kronberg. (German translation follows the 11th edition of the original Norwegian publication.)
Nagel, E. (1961). *The structure of science. Problems in the logic of scientific explanation*. London: Routledge and Paul Kegan.
Niggli, P. (1941). *Lehrbuch der Mineralogie*. Berlin: Borntraeger.
Nitecki, M. H., Lenke, J. H., Pullman, H. W. & Johnson, M. E. (1978). Acceptance of plate tectonic theory by geologists. *Geology*, **6**, 661.

Oehler, K. (1979). Idee und Grundriss der Peirceschen Semiotik. *Zeitschrift für Semiotik*, 1, 9ff.
Oeser, E. (1974). *System, Klassifikation, Evolution. Historische Analyse und Rekonstruction der wissenschaftstheoretischen Grundlagen der Biologie*. Vienna.
Ogorelec, B. & Rothe, P. (1979). Diagenetische Entwicklung und faziesabhängige Na-Verteilung in Karbonat-gesteinen Sloveniens. *Geologische Rundschau*, **68**, 965ff.
Paolo, D. J. & Johnson, R. W. (1979). Magma genesis in the New Britain Island arc. *Contributions to Mineralogy and Petrology*, **70**, 367ff.
Pawlowski, T. (1980). *Begriffsbildung und Definition*. Berlin: de Gruyter.
Peirce, C. S. (1931–58). *Collected papers*, **I–VI**, (1931–5). Ed. C. Hartshorne & P. Weiss. **VII–VIII**, (1958). Ed. A. W. Burks. Cambridge (Mass.): Harvard University Press.
Peirce, C. S. (1967–70). *Schriften*, ed. K.-O. Apel, **I** & **II**. Frankfurt: Suhrkamp/KNO.
Perelman, C. (1977). *L'empire rhétorique. Rhétorique et argumentation*. Paris. Transl. W. Kluback (1983). *The Realm of Rhetoric*. Illinois: University of Notre Dame Press.
Pettijohn, F. J. (1949). *Sedimentary rocks*. New York: Harper & Row.
Pittendrigh, C. (1958). *Behaviour and evolution*, ed. A. Roe and G. G. Simpson. New Haven: Yale University Press.
Pohl, J. & Angenheister, G. (1969). Die seismischen Messungen im Ries von 1948–69. *Geologica Bavarica*, **61**, 309ff.
Polanyi, M. (1964). *Personal knowledge*. London: Routledge & Paul Kegan.
Popper, K. (1934). *Logik der Forschung*. Vienna. Cited after 4th ed. (1971). Tübingen: J. C. B. Mohr. Transl. K. Popper & J. and L. Freed (1959). *The Logic of scientific discovery*. New York: Basic Books.
Popper, K. (1957). *The poverty of historicism*. London: Routledge & Paul Kegan.
Popper, K. (1963). *Conjectures and refutations. The growth of scientific knowledge*. London: Routledge & Paul Kegan.
Popper, K. (1972). *Objective knowledge*. Oxford: Clarendon press.
Pratt, V. J. F. (1972). Biological classification. *British Journal of the Philosophy of Science*, **23**, 305ff.
Proceedings of the lunar and planetary science conferences, (1970–80). **I–XI**. Houston.
Quine, W. O. V. (1951). *Two dogmas of empiricism*. Cited after reprint (1963), *W. O. V. Quine: From a logical point of view*, 20ff. New York: Harper & Row.
Quine, W. O. V. (1960). *Word and object*. Cited after the 1973 edition. Cambridge (Mass.): MIT Press.
Quine, W. O. V. (1969:114). *Ontological relativity and other essays*. New York: Columbia University Press.
Quintilian: *Institutio oratoria*.
Ramdohr, P., ed. (1954). *Klockmanns Lehrbuch der Mineralogie*. 14th edition. Stuttgart: F. Enke.
Rapaport, D. (1958). Various meanings of 'theory'. *American Political Science Review*, **52**, 927ff.
Ravetz, J. R. (1971). *Scientific knowledge and its social problems*. Oxford: Clarendon Press.
Reichenbach, J. (1928). *Philosophie der Raum-Zeit-Lehre*. Berlin: Viehweg. Transl. H. Reichenbach & J. Freund (1958). *The Philosophy of space and time*. New York: Dover Publications.
Reiner, M. (1949). *Deformation and flow*. London: Lewis.
Rescher, N. (1970). *Scientific explanation*. New York: Free Press.
Rescher, N. (1978). *Peirce's philosophy of science*. Notre Dame: University of Notre Dame Press.
Rosenbusch, J. (1910). *Elemente der Gesteinslehre*. Stuttgart: Schweizerbart.
Ruse, M. (1969). Definitions of species in biology. *British Journal of the philosophy of science*, 20, 97ff.

Rutte, E. (1974). Neue Befunde zu Astroblemen in der Schweifregion des Rieskraters. *Oberrheinischen geologischen Verein Abhandlung* 23, 97ff.

Satir, M. (1976). Rb-Sr- und K-Ar-Altersbestimmungen an Gesteinen und Mineralien des südlichen Ötztalkristalins und der westlichen Hohen Tauern. *Geologische Rundschau*, **65**, 394ff.

Savigny, E. von (1970). *Grundkurs im wissenschaftlichen Definieren*. Munich: dtv/KNO.

Schäfer, L. (1974). *Erfahrung und Konvention. Zum Theoriebegriff der empirischen Wissenschaften*. Stuttgart.

Schäfer, L. (1976). Der wissenschaftstheoretische Status synthetischer Urteile a priori. In *Von der Notwendigkeit der Philosophie in der Gegenwart*, ed. H. Kohlenberger & W. Lütterfelds, 267ff. Vienna: K. Ulmer.

Schalk, K. (1957). Geologische Untersuchungen im Ries. Das Gebiet des Blattes Bissingen. *Geologica Bavarica*, **31**.

Schieferdecker, A. A. G. (1959). *Geological nomenclature (English, Dutch, French, German)*. Gorinchem: Kluwer Group.

Schindewolf, O. (1950). *Grundzüge der Paläontologie*. Stuttgart: Schweizerbart.

Schlieben-Lange, B. (1975). *Linguistische Pragmatik*. Stuttgart: Kohlhammer.

Schnädelbach, H. (1977). *Reflexion und Diskurs*. Frankfurt: Suhrkamp/KNO.

Schubert, G., & Valastro, S. (1974). Late Pleistocene glaciation of Paramo de la Culata, north-central Venezuelan Andes. *Geologische Rundschau*, **63**, 516ff.

Schulz, W. (1972). *Philosophie in der veränderten Welt*. Pfullingen: G. Neske.

Schuster, M. (1926). Neues zum Problem des Rieses. In *Das Problem des Rieses*. Nördlingen: Oberrheinischen Geologischen Verein.

Schwarzbach, M. (1974). *Das Klima der Vorzeit*. Stuttgart: Ferdinand Enke.

Scrudato, R. J., & Estes, E. L. (1976). Clay-lead sorption relations. *Environmental Geology*, **1**, 167ff.

Shoemaker, E. M. (1963). Impact mechanics at Meteor Crater, Arizona. In *The Solar System*, **IV**, ed. B. M. Middlehurst & G. P. Kuiper, 301ff. Chicago: University of Chicago Press.

Shoemaker, E. M. & Chao, E. C. T. (1961). New evidence for the impact origin of the Ries Basin, Bavaria, Germany. *Journal of Geophysical Research*, **66**, 3371ff.

Siel, A. & Thein, J. (1978). Geochemische Trends in der Minette (Jura, Luxemburg, Lothringen). *Geologische Rundschau*, **67**, 1052.

Sneed, J. D. (1971). *The logical structure of mathematical physics*. Dordrecht: Kluwer Group.

Solla Price, D. J. de (1963). *Little science, big science*. New York: Columbia University Press.

Speck, J., ed. (1980). *Handbuch wissenschaftstheoretischer Begriffe*, **I–III**. Göttingen: UTB/BRO/Vandenhoeck & Ruprecht.

Spinner, H. (1974). *Pluralismus als Erkenntnismodell*. Frankfurt: Suhrkamp/KNO.

Staub, R. (1924). Der Bau der Alpen. In *Beiträge zur geologischen Karte der Schweiz*, Neue Folge, **52**.

Stegmüller, W. (1957). *Das Wahrheitsproblem und die Idee der Semantik*. Vienna: Springer Wien.

Stegmüller, W. (1969–73). *Probleme und Resultate der Wissenschaftstheorie und analytischen Philosophie*, **I**, **II**, & **IV**. Vol. I (1969), *Wissenschaftliche Erklärung und Begründung*. Vol. II (in two half-volumes 1970 & 1973), *Theorie und Erfahrung*. Vol. IV (in two half-volumes, 1973), *Personnelle und statistische Wahrscheinlichkeit*. Berlin: Springer.

Strawson, P. (1959). *Individuals*. London: Methuen.

Streckeisen, A. (1967). Classification and nomenclature of igneous rocks. *Neues Jahrbuch Mineralogische Abhandlungen*, 107, 144ff.

Streckeisen, A. (1980). Classification and nomenclature of volcanic rocks, lamprophyres, carbonatites and melilitic rocks. *Geologische Rundschau*, **69**, 194ff.
Ströker, E. (1973). *Einführung in die Wissenschaftstheorie*. Darmstadt: Wissenschaftliche Buchgesellschaft.
Ströker, E. (1974). Das Problem der Sprache in den exakten Wissenschaften. In *Aspekte und Probleme der Sprachphilosophie*, ed. J. Simon, 231ff. Freiburg: K. Alber.
Strunz, H. (1966). *Mineralogische Tabellen*. Leipzig: B. G. Teubner.
Suppe, F., ed. (1974). *The structure of scientific theories*. Urbana (Ill.): University of Illinois Press.
Toit, A. L. du (1927). *A geological comparison of South America with South Africa*. Carnegie Institute of Washington Publication 381.
Toit, A. L. du (1937). *Our wandering continents*. Edinburgh: Oliver & Boyd.
Tondl, L. (1973). *Scientific procedures*. Dordrecht: Riedel.
Toulmin, S. (1958). *The uses of argument*. Cambridge: Cambridge University Press.
Tricart, J. (1965). *Principes et méthodes de la géomorphologie*. Paris: Masson.
Tröger, W. E. (1967–71). *Optische Bestimmung der Gesteinsbildenden Minerale*. Part I (1971), ed. H. U. Bambauer & F. Taborsky & H. D. Trochim. Part II (1967) ed. O. Braitsch. Stuttgart: Nägel & Obermiller.
Upham, W. (1894). *American Geologist* **13**. (After Gilbert, 1896).
van Bemmelen, R. W. (1975). Kritik zur Plattentektonik. *Geologie en Mijnbouw*, **54**, 71ff.
van Bemmelen, R. W. (1976). Plate tectonics and the undation model: a comparison. *Tectonophysics*, **32**, 145ff.
Viallet, F.-A. (1958). *Zwischen Alpha und Omega. Das Weltbild Teilhard de Chardins*. Nürnberg: Glock und Lutz.
Wagenbreth, O. (1958). *Geologisches Kartenlesen und Profilzeichnen*. Leipzig: Teubner.
Wagenbreth, O. (1966). Bemerkungen zum Zeitbergriff in der historischen Geologie und zur Frage einer Unschärfebeziehung bei rhythmischer oder zyklischer Schichtengliederung. *Wissenschaftliche Zeitschrift der Hochschule für Architektur und Bauwesen Weimar*, 13, 617ff.
Waismann, F. (1965). *The principles of linguistic philosophy*, ed. R. Harré. London: Macmillan.
Wartofsky, M. W. (1968). *Conceptual foundations of scientific thought*. New York: Collier–Macmillan.
Watkins, J. (1970). Against normal science. In *Criticism and the growth of knowledge*, ed. I. Lakatos & A. Musgrave, 23ff. Cambridge: Cambridge University Press.
Wedepohl, K. H., ed. (1969–78). *Handbook of geochemistry*. Berlin: Springer Verlag.
Wedepohl, K. H., ed. (1979). Geochemische Aspekte der Diagenese von marinen Ton- und Karbonatgesteinen. *Geologische Rundschau*, **68**, 383.
Wegener, A. (1912a). Die Entstehung der Kontinente. *Geologische Rundschau*, **3**, 276ff.
Wegener, A. (1912b). Die Entstehung der Kontinente. *Petermanns Mitteilungen*, 185ff.
Wegener, A. (1915–29). *Die Entstehung der Kontinente und Ozeane*, 1st–4th edition. Braunschweig: Friedrich Viehweg.
Weingart, P., ed. (1972). *Wissenschaftssoziologie*, **I** & **II**. Frankfurt: Athenäum/VVA.
Weingart, P., ed. (1976). *Wissensproduktion und soziale Struktur*. Frankfurt: Suhrkamp/KNO.
Weingartner, P. (1971–6). *Wissenschaftstheorie* (in 2 vols). Vol. I (1971), *Einführung in die Hauptprobleme* (2nd improved edn. 1978). Vol. II, section 1 (1976), *Grundlagenprobleme der Logik und Mathematik*. Stuttgart: Frommann-Holzboog.
Weiskirchner, W. (1962). Untersuchungen und Überlegungen zur Entstehung des Rieses. *Jahresbericht und Mitteilungen des Oberrheinischen Geologischen Vereins*, **44**, 17ff.
Weizsäcker, C. F. von (1971). *Die Einheit der Natur*. Munich: dtv/KNO.
Wells, G., Brian, W. B. & Pearce, T. H. (1979). Comparative morphology of ancient and modern lavas. *Journal of Geology*, **87**, 427ff.

Werner, A. G. (1787). *Kurze Klassifikation und Beschreibung der verschiedenen Gebirgsarten*. Dresden.

Whewell, W. (1837). *History of inductive sciences*. **I–III**. London: J. W. Parker.

Wilson, J. T. (1968). Static or mobile Earth. The current scientific revolution. *American Philosophical Society Proceedings*, **112**, 309ff.

Wolff, R. (1971). *Die Sprache der Chemie. Zur Entwicklung und Struktur einer Fachsprache*. Bonn: F. Dümmler.

Wright, G. H. von (1971). *Explanation and understanding*. London: Routledge and Paul Kegan.

Wunderlich, H. G. (1974). Die Bedeutung der süddeutschen Grosscholle in der Geodynamik Westeuropas. *Geologische Rundschau*, **63**, 771ff.

Zimmerman, J. (1973). *Von der rationalistischen Sprachkritik zur Hermeneutik der Sprachspiele*. Dissertation, University of Tübingen.

Zimmerman, J. (1975). Psychische Realität. Aspekt einer hermeneutischen Grundlegung der Psychoanalyse. In *Die Beziehung zwischen Arzt und Patient. Zur psychoanalytischen Theorie und Praxis*, ed. S. Goeppert, 264ff. Munich.

Zimmermann, J. ed. (1978). *Sprache und Welterfahrung*. Munich: W. Fink.

Zittel, K. A. von (1899). *Geschichte der Geologie und Paläontologie*. Munich: Königliche Akademie der Wissenschaften. Transl. N. M. Ogilvie-Gordon (1962). *History of geology and palaeontology to the end of the nineteenth century*. New York: Hafner.

INDEX

Numbers in *italic* print refer to Figures
Numbers in **bold** print refer to Tables